T0336207

Liquid Crystal Displays

Wiley-SID Series in Display Technology

Series Editor:
Anthony C. Lowe

Consultant Editor:
Michael A. Kriss

Display Systems: Design and Applications
Lindsay W. MacDonald and Anthony C. Lowe (Eds.)

Electronic Display Measurement: Concepts, Techniques, and Instrumentation
Peter A. Keller

Reflective Liquid Crystal Displays
Shin-Tson Wu and Deng-Ke Yang

Colour Engineering: Achieving Device Independent Colour
Phil Green and Lindsay MacDonald (Eds.)

Display Interfaces: Fundamentals and Standards
Robert L. Myers

Digital Image Display: Algorithms and Implementation
Gheorghe Berbecel

Flexible Flat Panel Displays
Gregory Crawford (Ed.)

Polarization Engineering for LCD Projection
Michael G. Robinson, Jianmin Chen, and Gary D. Sharp

Fundamentals of Liquid Crystal Devices
Deng-Ke Yang and Shin-Tson Wu

Introduction to Microdisplays
David Armitage, Ian Underwood, and Shin-Tson Wu

Mobile Displays: Technology and Applications
Achintya K. Bhowmik, Zili Li, and Philip Bos (Eds.)

Photoalignment of Liquid Crystalline Materials: Physics and Applications
Vladimir G. Chigrinov, Vladimir M. Kozenkov and Hoi-Sing Kwok

Projection Displays, Second Edition
Matthew S. Brennesholtz and Edward H. Stupp

Introduction to Flat Panel Displays
Jiun-Haw Lee, David N. Liu and Shin-Tson Wu

LCD Backlights
Shunsuke Kobayashi, shigeo Mikoshiba and Sungkyoo Lim (Eds.)

Liquid Crystal Displays: Addressing Schemes and Electro-Optical Effects, Second Edition
Ernst Lueder

Transflective Liquid Crystal Displays
Zhibing Ge and Shin-Tson Wu

Liquid Crystal Displays: Fundamental Physics and Technology
Robert H. Chen

Liquid Crystal Displays

Fundamental Physics and Technology

Robert H. Chen

National Taiwan University

A John Wiley & Sons, Inc., Publication

Published by John Wiley & Sons, Inc., Hoboken, New Jersey
Published simultaneously in Canada

Library of Congress Cataloging-in-Publication Data:

Chen, Robert H., 1947-
Liquid crystal displays : fundamental physics & technology / Robert H. Chen.
p. cm.
Includes index.
ISBN 978-0-470-93087-8 (cloth)
1. Liquid crystal displays. 2. Liquid crystal devices. I. Title.
TK7872.L56C44 2011
621.3815'422–dc22

2010045220

Printed in the United States of America

Obook ISBN 978-1-118-08435-9
ePDF ISBN 978-1-118-08433-5
ePub ISBN 978-1-118-08434-2

10 9 8 7 6 5 4 3 2 1

Contents

Series Editor's Foreword

For once, I found it difficult to know how to begin. Writing forewords for the Wiley-SID series (this is my twentieth) is a demanding but extremely pleasurable task. In a qualitative sense this foreword is no different; it is the book that is different. Let me explain.

On reading two sample chapters of Robert Chen's manuscript, I realized that this would be like no other book in the series. Not only was its intended scope to cover the entirety of liquid crystal display (LCD) science and technology from the fundamentals of mathematics and physics to the production of products, but it was written by an author who has not only the academic background but also the experience as an executive in several major companies to provide first-hand insight and understanding of the global development of what is now a predominantly Asia-based industry.

The author has covered his subject matter with great proficiency and style. But there is more: the book is filled with interesting footnotes, often witty, of technical or historical relevance or a combination of all three. The most significant references are cited, but this is not a book where the reader will find a comprehensive list of all relevant publications. Other books in the series which address specific aspects of the technology provide that.

The unique feature of this book is that when discussing the global industrial development of the LCD industry, the author provides an account which is unprecedented—certainly in this series—in its level of detail, its understanding of cultural influences, and its degree of frankness. I believe that few will disagree with his arguments, but some will find it uncomfortable reading.

So, as the author aspires, this book may be read at several different levels. Anyone who reads it will find it rewarding as a technical introduction to the field replete with a sense of history. They will realize that this industry, which has made most of its growth in the last two decades, is built on the shoulders of scientific progress going back two centuries. Last, but certainly not least, I hope that they will find it a first-rate literary experience.

Anthony C. Lowe
Series Editor

Braishfield, UK

Preface

The liquid crystal display (LCD) has become the principal modern medium for visual information and image appreciation. It is now a pervasive and increasingly indispensable part of our everyday lives. Apart from its utility, this marvelous device relies on a science and technology that I believe makes the device all the more attractive and interesting.

This book is organized to highlight the basic physics, chemistry, and technology behind this intriguing product, and while describing the LCD, I attempt to provide some insight into that physics, chemistry, and technology. I believe that the history of the development of the LCD is equally intriguing, and thus I make excursions into tales of the principal contributors and their achievements and thinking in their research. Finally, the allure of liquid crystal television has made it a coveted symbol of modern life worldwide, and so apart from the technical descriptions, I also describe how the LCD business has become a global enterprise.

I attempt to describe the physics and technology in a clear and simple manner understandable to an educated reader. Further, I have endeavored to pay attention to literary exposition as far as I am able, in the hope that, in addition to describing the technology, the book may also provide some literary enjoyment. Of course whether I have succeeded here depends on the reader's assessment.

This book is written at an introductory level suitable for advanced undergraduates and first-year graduate students in physics and engineering, and as a reference for basic concepts for researchers. I also have tried to make the scientific and technical descriptions intuitively clear so that any educated person who has studied calculus can easily understand the exposition and thereby understand and appreciate liquid crystal displays and the science behind them.

Readers new to the field should read this book in chapter sequence to understand the gradual development of the LCD and the science and engineering involved; advanced researchers and practitioners can select the chapters and sections to find descriptions of the background of those selected topics.

Robert H. Chen

Taipei, Taiwan
June 2011

Acknowledgments

I would like to thank Professor Paul Nahin, for his books on mathematics and engineering from which I learned a great deal and borrowed liberally, and for his kind encouragement; Simone Taylor, Editorial Director at Wiley, who saw the potential of the manuscript and undertook the task of getting this book published while guiding me along the way; and most gratefully Dr. Anthony C. Lowe, the Editor of the Wiley-SID Series, who corrected mistakes and blocked metaphors (I am of course solely responsible for any that have gotten through). Further thanks are due to my wife Fonda, for her patient understanding; my daughter Chelsea, for cheerful enthusiasm; and my cat Amao, for accompanying me all the while. For my technical education, I would like to thank Dr. Hsu Chenjung, whose intelligence inspired me; Professor Andrew Nagy of Michigan and Professor Von Eshleman of Stanford, who supported me; and Chimei Optoelectronics Corporation where I learned about LCDs. Many of the drawings were done by Ingrid Hung at Chimei and Tsai Hsin-Huei of the National Taiwan University of Art.

About the Author

Robert Hsin Chen

Robert Hsin Chen is an adjunct professor at National Taiwan University and also teaches at Tsinghua and Jiaotong Universities in Taiwan. He was formerly a Senior Vice-President at Chimei Optoelectronics, a Director at Taiwan Semiconductor Manufacturing Company, Vice-President at Acer Corporation, and Of Counsel at the law firm Baker & McKenzie. Dr. Chen has a PhD from the University of Michigan (Space Physics Research Lab), a postdoctorate from Stanford University (Center for Radar Astronomy), and a JD from the University of California at Berkeley. He is a member of many scientific organizations, as well as the California Bar, and is a registered patent attorney; he has written many articles for international scientific and intellectual property journals, and is the author of *Made in Taiwan* (1997) and *Crystals, Physics, and Law* (in Chinese, 2010).

1

Double Refraction

The operation of liquid crystal displays is founded on the phenomenon of the double refraction of light as first recorded in Denmark by Erasmus Bartholinus in 1670. A piece of translucent calcite apparently divides incident light into two streams, producing a double image. This is depicted in Figure 1.1, as shown by the offset of the word "calcite." At about the same time in the Netherlands, Christian Huygens discovered that the light rays through the calcite could be extinguished by passing them through a second piece of calcite if that piece were rotated about the direction of the ray; this is depicted in Figure 1.2. This may be observed by taking two pairs of polarizing sunglasses and rotating them relative to each other.

One hundred and thirty-eight years later, in 1808, a protégé of the famous French mathematician Fourier, Etienne Louis Malus, observed that light reflected from a window, when passing through a piece of calcite also would change intensity as the calcite was rotated, apparently showing that reflected light was also altered in some way. The intensity of the light changed in both cases because the molecules of calcite have a crystal order that affects the light in an intricate but very understandable way called *polarization*.

Liquid Crystal Displays, First Edition. Robert H. Chen.

Figure 1.1 Double refraction in calcite. From http://www.physics.gatech.edu/gcuo/lectures.

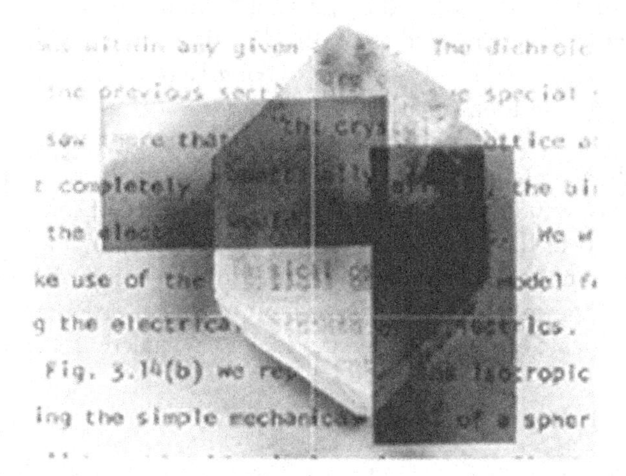

Figure 1.2 Two pieces of calcite at an angle. From http://www.physics.gatech.edu/gcuo/lectures.

It would be another 80 years later in Austria that double refraction, also called *birefringence*, and light polarization would be observed, not in crystalline rocks, but in a viscous liquid, later to be called a "liquid crystal." Although no doubt intriguing to natural scientists, intensive investigation of liquid crystals had to wait for yet another 80 years, when commercial interests provided the impetus for further study.

Briefly, a liquid crystal display can reproduce an image of a scene through the use of a video camera that, upon receiving the light reflected from the scene through its lens, in accord with the photoelectric effect first explained by Einstein, an electric current is generated in a metal when struck by light of sufficient energy, the current being proportional to the intensity of that light. That current is then transmitted to transistors that control an analog voltage that is applied to a pair of transparent electrode plates. Those plates enclose a thin layer of liquid crystal between them, and the voltage on the plates generates an electric field that is used to control the orientation of the electric dipole moment of the liquid crystal molecules, causing them to turn. Then light from a light source placed behind the liquid crystal layer, after being linearly polarized by a polarizer, will have its polarization states altered by the different orientations of the liquid crystal molecules, in accord with the liquid crystal's degree of birefringence. The beauty of the liquid crystal display is that the birefringence effected by a liquid crystal is precisely controllable by that electric field. The different polarization states of the light in conjunction with a second polarizer changes the brightness of the light emanating from the backlight source, and that modulated brightness can represent the light intensity of the original scene; the millions of picture elements so produced then combine to form an image that replicates the original scene.

Liquid crystal displays thus are based on an optical phenomena of electrically controlled birefringence and polarization, which can only be understood through knowledge of the interaction of light and matter.

However, light may be familiar to everybody, but Samuel Johnson succinctly observed that [1]*

> We all know what light is, but it is not easy to tell what it is.

* Samuel Johnson (1709–1784), English lexicographer, critic, poet, and moralist who completed the *Dictionary of the English Language* in 1755; Johnson is one of the preeminent authorities on the English language.

The understanding of light can gainfully begin at the outset with an appreciation of light as described by the Maxwell equations.

Reference

[1] Johnson, S. 1755. *Boswell's Life; Dictionary of the English Language*; quoted in Clegg, B. 2001. *Light Years*. Piatkus, London.

2

Electromagnetism

The scientific study of light has more than 1500 years of illustrious history. Beginning with Euclid and his geometrical study of light beams, the list of luminaries includes the great scientist/mathematicians Descartes, Galileo, Snell, Fermat, Boyle, Hooke, Newton, Euler, Fourier, Bartholinus, Huygens, Malus, Gauss, Laplace, Fresnel, Hamilton, Cauchy, Poisson, Faraday, and Maxwell. From those classical beginnings, the theories have evolved into atomic and quantum mechanical theories of light, developed by the great physicists Planck, Bohr, Heisenberg, Schrodinger, Born, Dirac, and Einstein. With such brainpower as driving force, the subsequent profound understanding of light should not have been unexpected.

The first mathematical treatments of light however quickly became mired in an ineluctable *æther*; that is, the early physical theory of *action at a distance* required the presence of an all-pervasive, elastic, and very subtle material to serve as the medium through which forces could transfer their effect. Simply put, although often not easy to apply, the interaction between two separate bodies is determined by a mechanical transfer of force acting along a line connecting the bodies, that force weakening with the distance between the bodies. The action at a distance theory could successfully describe many observations in common experience, the most cogent example being sea waves. But this "æthereal" view of Nature confounded even its proponents

Liquid Crystal Displays, First Edition. Robert H. Chen.

when faced with the equally naturally observed electromagnetic phenomena, such as the effects of a magnet on a current-carrying wire and the invisible transfer of electromagnetic forces through a vacuum.

The great mathematical physicist Maxwell too was caught up in the æthereal action at a distance and a physics based on mechanics and fluid dynamics, so his initial efforts to mathematically describe the observed electromagnetic phenomena were based on such conceptualizations of electrical energy as the stored energy in a spring, and magnetic energy as the kinetic energy of a flywheel, and of course, electric current was seen as flowing water (an analogy nonetheless still used today). When Maxwell faced the *interaction* between electricity and magnetism, however, he was confounded: how would an electric current in a wire produce a concentric circulating magnetic force, and how would a moving magnet near a wire coil produce an electric current in that coil? The description of all the parts and the mutual interactions among them using purely mechanistic and fluid formulations would result in some strange machines [1].

For example, the *deus ex machina* sketched in Figure 2.1 consisting of an interconnected contraption of balls, wheels, gears, and tubes. Solidly ingenious as it was, in order to explain what the experimentalists Oersted, Ampere, and Faraday had observed in Nature and experiments, it also was clear that this mechanical beast was going to be very difficult to tame.

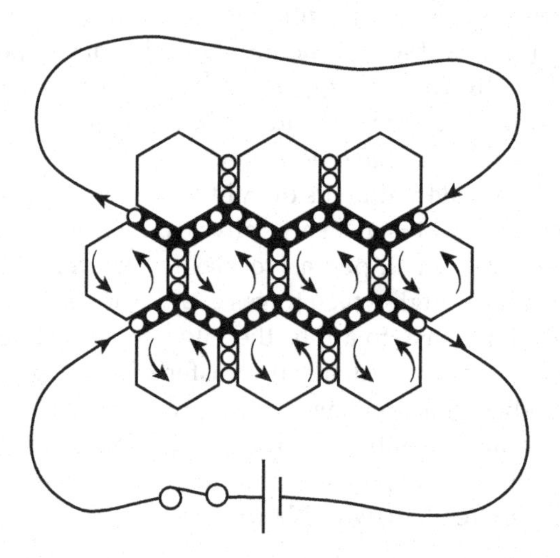

Figure 2.1 Maxwell's electromagnetic mechanical machine. From Mahon, B. 2003. *The Man Who Changed Everything, The Life and Times of James Clerk Maxwell*, Wiley, p. 100.

Indeed, the intricate *pas de deux* of electromagnetic forces at the time was clearly observed but only murkily understood. One force apparently engenders another force, but the generation clearly was not acting in the line through the distance between the bodies producing those forces. In a vain attempt to tie the forces, the construction of springs, flywheels, balls, and interconnecting water pipes, ropes, pulleys, and gears became just too complicated and contrived to attain the simplicity and elegance sought by a mathematical theoretician like Maxwell. But worst of all, the bits and pieces could not hope to operate to produce electromagnetic forces in a vacuum; the mechanical theory still relied upon the ethereal yet ubiquitous *æther* and all its attendant mystery. If Maxwell was to overcome the *æther's* dark art, he needed the power of mathematical physics to smite that *ævil* witch.

Faraday's Intuitive Field

The untenable complications wrought by the purely mechanistic and æther-laden action at a distance were unraveled by the great experimentalist Faraday. Having had little formal education, Faraday was not equipped to use mathematical physics to describe what he observed; instead he depended on his (considerable) powers of intuition.

To start off, a point charge (q) acted upon by an electric force (E) will experience a mechanical force (F) described simply by the equation $F = qE$, where the force is directly proportional to, and in the same direction as, the electric force. But Faraday observed that the effect that a magnet has on a current-carrying wire is to move it, as shown in the schematic drawing in Figure 2.2 as the dashed line. That is, when the current is turned on, the wire near the magnet will move horizontally in a direction perpendicular to the direction of the South to North poles of the magnet, so mathematically the magnetic force emanating from the magnet produces a mechanical force F that can be described by the vector equation $F = qv \times B$. The equation says that a point charge traveling in the wire at a velocity (v) will be subject to a force (F) that is proportional and perpendicular to both v and B (the *cross product* in the vector calculus). The electric and magnetic forces combined in a single equation is the well-known Lorentz force,

$$F = q(E + v \times B).$$

From the above equation, it is clear that while there is a force (E) associated directly with an electric charge (q), a magnetic force requires motion (v) to act.

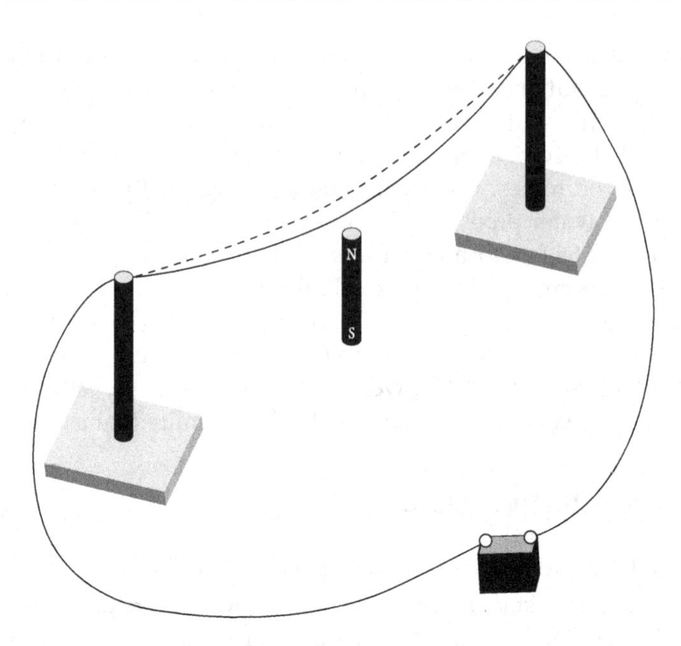

Figure 2.2 Magnetic force acting on a current in a wire moves the wire.

Other observations were not so simply describable, however; for example, the subsequent mutual interaction of the current, the generated magnetic force, and the magnet's magnetic force.

To help matters along, Faraday here visualized the force effect of the magnet as a pattern of *lines of force*, the grouping together of which constituted a *flux* of force lines, the number and closeness of the lines representing the density of the flux, and that flux density indicating the strength or intensity of the magnetic force. Faraday's own drawing of the lines of force emanating from a bar magnet is shown in Figure 2.3, where he described the flux lines as a *field*. This then was the basic idea of a field to intuitively conceptualize electromagnetism.

The idea of lines of force constituting a field to describe the electric and magnetic effects was not the only pivotal concept invented by Faraday; another critical idea was that the field lines could be superimposed to describe the cumulative effect from many different sources of electric and magnetic force. This *principle of superposition* can reduce a very complicated collection of electromagnetic sources of force into a simple addition of the different contributions from the various sources. That is, at a given point in space, no matter what the distribution of the other electric or magnetic

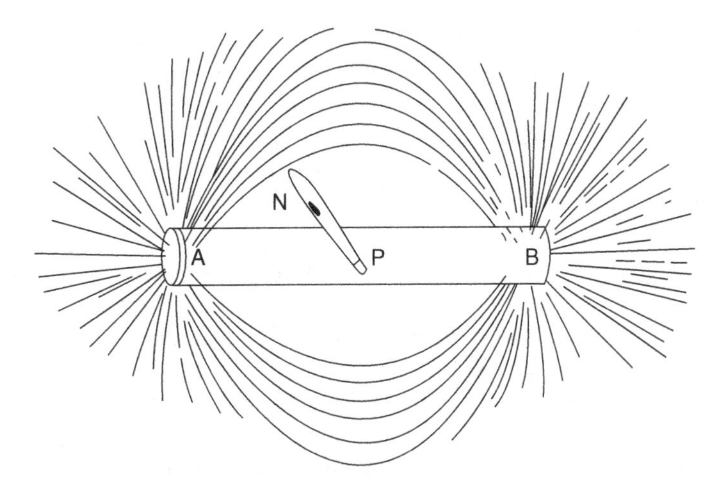

Figure 2.3 Faraday's own drawing of the magnetic lines of flux.

charges happens to be, each contribution to the field at that point can be simply added to derive their total cumulative effect. No matter how the electric or magnetic field was produced, whatever its character, or what the distribution of sources is, the cumulative field by itself only, will determine the influence on anything at the point in space in question. As a consequence, *field theory* can resolve a physically very complicated electromagnetic situation into something amenable to a simple additive analysis.

Maxwell's Equations

Physics is based on observed phenomena, and any theory must of course comport with what is observed in Nature. The intricate interplay between electric and magnetic phenomena was unambiguously observed—a moving charge feels the magnetic force and a moving charge can generate a magnetic force but not in the same direction. That is, one moves, the other is created but not directly in-line; one dies, the other then expires; one again stirs, the other is reborn, but again not in-line; the problem was how to describe that in a coherent and self-consistent mathematical theory.

The famous Maxwell equations evolved from a compendium of ideas gleaned primarily from the mathematical theories of Gauss, Laplace, and Poisson, and the observations of the experimentalists Coulomb, Oersted, Ampere, and Faraday. Maxwell himself admonished those at the Cambridge Philosophical Society that in particular, "the understanding of electromagnetism requires . . . carefully reading Faraday's *Experimental Researches in Electricity*" [2]. Within that book lay not only experimental data and analysis, but also the critical concept of the intuitive field and principle of superposition. Leaning heavily on Faraday's book, Maxwell set about to combine all the then current observations of electromagnetic phenomena into a concise set of mathematical equations.

The *Maxwell equations* are beautiful not only because they succinctly describe the complex electromagnetic phenomena observed in Nature, they also beget a new physical reality. That is, the Maxwell equations, in bringing to bear the knowledge and techniques of mathematical physics to the study of light, have not only succinctly explained electromagnetism, but also created an entirely new reality, the fruits of which are the many wondrous electronic products of our modern technological society, including of course the marvelous liquid crystal display. Viewed in this way, the significance of the Maxwell equations cannot be overestimated.

In Maxwell's theory, the electric and magnetic fields can be represented by two vectors, an electric vector E and a magnetic induction vector B (the word "induction" indicating that one is induced by the other), and a current density J resulting from the interaction of the electric and magnetic fields in the medium. Maxwell's famous four simultaneous partial differential equations describing the intricate interactions among those vectors are first set forth here for visual appreciation; they will be derived in turn below.

$$\nabla \circ E = \frac{\rho}{\varepsilon}$$

$$\nabla \circ B = 0$$

$$\nabla \times E = -\frac{\partial B}{\partial t}$$

$$\nabla \times B = \varepsilon\mu \frac{\partial E}{\partial t} + \mu J.$$

In the equations, t is the parameter of time and ρ is the charge density, while the material coefficients, dielectric permittivity (ε, also called the *dielectric constant*), and magnetic permeability (μ) describe the electromagnetic character of a material body that is subjected to electric and magnetic forces. The significance of particularly the dielectric constant in liquid crystal materials will become exceedingly clear through the descriptions of liquid crystal display technology to be presented throughout this book.

Having had no formal scientific training, Faraday relied on experiments and intuition to comprehend electromagnetic phenomena, the scholarly Maxwell began with the mechanistic view in accord with his classical training, and the different approaches together produced the theory. Maxwell's purely mechanical constructs at first lead to a hopeless complexity, but they contributed clearly analogical mathematical wherewithal, such as the vector calculus of fluid mechanics, to attack the electromagnetic problem, providing not only help in understanding, but also a mathematical formalism propelled by the powerful analytical tools of mathematical physics. After looking at Maxwell's equations long and hard enough, hopefully the meaning of the eminent mathematical physicist Paul Dirac's famous quote for physical equations will be realized [3]:

> I understand what an equation means if I have a way of figuring out the characteristics of its solution without actually solving it.

First off in pursuing this hoped for understanding, the electric and magnetic parts of the equations can be separated by simply not considering changes with time; that is, the static case is presented, wherein there are no changes of the electric and magnetic fields with time ($\partial / \partial t = 0$), this effectively decouples Maxwell's equations, becoming for the static case,

$$\nabla \circ E = \frac{\rho}{\varepsilon}$$

$$\nabla \times E = 0,$$

and

$$\nabla \circ B = 0$$

$$\nabla \times B = \mu J.$$

In the static case, electric and magnetic effects are independently described by the equations, and since no equation contains both E and B together, the change in their values with time is revealed as the critical link to their basic interdependent nature, that is, they mutually create one another by dint of their change with time.

Now further understanding of the equations requires their derivation, and the procedures employed in the derivation hopefully will provide further insight into the full dynamic character and meaning of the equations [4].*

* Maxwell's equations are set forth here in the meter–kilogram–second unit system (mks).

The Derivation of $\nabla \circ E = \frac{\rho}{\varepsilon}$

First, the $\nabla \circ E$ term in Maxwell's first equation is what is known in the vector calculus as *flux divergence*, and as the name implies, it is a measure of the spreading out of the flux of the electric field vector E. The equation simply states, reasonably enough, that this spreading out depends directly on the density of electric charge (ρ) emanating from a point charge, and inversely as the permittivity (ε) of the medium, which is a measure of its concordance with electricity, or how well it "permits" electricity to course through it (ε has been variously also called the *dielectric coefficient*, the *dielectric constant*, and the *electric permeability*). As shown in Figure 2.4, a point charge (at left) and a spherically distributed charge (at right) will produce a spherical electrical flux diverging from the center.

Faraday's physical intuition regarding the divergence of flux is manifested in the mathematical Gauss's law, which will be seen shortly to be the starting point of the derivation of Maxwell's first equation. Gauss' law is

$$\oiint E \circ \hat{n} dA = q / \varepsilon.$$

The closed surface integral (the meaning of $\oiint$) of the electric field vector (E) in the direction perpendicular to the surface of the enclosed area, with the latter being denoted by the unit vector $\hat{n}$, wherein $E \circ \hat{n}$ is simply the projection of E onto $\hat{n}$, or in other terms, $E \cos\theta$, where θ is the angle between E and $\hat{n}$. For example, in the simple spherical charge distribution shown above with the point charge becoming the density of charge over the surface of the sphere, $q \rightarrow \rho$, then

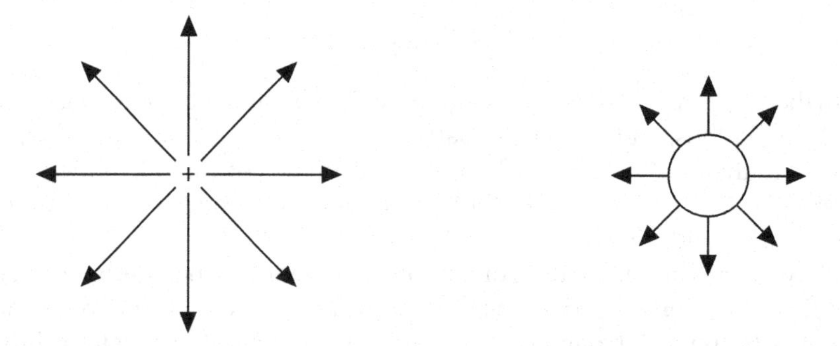

Figure 2.4 A point charge (left) and a spherical charge (right) produce an emanating electrical flux.

$$\oint\!\!\oint E \circ \hat{n} dA = \rho / \varepsilon,$$

where again ρ is the charge density within the closed surface, and ε is the electric permittivity of the medium. Clearly, the greater the charge density, the more electric flux emanating through the closed surface, with the dielectric constant modulating that electric flux, and thus the latter is indeed a measure of the material medium's concordance, or lack thereof, with electric fields. This dielectric constant plays a central role in describing the liquid crystals used in displays, as will be seen in later chapters.

Gauss' law should be easy to understand: it states simply that electric flux emanating perpendicularly from a closed surface is produced by the charge within and modulated by the permittivity of the medium in which the charge is placed. This then is just the *net flux* emanating through the surface, because those flux lines that are not perpendicular to the surface will cancel each other out on their journey outwards through that surface.

The divergence theorem of the vector calculus states that the flux emanating through a closed surface must be equal to the divergence of that flux within the volume of the closed surface, which if one thinks about it, is eminently understandable. In mathematical terms, the net flux through an arbitrary closed surface is equal to a volume of diverging flux within that surface:

$$\oint\!\!\oint E \circ \hat{n} dA = \oint\!\!\oint\!\!\oint (\nabla \circ E) dV.$$

The forbidding-looking double and triple integrals may give one pause, but they are only used to define the closed surface and the closed volume, respectively. The divergence theorem is very useful in mathematics and physics, and later on it will be used in developing the theories of liquid crystals.

Returning now to the derivation of Maxwell's first equation, for any arbitrary distribution of charge, the total charge can be simply written as the volume integral of the charge density $\oint\!\!\oint\!\!\oint \rho dV$, and therefore the most general form of Gauss' law now can be written as

$$\oint\!\!\oint E \circ \hat{n} dA = \oint\!\!\oint\!\!\oint \frac{\rho}{\varepsilon} dV,$$

then using the divergence theorem, Gauss' law becomes

$$\oint\!\!\oint E \circ \hat{n} dA = \oint\!\!\oint\!\!\oint (\nabla \circ E) dV = \oint\!\!\oint\!\!\oint \frac{\rho}{\varepsilon} dV.$$

Because the surfaces and volumes were chosen arbitrarily, the integrands of the volume integrals are equal, and so

$$\nabla \circ E = \frac{\rho}{\varepsilon},$$

which is just Maxwell's first equation which derivation is now complete. The equation says that the divergence of the electric flux is just equal to the charge density producing that flux, modulated by the electric permittivity of the material through which the flux passes.

The Derivation of $\nabla \circ \boldsymbol{B} = 0$

Maxwell's second equation ($\nabla \circ B = 0$), as now can be explained using the vector calculus, means that the divergence of an induced magnetic field is zero, which further means that a "magnetic point charge" generates no magnetic flux. That is because so far, no such thing as a point magnet has been observed in the Universe. Of course, just because something has not been observed does not mean it does not exist, for example, the "magnetic monopole" may just be too small to observe, or present-day instrumentation is simply not sophisticated enough to find it [5].* At any rate, right now the understanding is that Maxwell's second equation holds true (at least for the macroscopic world), and there is no magnetic flux emanating from a point because a magnetic point does not exist. Incidentally, such a simple equation will be predictably easy to derive. If Gauss' law and the divergence theorem are combined into a single equation, this time using the magnetic induction vector B, and accepting that there is no such thing as a magnetic charge,

$$\oiint B \circ \hat{n} dA = \oiiint (\nabla \circ B) dV = 0,$$

first, clearly (for later use),

$$\oiint B \circ \hat{n} dA = 0,$$

and also since the closed volume integral of the second term is arbitrarily chosen and is equal to zero (physicists like to say it "vanishes"), its integrand must also be zero, so

* The new supercolliding accelerator at the *Conseil European pour la Recherche Nucleaire,* (CERN) in Geneva may find in its search for the missing mass in the Universe, presumably hidden in the elusive Higgs Boson, some indications of the existence of magnetic monopoles.

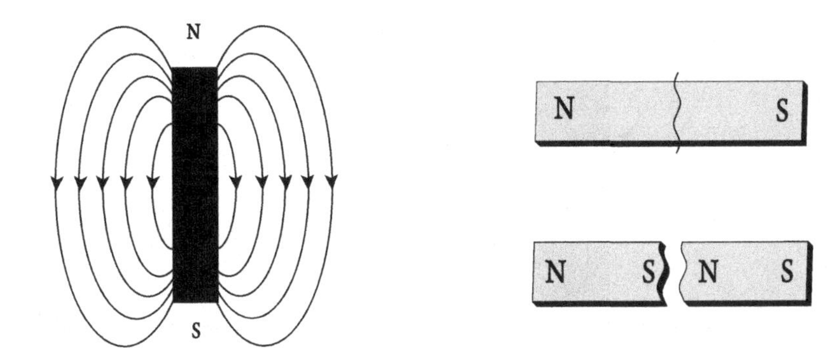

Figure 2.5 Cutting a magnetic dipole results in another dipole, not a monopole.

$$\nabla \circ B = 0,$$

which is just Maxwell's second equation, very easily derived. The message is clear for now that the flux divergence of a magnetic field is zero.

Now the physical significance of this equation is the non-existence of magnetic charges or monopoles. Indeed every schoolchild knows that a magnet has a north and a south pole (a magnetic dipole, as shown in Fig. 2.5 at left), and if a magnet is cut in half to try to get a single magnetic pole, the result will be that no matter how finely it is cut, there are always two poles, as shown schematically in Figure 2.5 at right. Now the idea that the magnetic monopole perhaps being too small to observe is revealed as a real possibility; that is, perhaps the magnet just can not be cut thin enough with the tools presently available to separate the poles.

The Derivation of $\nabla \times \boldsymbol{E} = -\frac{\partial \boldsymbol{B}}{\partial t}$

Maxwell's third equation is a little more difficult to derive, but just looking at it, the equation seems to say that a change of the magnetic induction vector ($\partial B / \partial t$) with time produces a circulating electric field, as denoted by the so-called *curl* $\nabla \times E$, and, vice-versa, the curl of E instigates a change in the magnetic induction vector B. Now this is in accord with what is observed experimentally, so the mathematical formulation above can describe the phenomenon observed, and a starting point is the Lorentz force.

Recalling that the Lorentz force equation $F = q(E + v \times B)$ says that a force on a charge is due to the electric field and the motion of the charge in a magnetic field, the derivation of Maxwell's third equation can begin with the observation of a force exerted on the electron current in a wire by a magnet.

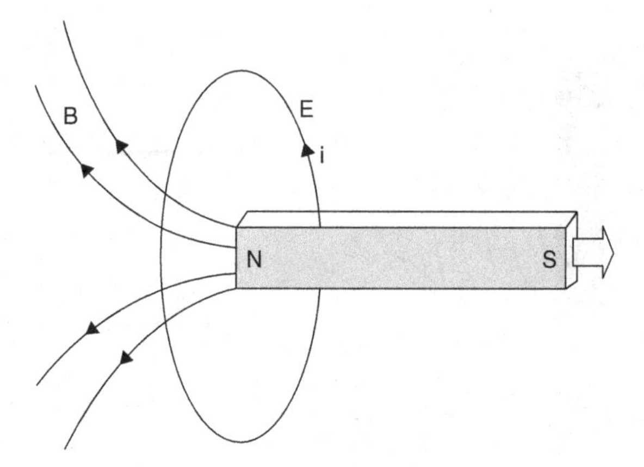

Figure 2.6 A moving magnet generates an electric field and current in the wire.

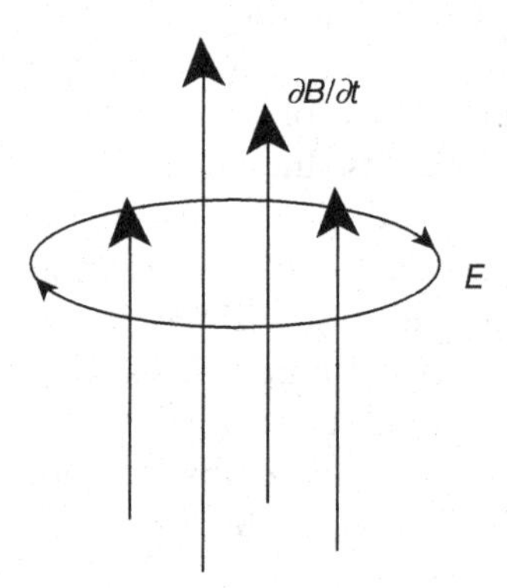

Figure 2.7 A changing magnetic field generates an electric field.

A schematic drawing of a magnet being thrust through a wire coil is shown in Figure 2.6; the changing magnetic field of the magnet (B) generates an electric field (E) in the wire through the Lorentz force; E then pushes the electric charges with a force given by the first term $F = qE$, constituting an electromotive force (*emf*), but from the second term $F = q(v \times B)$, the force exerted on the moving electron (v) should be perpendicular to the magnetic field, thereby producing a circulating electric field that pushes the electrons forming a current (i) (Fig. 2.7). If the magnet is thrust in the opposite direction, from the Lorentz force equation, the current will be in the opposite direction, so that a magnet moving back and forth will generate an alternating current. Conversely, if there is an alternating current in the wire, a freely hanging magnet will move back and forth, and by cleverly postitioning the wire, the magnet can be made to rotate; this is just a dynamo which is the basis of an electric motor.

So a change in the magnetic field causes a circulating electric current, and thus it is reasonable that the electromotive force can be given by the time rate of change of the net magnetic flux passing through the surface formed by the wire (the surface helping to determine just how the magnetic flux is changing),

$$emf = -\frac{d}{dt}\iint (B \circ n)dA.$$

Similarly, the wire can be seen as a closed line integral of the E field, which is defined as the *circulation* of the electric field, and that circulation can be related to the time rate of change of the net magnetic flux through the surface formed by the wire (this time taking the derivative inside the surface integral, which can be done since none of the other terms change with time or over space, and changing it to a partial derivative with t since B can also change with space as well as time),*

$$\oint E \circ ds = -\iint \left(\frac{\partial B}{\partial t} \circ \hat{n} \right) dA.$$

To sum up, the first equation above says that an electromotive force is generated by the change of the magnetic induction flux emanating perpendicular to the surface formed by the wire. The second equation says that the line integral of the electric field in a closed loop (the circulation) is equal to the surface integral of the change of the magnetic induction flux going through the surface.

Experimentally, if a magnet is moved to produce a change in the magnetic flux through the plane of the wire loop, an electric current will be generated in the wire.

Now the right hand side of the first equation is

$$-\frac{d}{dt}\iint (B \circ n)dA.$$

* The cavalier transposing of the integral and differential operators might not be acceptable to mathematicians, but physicists can rationalize it because the transpositions are in accord with physical reality, as proven by the many electronic devices that obey the Maxwell equations.

However, recall that in the derivation of Maxwell's second equation in the previous section, the net magnetic flux was equal to zero ($\oiint B \circ \hat{n} dA = 0$), why does it not now vanish as well, leaving nothing to differentiate in time? The difference is that the earlier derivation used Gauss' law, which holds true only for *closed* surfaces, and for the present derivation, the surface is not closed on itself, but merely defined by the wire (so there is no circle through the integral signs).

The plane surface of the wire may not be closed, but the line of the wire *is* closed, and so the left-hand side of the second equation above is a so-called closed line integral. In electromagnetic theory, a closed line integral of an electric field vector describes the work done by an electric field pushing electric charges in the line integral to become an *electric circulation.*

The integral equations above later came to be known collectively as "Faraday's law," but the typical exposition of the Maxwell equations is in derivative forms, which are more easily comprehensible by inspection than the integral forms. In order to derive the derivative form, another well-used vector calculus theorem, namely *Stokes' theorem,* will be employed,

$$\oint E \circ ds = \oiint (\nabla \times E) \circ \hat{n} dA.$$

Just a glance at Stokes' theorem indicates that it is eminently suitable for the task at hand, since it relates the circulation to the curl, so inserting the right hand side of the circulation equation above into Stokes' theorem,

$$\oint E \circ ds = -\iint \left(\frac{\partial B}{\partial t} \circ \hat{n} \right) dA = \oiint (\nabla \times E) \circ \hat{n} dA,$$

and again because the surfaces are chosen arbitrarily, the integrands of the two surface integrals should be equal, so

$$\nabla \times E = -\frac{\partial B}{\partial t}.$$

The above equation is just Maxwell's third equation. It states that the swirling of the electric field vector ($\nabla \times E$), for which Maxwell himself coined the term "curl," describes a rotation of the electric field vector. So when the magnetic field changes with time, just like water in a bathtub swirling down the drain, the electric field swirls in a curling motion as shown schematically in Figure 2.8. This is a good example of how the techniques of fluid dynamics can be profitably used in the study of electromagnetics.

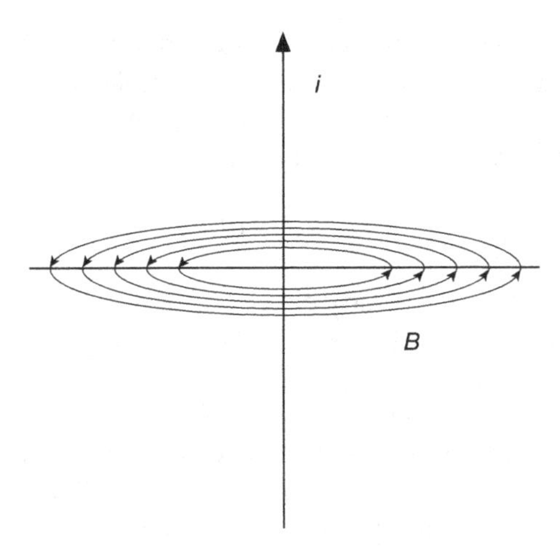

Figure 2.8 A steady current generates a magnetic field, but only a changing magnetic field generates a current.

It should be noted that the minus signs in the equations above mean that whatever force is generated must oppose the change in flux (Lenz's law), otherwise an infinitely increasing perpetual motion would result, which Nature will not accept.

This electric and magnetic mutually induced interaction equation is known formally as "Faraday's Law of Electromagnetic Induction," and it is verily the theoretical source of all electromagnetically driven devices, such as electric motors, generators, telephones, and many other modern electronic products.

The Derivation of $\nabla \times \boldsymbol{B} = \varepsilon\mu \frac{\partial \boldsymbol{E}}{\partial t} + \mu \boldsymbol{J}$

From the creation of the Universe, there was magnetism from the motion of charged particles in the primordial plasma, and consequently magnetism came to reside in planetary bodies. At the very beginning of the first millennium (4 AD) in the time of the Western Han Dynasty, the magnetic compass apparently was invented in China [6]. Its navigational uses of course opened the way for world exploration, but the scientific investigation of magnetism did not progress much further until near the end of the second millennium, when the Danish scientist Oersted noticed that a current-carrying wire when placed near a compass would cause the magnetic compass needle to move. Shortly thereafter, Ampere in France discovered that a steady electric current would produce a concentrically circulating magnetic field, and vice-versa, the circulating magnetic field would produce an electric current. The mathematical representation of this phenomenon using the closed line integral circulation introduced above, came to be known as "Ampere's law,"

$$\oint B \circ ds = \mu I.$$

As shown in Figure 2.8, a steady electric current (i) generates a concentric magnetic field (B), but it must be emphasized that only a *changing* magnetic field generates an electric current.

Ampere's discovery, decisive as it was, however, could only describe the effects of a static, unchanging electric current. While trying to portray Ampere's law in his mechanical *dæmon*, Maxwell inveighed a whirlpool-like mechanism in an attempt to represent the magnetic forces pushing the electric particles and the concomitant converse of the moving electric particles creating the magnetic forces, whereupon Maxwell found that the whirlpool then required an elasticity, and only then would Ampere's law be able to describe dynamic changes in the magnetic field and their effect on electric particles. So Maxwell postulated an addition to Ampere's law, and called it the *electric displacement current* (later to appear in his field equations as the *electric displacement vector D*), which was manifested by a change in the net flux of the electric field. This electric displacement current would provide the magnetic field elasticity and the total electric current needed to comport with experiment.

$$I_{\text{displacement}} = \varepsilon \frac{d}{dt} \iint (E \circ \hat{n})\, dA.$$

With the addition of the displacement current, the magnetic circulation now becomes

$$\oint B \circ ds = \mu (I + I_{\text{displacement}}),$$

and inserting the expression for the displacement current given above, Ampere's law then becomes what later came to known as the "Ampere–Maxwell law",

$$\oint B \circ ds = \mu \left(I + \varepsilon \frac{d}{dt} \iint (E \circ \hat{n})\, dA \right).$$

The derivative form of the above Ampere–Maxwell law is just Maxwell's fourth equation, and as might be expected, the derivation again uses Stokes' theorem, this time for the magnetic induction vector (B),

$$\oint B \circ ds = \oiint (\nabla \times B) \circ \hat{n} dA.$$

Substituting the right hand term of Stokes' theorem into the Ampere–Maxwell law equation gives,

$$\oiint (\nabla \times B) \circ \hat{n} dA = \mu\left(I + \varepsilon \frac{d}{dt} \iint (E \circ \hat{n}) dA \right).$$

Because electromagnetic fluxes are generally continuous and differentiable in the mathematical curve sense, the derivative operator can be brought into the integrand and the surface integral brought outside, so

$$\oiint (\nabla \times B) \circ \hat{n} dA = \mu \iint \left[I + \varepsilon \frac{d}{dt} (E \circ \hat{n}) \right] dA,$$

and since the definition of electric current density is

$$J = \iint I dA,$$

the first surface integral on the right-hand side is just J, and again, since the surfaces are arbitrarily chosen, the integrands are equal, but because the spatial change of the electric field vector E must be considered, the single-variable differential operator d/dt becomes the partial differential operator $\partial/\partial t$, so

$$\nabla \times B = \varepsilon\mu \frac{\partial E}{\partial t} + \mu J.$$

The equation above is just Maxwell's fourth equation in differential form. The continuously self-circulating magnetic field is described by its curl, and the change in the electric field ($\partial E/\partial t$) produces a current denoted by its current density J; and conversely, the electric current and the change in the electric field produce the concentric curling magnetic field. An electric charge can only produce a magnetic field if the charge is moving, and electric charges are not affected by a magnetic field unless the charge is moving. The interdependent life and death, flourish and decay dance of the swirling electric and magnetic fields is manifest in Maxwell's fourth equation.

To say that the Maxwell equations merely describe electromagnetic phenomena ill serves their greatness. It should be understood that the reality of the electromagnetic fields themselves depends on the verity of the Maxwell equations. That is, the existence of the electromagnetic field and the truth of the Maxwell equations are inexorably intertwined and mutually dependent, and which begets the other is a question for epistemological discourse.

Vector Analysis

The point of departure of the exposition of the Maxwell equations, was to describe physical reality gleaned from observations of Nature and controlled experimentation. Further elucidation of the electromagnetic theory requires a brief foray into the realm of the vector calculus, a mathematical formalism the bases of which lay in theories of fluid mechanics and electromagnetics, but nonetheless also stands independently as a mathematical discipline in and of itself.

To start, it is necessary to establish a spatial coordinate system (x,y,z) upon which the respective unit vectors **i,j,k** lie collinearly, and accordingly the electric and magnetic field vectors can be resolved into their spatial coordinates lying along the spatial unit vector directions, so that the ***E*** and ***B*** fields can be written as,

$$\mathbf{E} = \mathbf{i}E_x + \mathbf{j}E_y + \mathbf{k}E_z$$

$$\mathbf{B} = \mathbf{i}B_x + \mathbf{j}B_y + \mathbf{k}B_z.$$

A basic vector calculus operator ∇ can transform a scalar into a gradient vector, by means of an operator called the *gradient*

$$grad = \nabla = \left(\mathbf{i}\frac{\partial}{\partial x} + \mathbf{j}\frac{\partial}{\partial y} + \mathbf{k}\frac{\partial}{\partial z}\right),$$

and from that operation, one can see that the effect of the gradient operator is to display a scalar function in all its directional glory, and in particular the gradient will indicate the direction of the greatest slope of the function (that is why it is called a "gradient").

The divergence ($divB = \nabla \circ B$), recalling from the derivation of the Maxwell equations above, when acting upon a vector, produces the *scalar product*, as for example acting on ***B***,

$$div\mathbf{B} = \nabla \circ \mathbf{B} = \left(\mathbf{i}\frac{\partial}{\partial x} + \mathbf{j}\frac{\partial}{\partial y} + \mathbf{k}\frac{\partial}{\partial z}\right) \circ (\mathbf{i}B_x + \mathbf{j}B_y + \mathbf{k}B_z) = \frac{\partial B_x}{\partial x} + \frac{\partial B_y}{\partial y} + \frac{\partial B_z}{\partial z}$$

From the above equation, one can see that the divergence operator reveals the vector's overall change with respect to the $x,y.z$ coordinate directions.

Further, if the ∇ operator operates on itself in a scalar product fashion, a new second derivative operator ∇^2 is formed, called the divergence of the gradient,

$$divgrad = \nabla \circ \nabla = \nabla^2 = \frac{\partial^2}{\partial x^2} + \frac{\partial^2}{\partial y^2} + \frac{\partial^2}{\partial z^2}.$$

The second derivative in physics is "the change of a change," for example, if a time rate of change, then it is the rate of change of velocity, which is acceleration; if a spatial rate of change, then it is the change of slope of a curve. Because it very often appears throughout mathematical physics, ∇^2 has its own name, *the Laplacian,* named of course in honour of the great French mathematician Laplace. The Laplacian not only can operate on a scalar, it can also operate on a vector, thusly

$$\nabla^2 B = \frac{\partial^2 B_x}{\partial x^2} + \frac{\partial^2 B_y}{\partial y^2} + \frac{\partial^2 B_z}{\partial z^2}.$$

So far the vector calculus is fairly straightforward, going up or down gradients and spreading out in space as it were. Now the turns become trickier with the curl (e.g., $\nabla \times E$ for the electric vector), which is the ∇ operator operating as a vector product. Its definition is the circulation of a vector (here B) in the limit of a unit area going to a point,

$$\nabla \times B = \lim_{\Delta A \to 0} \frac{1}{\Delta A} \oint B \circ ds.$$

This definition provides every point in space with a curl (since the area ΔA goes to zero to form a point). For example in fluid flow, the value of the curl at a point depends on whether the flow on the sides around that point are moving in the same or different directions, or moving at the same or different speeds. This can be visualized by imagining a small paddlewheel (Fig. 2.9) placed at a point in a river. If the water is flowing at about the same speed in about the same direction (for instance in the middle of the river), the paddlewheel will not turn, that is, there is zero curl at that point in the river. If, however, the paddlewheel is placed near the bank of the river where the water flow is slower compared with that towards the middle of the river, it may rotate, and that point in the river has curl. If the flows on opposite sides of the paddlewheel are in different directions and/or of different speeds, the paddlewheel will turn, and there will be curl. Of course, if the paddlewheel is placed in a whirlpool in the river, it will definitely show the curl.

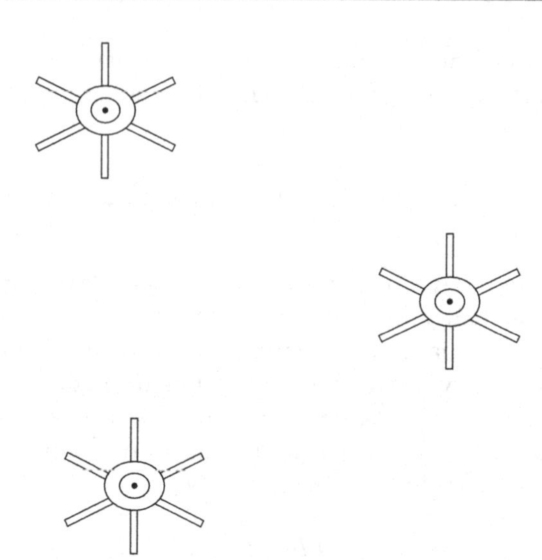

Figure 2.9 Paddle wheels to indicate curl.

Since the curl is a vector, it must have a direction, and by convention, the right-hand rule is used; that is, with fingers of the right hand curling in the direction of rotation, the raised thumb gives the vectorial direction or the curl. By taking the limit of an infinitesimally small paddlewheel, the curl at every point in space can be mathematically represented.

The calculation of the curl may be performed very conveniently using a determinant

$$curlH = \nabla \times B = \begin{vmatrix} \mathbf{i} & \mathbf{j} & \mathbf{k} \\ \dfrac{\partial}{\partial x} & \dfrac{\partial}{\partial y} & \dfrac{\partial}{\partial z} \\ B_x & B_y & B_z \end{vmatrix} = \mathbf{i}\left(\frac{\partial B_z}{\partial y} - \frac{\partial B_y}{\partial z}\right) + \mathbf{j}\left(\frac{\partial B_x}{\partial z} - \frac{\partial B_z}{\partial x}\right) + \mathbf{k}\left(\frac{\partial B_y}{\partial x} - \frac{\partial B_x}{\partial y}\right).$$

Looking at the equation as a whole, it is not easy to discern the curling action, but a closer examination of each term will reveal the curl. Each spatial direction (signified by **i,j,k**) has in the parenthesis after it the difference of the changes of the magnetic field components in the plane formed by the values of the changes in the other two directions. For example, the ***i*** unit vector is in the x direction, and the rotation is in the yz-plane. The relative difference between the variation of B_z with y ($\partial B_z / \partial y$) and the variation of B_y with z ($\partial B_y / \partial z$) determines the x-component of the contribution to the curl. It can thus be seen that there must be a relative difference, otherwise that component of the curl will be zero and will have no contribution to the overall curl. It might be helpful again to think of the different speeds and directions of water flow around the paddlewheel for one plane. Similarly, the y-component of the curl is calculated by subtracting the changes in the magnetic vector components in the zx-plane, and the z-component of the curl is calculated by subtracting the changes in the magnetic vector components in the xy-plane. Thus all the values of the changes in magnetic field components are reflected in their relative changes, together constituting a three-dimensional curling motion depending on the relative values of the changes in the magnetic vector components. The calculation itself is performed efficaciously by simple Gaussian elimination in a system of simultaneous equations utilizing the convenient determinant as the operational scheme.

The definition of the curl given above as a limit in space illustrates how Maxwell "mathematicized" Faraday's intuitive field, that is, wire coils and magnets all can be taken to their infinitesimal limit at a point in space, and all the points in space constitute a mathematical field. So the substantive constituents of electromagnetism were abstractly reduced to their mathematical essence by Maxwell's equations.

Finally, an intuitive understanding of the divergence and curl can be given in the well-known and often-used vector relation

$$\nabla \circ (\nabla \times B) = 0.$$

This is simply stating that a curling point has no divergence, which when one thinks about it, makes eminent sense, since a curling field apparently can not have a radially emanating flux perpendicular to its curling motion.

Light Is an Electromagnetic Wave

Maxwell's equations can be said to have provided the mathematical formulation of Faraday's intuitive field and successfully described the observed electromagnetic phenomenon of the time; that alone is a great achievement. But before the Maxwell equations, light and electromagnetism were considered entirely separate things. Maxwell's mathematical formulation of electromagnetics showed that light and electromagnetism were actually one and the same, and thus the power of mathematical physics could be brought to bear in the study of light, in particular the polarization of light, which will come to be seen as the very essence of the operation of liquid crystal displays.

Starting with the differential form of the Maxwell equations above, it is possible to derive the interaction of electric and magnetic fields that comprise an "electromagnetic wave," and forthwith show that light is itself an electromagnetic wave. This demonstration begins by taking the curl of both sides of Maxwell's third equation (Faraday's law) as follows,

$$\nabla\times(\nabla\times E)=\nabla\times\left(-\frac{\partial B}{\partial t}\right).$$

Because electromagnetic phenomena in Nature generally proceed smoothly and continuously, the mathematical functions and their derivatives that describe them are also smooth and continuous, and so the partial derivative operator on the right-hand side may slide out to the front of the curl without too much offense,

$$\nabla\times\left(-\frac{\partial B}{\partial t}\right)=-\frac{\partial(\nabla\times B)}{\partial t}.$$

A well-known identity in vector analysis is

$$\nabla\times(\nabla\times E)=\nabla(\nabla\circ E)-\nabla^2 E,$$

then using the curl of Faraday's law given above and the partial derivative slide,

$$\nabla\times(\nabla\times E)=\nabla(\nabla\circ E)-\nabla^2 E=-\frac{\partial(\nabla\times B)}{\partial t}.$$

From the Ampere–Maxwell law,

$$\nabla\times B=\varepsilon\mu\frac{\partial E}{\partial t}+\mu J,$$

therefore

$$\nabla\times(\nabla\times E)=\nabla(\nabla\circ E)-\nabla^2 E=-\frac{\partial\left(\varepsilon\mu\frac{\partial E}{\partial t}+\mu J\right)}{\partial t}=-\varepsilon\mu\frac{\partial^2 E}{\partial t^2}-\mu\frac{\partial J}{\partial t}.$$

Now, using Gauss' law

$$\nabla\circ E=\frac{\rho}{\varepsilon},$$

then

$$\nabla\times(\nabla\times E)=\nabla\left(\frac{\rho}{\varepsilon}\right)-\nabla^2E=-\varepsilon\mu\frac{\partial^2E}{\partial t^2}-\mu\frac{\partial J}{\partial t},$$

and rearranging,

$$\nabla^2E-\varepsilon\mu\frac{\partial^2E}{\partial t^2}=\nabla\left(\frac{\rho}{\varepsilon}\right)+\mu\frac{\partial J}{\partial t}.$$

In the vacuum of free space, J is zero, and there is no current density, so $\rho = 0$ as well, so

$$\nabla^2E=\varepsilon\mu\frac{\partial^2E}{\partial t^2}.$$

In exactly the same way, an equation of the same form for the magnetic field vector may be derived,

$$\nabla^2B=\varepsilon\mu\frac{d^2B}{dt^2}.$$

The two equations above, beautifully symmetric, are also of exactly the same form as a mechanical wave in air (sound) or in water (sea wave), that is, the classical equation for a mechanical wave propagating through a medium with velocity v is

$$\nabla^2\Psi=\frac{1}{v^2}\frac{d^2\Psi}{dt^2}.$$

From this similarity, not only can the concept of an "electromagnetic wave" be established, but from the material coefficient factors in the expressions, it can also be seen that

$$\varepsilon\mu=\frac{1}{v^2},$$

thus the speed of the electromagnetic wave is

$$v=\frac{1}{\sqrt{\varepsilon\mu}}.$$

The German experimentalists Kohlrausch and Weber in 1856 [7], based on the Faraday law of electromagnetic induction equation previously described, in experiments utilizing charges in a condenser, were the first to measure the values of ε and μ as (where ε_o and μ_o are their values in vacuum),

$$\varepsilon = \varepsilon_o = 8.8541878 C^2 \sec^2 / \mathrm{kg} \times \mathrm{m}^2,$$

and

$$\mu = \mu_o = 4\pi \times 10^{-7} \mathrm{m} \times \mathrm{kg} / C^2,$$

so that

$$v = \frac{1}{\sqrt{\varepsilon_o \mu_o}} = \frac{1}{\sqrt{8.8541878 C^2 \sec^2 / \mathrm{kg} \times \mathrm{m}^2 \times 4\pi \times 10^{-7} \mathrm{m} \times \mathrm{kg} / C^2}}$$
$$= 299{,}792{,}458 \ \mathrm{m/s}.$$

That number, as many know, is exactly the speed of light in vacuum [8]. So the speed of the electromagnetic wave is exactly the same as the speed of light, that is $v = c$. This astounding result surprised even the trenchant Maxwell. It was an indisputable mathematical physics demonstration that an electromagnetic wave travels in space at the speed of light, and this led to the unavoidable conclusion that *light is an electromagnetic wave.*

As the electromagnetic field, personified as it is by the space and time second derivatives of the electric and magnetic vectors, can travel through space, so action at a distance was possible through the medium-less vacuum of space, and thus the conundrum of the ineluctable electromagnetic *æther* was at last resolved, Maxwell did indeed smite the *ævil* witch.

In 1888 Heinrich Hertz performed his famous experiment proving the existence of electromagnetic waves. An induction coil* acted as an oscillating dipole, and some distance away were placed small wire coils with tiny gaps. When the induction coil was turned on, sparks appeared in the gaps of the distant wire coils, showing the reception of electromagnetic signals from the oscillating dipole; and when the small wire coils were turned 90°, there were no sparks, demonstrating that the oscillating dipole generated signals with a transverse magnetic field that induced the voltage that generated the sparks, thereby proving that the signal was a *transverse* electromagnetic wave

* An induction coil is a step-up transformer with an open core and a vibrator-interrupter in series with the primary winding. A battery provides low-voltage dc to the primary winding and the vibrator-interrupter (vibrating armature that alternately makes and breaks contacts) breaks up the current into short pulses to generate through the secondary winding a high-voltage ac which produces an oscillating dipole signal.

(meaning that the electric and magnetic field vectors were mutually perpendicular, and both perpendicular to the direction of travel of the electromagnetic wave).

Hertz thereby established once and for all that Maxwell's promulgation in 1862 of his *Theory of Electromagnetism* was the new truth for a new age. And not insignificantly, the understanding that light is an electromagnetic wave, and an electromagnetic wave is light makes possible, among all the other wondrous new electronic and optical products of today, the existence of the liquid crystal display.

It is no overestimation to say that Maxwell's transformational discovery is a milestone in the intellectual history of mankind; indeed, Einstein himself said [9],

> Faraday and Maxwell's electromagnetic theory is the greatest alteration . . . in our conception of the structure of reality since the foundation of theoretical physics by Newton.

and

> The theoretical discovery of an electromagnetic wave spreading with the speed of light is one of the greatest achievements in the history of science.
>
> Albert Einstein

Today, radio, television, computers, mobile phones, and all the other now indispensable items of our electronic age depend on Maxwell's electromagnetic waves to transmit information and images. Many of the later theories in physics also take as their point of departure the Maxwell equations.

Faraday's intuition and Maxwell's mathematics together formed the new mathematical physics construct of a continuous *field theory*. Today Einstein's *general theory of relativity* describing the gravitational force pervading the Universe, and the relativistic quantum electrodynamics describing the interaction of electromagnetism with atomic matter are well-established but (so-far) separate field theories. At the sub-atomic scale, there are the strong and weak forces in the atomic nucleus, but so far only the electromagnetic and weak nuclear forces have been unified in the so-called *electroweak model* with the addition of nuclear strong force added in the current *standard model*, which are discrete (not continuous) *quantum field theories*. The unification of all these forces with Einstein's gravity is the objective of the *grand unified field theory*, the Holy Grail of fundamental physics, and if achieved indeed will be the Mother of All Knowledge. The ultimate theory of matter will be a quantum field theory, and the modern field concept can be said to have began with Faraday's electromagnetic fields.*

The Light Wave

The solution to the electromagnetic wave differential equation can be improbably found by simply guessing at answers and substituting them back into the differential equation to see if they work. For second-order differential equations such as the wave equation, a good guess at a solution would be sinusoidal functions, since when differentiated twice, they return to the original function; thus substituting a sine function back into the differential equation likely would give an equality, thereby satisfying the differential equation. And so it is that a sine or cosine function does indeed satisfy the electromagnetic wave equation, and from convention, a cosine function is usually chosen because it constitutes the real part of the complex notation of a wave motion (to be discussed below); that is, the solution to the wave equation can have the form of the x- and y-components of the light's electric field vector,

$$E_x = a_x \cos(\omega \cdot t + \delta_x)$$

$$E_y = a_y \cos(\omega \cdot t + \delta_y),$$

where ω is the angular velocity of the cosine function, also known as its angular frequency (ω depends on the frequency of an oscillating charge source, such as that used by Hertz to experimentally demonstrate the wave nature of light), and thus is also dependent on the nature of the material in which the wave is vibrating; that is, it is related to ε and μ; a_x and a_y, are the amplitudes of the cosine function, and the "phase difference" between the electric field vector components E_x and E_y is $\delta = \delta_y - \delta_x$. The four numbers a_x, a_y, δ_y, and δ_x are constants of integration derived from solving the differential equation. The idea of a phase difference will become very clear as the pursuit of the understanding of liquid crystal displays proceeds; for now, only a short introduction to its mathematical character will be given. It should be

* The electromagnetic, strong, and weak nuclear forces are united in the standard model in terms of a gauge theory, but the gravitational force is not included (cf. Chapter 7), although some physicists say that gravity does not have to be included because *relative* to the other forces, gravity is negligible (for example, the electromagnetic force between two protons is 10^{35} times stronger than the gravitational force).

noted at the outset that the phenomenon of polarization depends on the fact that the electric field of light can be characterized as a *vector*, that is, it can be resolved into the components E_x and E_y.

When $\delta = 0$ or π, or integer multiples of π, the light is "linearly polarized"; when $a_x = a_y$ and $\delta = \pm\pi/2$ (or integer multiples of $\pm\pi/2$), then the light is "circularly polarized," and the + sign means that it is left-hand circularly polarized (clockwise rotation in the x,y-plane), and the – sign means that it is right-hand circularly polarized (anticlockwise rotation).* If the $\omega \cdot t$ is cancelled in the E_x and E_y equations above, some simple algebra will produce the equation of a conic section,

$$\left(\frac{E_x}{a_x}\right)^2 + \left(\frac{E_y}{a_y}\right)^2 - 2\frac{\cos\delta}{a_x a_y}E_x E_y = \sin^2\delta.$$

If the principal axes of the conic section are rotated to align with the x,y-coordinate system, the conic section will take the form of an ellipse,

$$\left(\frac{E_{x'}}{a_{x'}}\right)^2 + \left(\frac{E_{y'}}{a_{y'}}\right)^2 = 1,$$

where the primes indicate the coordinate system where the xy-axes is taken along the principal axes of the ellipse, and the $a_{x'}$ and $a_{y'}$ are the semi-major and semi-minor elliptic axes. The *ellipticity* (e) of the ellipse is defined as

$$e = \pm\frac{a_{y'}}{a_{x'}},$$

where the positive values are for right-hand rotations and the negative for left-hand rotations.

Because there is no correlation between the amplitudes or phase of the light's electric vector components, this elliptical polarization is the most general state of the polarization of the light.

Another solution to the electromagnetic wave equation is the exponential function e^x, which indeed upon differentiation remains the same, thus making it the ideal candidate for solutions of differential equations. Further, if the argument of the exponential is a complex variable, it is more compact and ultimately more useful because of the mathematical attributes of complex

* This is the quantum electrodynamics convention for the naming of the left- and right-hand circular polarization; in some optics books, the definitions of the directions are just the opposite of those given here.

variables. Still further, the well-known Euler formula ties the complex exponential function to the sinusoidal functions as (where incidentally it can be seen that the cosine function is indeed the real part of the function),

$$e^{-i(\omega t+\delta)} = \cos(\omega \cdot t + \delta) - i\sin(\omega \cdot t + \delta),$$

then the electromagnetic wave can be written conveniently as (where A is the amplitude of the wave):

$$E = Ae^{-i(\omega t+\delta)}.$$

From the character of the sinusoidal or complex exponential functions, the purely mathematical derivation of the equations governing the journey of light passing through a material medium (such as a liquid crystal) may be performed, and the resulting phase difference arising naturally from the mathematics turns out to be the key to the operation of liquid crystal displays. Most optics textbooks introduce the phase difference purely in terms of the sinusoidal and complex exponential functions, but a physical understanding of phase difference requires a more physically intuitive approach, which will be presented in Chapter 4.

A summary of the principal equations of this chapter:

Gauss' Law (Electric Field)

$$\oint\!\!\oint E \circ \hat{n} dA = q/\varepsilon \rightarrow \text{Divergence Theroem} \rightarrow \nabla \circ E = \frac{\rho}{\varepsilon}$$

Gauss' Law (Magnetic Field)

$$\oint\!\!\oint B \circ \hat{n} dA = 0 \rightarrow \text{Divergence Theorem} \rightarrow \nabla \circ B = 0$$

Faraday's Law

$$\oint E \circ ds = -\frac{d}{dt}\iint (B \circ \hat{n}) dA \rightarrow \text{Stokes' Theorem} \rightarrow \nabla \times E = -\frac{\partial B}{\partial t}$$

Ampere–Maxwell Law

$$\oint B \circ ds = \mu(I + \varepsilon\frac{d}{dt}\iint (E \circ \hat{n}) dA \rightarrow \text{Stokes Theorem} \rightarrow \nabla \times B = \varepsilon\mu\frac{\partial E}{\partial t} + \mu J$$

Wave Equation

$$\nabla^2 E = \varepsilon\mu\frac{\partial^2 E}{\partial t^2}$$

$$\nabla^2 B = \varepsilon\mu \frac{d^2 B}{dt^2}$$

Wave

$$E_x = a_x \cos(\omega \cdot t + \delta_x)$$

$$E_y = a_y \cos(\omega \cdot t + \delta_y)$$

The eminent theoretician P.A.M. Dirac once wrote that*

Physical laws should have mathematical beauty.

Are the Maxwell equations beautiful? Because "beauty" is often subjective and hardly immutable, a certain answer is clearly difficult. However, focusing on the apparently endurable aspects of beauty in physics might be helpful, for example, the title of the biography of China's first Nobel Prize winner, C.N. Yang, is *The Beauty of Order and Symmetry* (translated). The thesis is that there is beautiful symmetry and order in the Universe yet to be discovered. So just as a symmetric visage and figure are pleasing to the eye, so an equation possessing and describing a symmetry in Nature may manifest enduring qualities of physical beauty. But perhaps most important, there is the inner beauty of human intelligence and depth; and so do the equations, aside from their symmetry, further give order to apparent chaos to reach far, wide, and deep to explain a sublime reality?†

* Dirac wrote this on a blackboard during a seminar at Moscow University in 1955 in response to a request to summarize his philosophy of physics, quoted in Farmelo, G. (ed.). (2002). *It Must be Beautiful, Great Equations of Modern Science*, Granta, p. xiii.

† Irrespective of their monumental contributions to science and human progress, especially Faraday's humility and Maxwell's civility are worthy of praise. Michael Faraday developed his field concepts while performing experiments in his basement laboratory at London's Royal Institution; often, and particularly at Christmastime, Faraday would talk about and give demonstrations of the scientific discoveries of the day to an audience that especially included many children. No doubt a considerable portion of his lectures was about his electromagnetism discoveries, but Faraday very early also revealed a facet of the conscientious scientist, being among the few well-known personages in those early industrializing times to evince concern about man's effect on the environment. Faraday himself in fact led a movement to clean up the terribly polluted Thames River that wended its dark way through London at the time, spewing noxious smells and spreading pollutants in its path. Maxwell not only formulated electromagnetic theory, he contributed in areas as diverse as the theories of heat and color, but as far-ranging as the rings of Saturn, of which he postulated that, based on Kepler's laws, they must be composed of many, many particles of ice rather than the solid disk commonly believed at the time. Paul Dirac is the mathematician/physicist who formulated quantum mechanics, and postulated the positron and ultimately antimatter, among many other achievements.

References

[1] Mahon, B. 2003. *The Man Who Changed Everything: The Life and Times of James Clerk Maxwell*. Wiley, Chicester.

[2] Faraday, M. 1832. *Experimental Researches in Electricity*. London; Maxwell, J.C. 1856; On Faraday's lines of force, Cambridge Philosophical Society, *Trans. Cambridge Phil. Soc.*, **10**.

[3] Feynman, R., Leighton, R.B., and Sands, M. 1962. *Lectures on Physics*, Vol. II-2-1, Addison-Wesley, Reading.

[4] Fleisch, D. 2008. *A Student's Guide to Maxwell's Equations*. Cambridge. Maxwell, J.C. 1891/1954; *A Treatise of Electricity and Magnetism*, Vols. 1 and 2, Clarendon and Dover, New York; Feynman, R., Leighton, R.B., and Sands, M. 1962. *Lectures on Physics*, Vol. II, 11-1, 3, Addison-Wesley, Reading; Born, M., and Wolf, E. 1999. *Principles of Optics*, 7th edition. Cambridge; and Slater, J.C., and Frank, N.H. 1969. *Electromagnetism*. Dover, New York.

[5] Dirac, P.A.M. 1948. The theory of magnetic monopoles, *Phys. Rev.*, **74**, 48, 817–830; and Pais, A., et al., *Paul Dirac: The Man and His Work*, Cambridge, 1988.

[6] Needham, J. 1980. *Science and Civilization in China*. The shorter, C.A. Ronan abridgement, Cambridge, p. 38.

[7] Kohlrausch, R., and Weber, W. 1856. *Pogg. Ann. Physic u. Chem.(2), 99*, 10.

[8] *Conference Generale des Poids et Mesures, XV, Paris*, 1975.

[9] Einstein, A., and Infeld, L. 1938. *The Evolution of Physics*. Touchstone Simon & Schuster, New York.

3

Light in Matter

In order to describe light waves in a material medium (i.e., not in a vacuum), some more vectors are required. Besides the electric and magnetic fields E and B and the current density J already introduced, the magnetic vector H and the electric displacement vector D are used to describe the electromagnetic wave in matter. In stable isotropic materials, the five vectors (E, B, D, H, J), together with the electric permittivity ε and the magnetic permeability μ already mentioned, plus the conductivity σ, are the material coefficients that manifest the effect of matter on the electromagnetic wave. In their simplest forms, the relations are

$$\mathbf{B} = \mu \cdot \mathbf{H}$$

$$J = \sigma \cdot \mathbf{E}$$

$$\mathbf{D} = \varepsilon \cdot \mathbf{E}.$$

If the coefficients σ, ε, and μ are taken just as scalars, these purely multiplicative relations are at best an approximation of the complicated interaction of matter with electromagnetic waves, assuming as they do a directional isotropy of the material. An electromagnetic wave passing through a gaseous,

Liquid Crystal Displays, First Edition. Robert H. Chen.

solid, or liquid material of course will be affected differently, and further different gases, solids, and liquids will also have different effects on the wave depending on their gross structure and their molecular structure, and in particular whether they are directionally the same (isotropic) or not (anisotropic) insofar as the electromagnetic wave is concerned. The scalar material coefficients provide a means to differentiate the effects in only an approximate way; below, the anisotropy of materials and their effect on light will be addressed by investigating the charge polarization of the material, and later in Chapter 13, the dielectric anisotropy will be addressed in more general terms using tensors.

An electric field E can generate an electric current density J, the character of which will be determined by the conductivity σ of the material medium; that is, a very small value of conductivity means the material is likely an insulator, and larger values for the conductivity mean that the material is a conductor. The value of the electric field vector E changes in a material depending on the latter's character, and the changes are described by the historically named *dielectric constant*, but better termed the *dielectric coefficient* or *electric permittivity*. This dielectric constant in the course of the study of liquid crystal displays will become very familiar. The difference between the electric field inside and outside of a material is described by D, the electric displacement vector, which was introduced by Maxwell to adjust Ampere's law to comprise a certain electromagnetic elasticity, as described in Chapter 2. The dielectric constant in an anisotropic medium may have different values for different directions in the material (the meaning of "anisotropy"), and in fact, it is just this dielectric anisotropy that makes the operation of liquid crystal displays possible, as will be described in detail in chapters to follow.

The magnetic induction vector B describes how the magnetic vector H behaves within a material as characterized by the magnetic permeability of that material, which name is an indicator of just how "permeable" to magnetic force the material may be.

As Maxwell has shown that light is an electromagnetic wave, it is now possible to use electromagnetic theory to divine the character of light. For instance, as derived above, the speed of the electromagnetic wave is related to the speed of light in a medium divided by the square root of the dielectric constant and the magnetic permeability,

$$v = \frac{c}{\sqrt{\varepsilon\mu}}.$$

This first of all says that the speed of the electromagnetic wave in a medium will be less than the speed of light in vacuum provided that ε and

μ are both greater than unity (which so far is true for all known matter). From the point of view of light in the medium, the dielectric constant's counterpart is the *index of refraction*, and traditional measurements of the speed of light through a medium revealed that the medium may be characterized by its effect on the speed of light simply as follows,

$$n = \frac{c}{v},$$

then from the relation immediately preceding, the relation between the macroscopic index of refraction characterization of a material's interaction with light and the microscopic atomic-level dielectric constant and the magnetic permeability is just

$$n = \sqrt{\varepsilon\mu},$$

Experimental evidence shows that $\varepsilon > 1$ and $\mu \sim 1$ for the almost non-magnetic liquid crystals used in displays, so the relation between the index of refraction, a macroscopic measure of the effect of a material on light, and the dielectric constant, describing the atomic-level interaction of matter with an electromagnetic wave, for liquid crystals is simply

$$n = \sqrt{\varepsilon}.$$

In deriving the equations commonly used in liquid crystal research, the Gaussian centimeter–gram–second (*cgs*) system of units historically has been used, and the Maxwell equations in Gaussian units are slightly different from the equations derived in the modern meter–kilogram–second (*mks*) or *Systeme Internationale* (*SI*) units. The different units systems unavoidably result in some confusion and unnecessary consternation, but in their physical essence, of course the Maxwell equations are the same in either system. The Gaussian unit system form of the Maxwell equations are

$$\nabla \circ D = 4\pi\rho$$

$$\nabla \circ B = 0$$

$$\nabla \times E = -\frac{1}{c}\frac{dB}{dt}$$

$$\nabla \times H = \frac{1}{c}\frac{dD}{dt} + \frac{4\pi}{c}J.$$

The 4π factor is a consequence of the Gaussian unit system particularly utilizing charged spheres as a basis for development of the electromagnetic theory.

The Electric Dipole Moment

The simple expression for the relation between index of refraction (the material's effect on light) and the dielectric constant (the property of the material) given above are sufficient for describing simple systems, that is, non-magnetic, dielectrically isotropic materials, but liquid crystals in particular are anisotropic, the anisotropy in effect being the cause of its birefringence. A more complete description of the dipole moment would require a method of quantifying the effects in different directions in the material; this can be done by the *tensor analysis* which will be introduced in Chapter 11 for treating elastic deformations of the liquid crystal, but for the dipole moment (intrinsic and induced), the average dipole moment will be used to finesse the problem, as to be described below.

To treat liquid crystals interaction with light at the atomic level requires delving into quantum electrodynamics, and since light is involved, the theory should also be relativistic, but the relativistic quantum electrodynamics theoretical approach for a complex macroscopic system of photon interactions with the multi-atom macromolecules constituting a liquid crystal will, as might be expected, quickly devolve into systems of equations of overwhelming complexity.

Pulling back from the relativistic quantum mechanical abyss, an orderly retreat from the interaction of light at the atomic structure level to that of a gross molecular structure level can be achieved through the utilization of the concept of an electric dipole moment. That is, instead of the individual atoms, the system may be modeled to good approximation (meaning that it works for meaningful situations) by treating the interaction of light and the electric field produced by the charge separation intrinsic to and induced by external fields in polarized molecules, the latter being represented by a dipole.

As early as 1876, Kerr found that certain liquids, when placed in an electric field, would behave like a double refracting crystal. The molecules in those liquids were long and dipolar, but in the absence of an applied electric field, the orientations of the molecules in the liquid are randomly distributed, and there was no net charge polarization and thus no net polar structure to produce the birefringence of light passing through. However, when an electric field was applied, the now net-polarized molecules would tend to align with the field direction, and the liquid would exhibit birefringence. This phenomenon became known as the Kerr effect, and the Kerr cell could be said to be the forerunner of the liquid crystal display.

Liquid crystal molecules are composed of atoms wherein congregated electrical charges are structurally separated, what is usually called *permanent* polarization, but better termed *intrinsic* polarization since it is a characteristic of the particular molecular structure. Because of the charge separation, the molecule in electrical terms is a *dipole* (literally meaning "having two poles, one negative and one positive"). Molecules that are not intrinsically polarized may be charge-separated by the Coulomb force exerted by an external electric field, to form a structure that is called an *induced* molecular charge polarization; molecules that are already intrinsically polarized may have their charge separation enhanced by the external electric field. In both cases, the polarized molecules may be turned about a point by the force from the electric field; for any force, this is called the *moment*, for an electric dipole, this is called the *dipole moment*. That is, since opposite charges attract, and the charges in the liquid crystal molecule are separated, the positive side of the applied electric field will attract the negatively-charged end of the molecule, and the negative side of the electric field the positively-charged end of the molecule. This results in a turning force acting through the lever between charges manifesting the liquid crystal's dipole moment.

As in the Kerr effect, when there is no external electric field, the orientation of the albeit intrinsically polarized liquid crystal molecules is random, and there is no net polarization of the liquid crystal as a whole; that is, the liquid crystal is *non-polar*. When an electric field is applied, however, the intrinsically polarized liquid crystal molecules will turn to align with the direction of an applied electric field because of their dipole moment. The alignment will render the previously non-polar liquid crystal, *polar*.

When the polarized molecules re-orient in the electric field, the interaction of the molecules with the electric field vector of the incident light also changes, and there will be produced a phase difference of that light, and so the polarization state of that light will be altered. That change in polarization state in turn changes the intensity of that light when passed through a polarizer (just as in the case of the two rotated polarizing sun glass lenses). So the electric dipole moment of the liquid crystal molecule is the conceptual tool that allows an external electric field to control the intensity of light passing through to convert a conglomerate of liquid crystal molecules into a display. This is the essence of the operation of a liquid crystal display.

As an aside, historically, the word "polarization" unfortunately has two totally different meanings in relation to matter on the one hand and electromagnetic radiation (and light) on the other, although they are mutually influential. In materials, "polarization" means the separation of charge in the structure of a molecule, the molecule being "polarized." For light, "polarization" means the state produced by the phase differences between the electric field vector components. To differentiate, the term "charge polarization" will indicate the charge separation in molecules, and the single word "polarization" will be reserved for light whenever there is danger of confusion. As the story of liquid crystals unfolds, understanding of the two phenomena will increase, and the polarization in question will be clear in context, rendering particularization unnecessary.

As described above, a liquid crystal's molecules have a dipolar intrinsic charge polarization structure that can be described by a *dipole,* which can be described mathematically as just the separation of positive and negative charge (q) times the distance and direction between the charges expressed by a vector lever arm ($\boldsymbol{l}$); this is the *intrinsic electric dipole moment* $\boldsymbol{p} = q\boldsymbol{l}$, as shown schematically in Figures 3.1 and 3.2 for between the charges and farther away from the charges respectively.

When no external electric field is applied, the intrinsic electric dipole moments in the liquid crystal are randomly oriented with no particular orientational order, and thus cancel each other out to have no net gross effect (the liquid crystal is *non-polar*). But applying an electric field will cause the electric dipole moments to tend to align with the field to produce a gross *electric dipole field,* that from a large distance ("large" meaning compared with

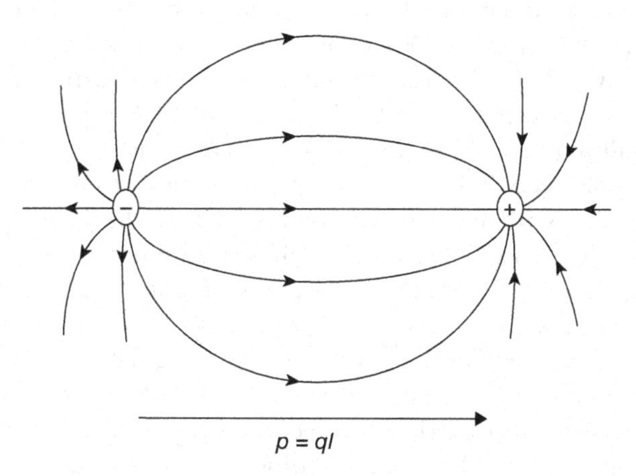

Figure 3.1 Dipole field between the charges.

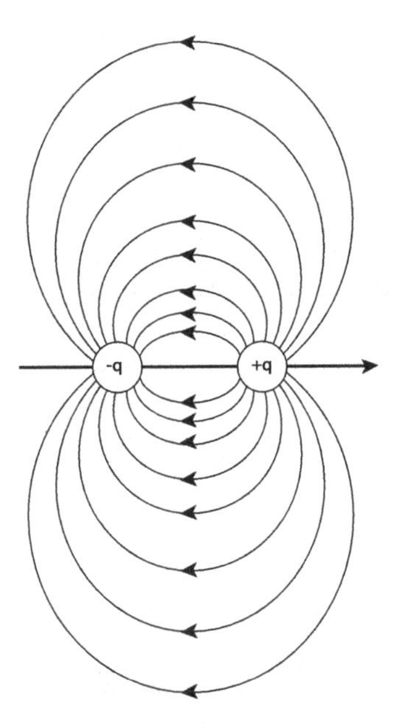

Figure 3.2 Dipole field away from the charges.

the length of the dipole's lever arm) manifests the response of the liquid crystal to the external electric field. In truth, since the human eye cannot perceive the microscopic individual dipoles, the gross average effect is the phenomenon that is observed anyway. So the molecules' dipole moments produce an average dipole moment field that manifests the liquid crystal's structural anisotropy. That *dielectric anisotropy* characterizes the response of the specific liquid crystal in question to light.

Relating the liquid crystal's susceptibility to charge polarization with its dielectric anisotropy and the index of refraction to describe the interaction with incident light thus would be extremely useful for designing liquid crystal displays. Mean electric dipole field theories for liquid crystals will be described in later chapters; here the useful equation that links Maxwell's electromagnetic theory to the molecular structure of matter will be derived.

The Lorentz-Lorenz Equation

Complicated molecular structures such as liquid crystals can be represented electrically by an average electric dipole moment field that can be described by the charge polarization (p) produced by the separation of the positive and negative charges in the molecular structure. If there are N molecules in the molecular array, then a unit volume average dipole moment is given by

$$P = Np,$$

where P is an average charge polarization vector. If the charge polarization is more or less uniform within the material, there will be only so-called "bound charges" (bound by the charge polarization) on the surface of the material. since the other charges will be averaged out within the material. But if the charge polarization is not uniform, then the bound charges may aggregate within the material at certain places, and this non-uniform distribution of polarization charge can be described by the divergence of the polarization vector, as

$$\rho_{\text{bound}} = -\nabla \circ P,$$

where ρ_{bound} is the density of that bound charge that migrates because of the applied electric field, but is not otherwise free to move within the material (the minus sign denotes the aggregation as opposed to divergence). Because of the existence of bound charges in intrinsically polarized material, the Maxwell equation introduced in Chapter 2 (Gauss' law) for that material is now

$$\nabla \circ E = \frac{1}{\varepsilon}(\rho_{\text{free}} + \rho_{\text{bound}}),$$

where the ρ_{free} is the density of the so-called "free electrons," which can move throughout the material, and ρ_{bound} is the density of the bound electrons just described.

If the electric divergence equation is substituted into the above ρ_{bound} equation then

$$\nabla \circ \varepsilon E = \rho_{\text{free}} - \nabla \circ P.$$

Collecting the divergence operators to the right-hand side (which can be done because the sum of divergences equals the divergence of the sums) gives

$$\nabla \circ (\varepsilon E + P) = \rho_{\text{free}}.$$

The divergence operand $(\varepsilon E + P)$ is called the *electric displacement vector;* and this is the genesis of the displacement current described previously as introduced by Maxwell in his equations, so

$$D = \varepsilon E + P.$$

With the electric displacement vector D, the Maxwell equations can be written in the form commonly used in textbooks as

$$\nabla \circ D = \rho_{\text{free}}$$

and Gauss' law (formerly for E) now can be written without the ε as

$$\oiint (D \circ \hat{n})\, dA = q_{\text{free}},$$

and the Maxwell equations can take into account light traveling through a charge-polarized material and the effect of the light electric field vector on that material and vice-versa with a single vector D.

From the above equations, the meaning of the electric displacement becomes clear: In isotropic materials, the simple multiplicative dielectric constant ε is sufficient to describe the influence of the material structure on light passing through it, that is, the direction of the light's electric field vector will not change in isotropic materials, but in charge-polarized anisotropic materials, such as liquid crystals, the simple multiplicative relationship must give way to a vector-additive relationship that considers direction, as well as magnitude. The charge-polarized molecules of liquid crystals constitute just that anisotropic structure, which polarizes incident light and renders liquid crystal materials birefringent. The electric displacement vector not only generalizes the Maxwell equations to describe anisotropic material bodies, it also simplifies calculations by eliminating the need to include the (very difficult to assess) distribution of bound charge, the free charge distribution now sufficing.

For purposes of deriving a useful dielectric constant and charge polarization relation, it is necessary to introduce a *dielectric susceptibility coefficient* η, which provides the direct relationship between the polarization vector and the electric field,

$$P = \eta \cdot E.$$

Now the Lorentz–Lorenz equation can be derived, and while not straightforward, armed with the physical concepts and mathematical wherewithal introduced above, the derivation can be instructive as to how simple models of charge distribution are useful in theoretical physics, so odd as the constructs may sometimes appear, they are common in the mathematical derivations of physics [1].

An electric field acting on molecules can be expeditiously described by an *effective field* E', wherein the difference with the average electric field E is due to the space between the molecules (i.e., the average electric field does not consider space between the molecules), and therefore the difference should be related to the number of molecules. Assuming that a given molecule is surrounded by a sphere, outside the sphere one can then neglect any effect due to the structure of the molecule and construct a continuous charge sphere as the unit to model the system. Then also assuming that the charge polarization vector P outside the sphere is a constant, inside the sphere the molecules' structure will have no effect and only affect the electric field outside the sphere. The question then becomes how does the polarization P change from the zero value inside the sphere to its substantive value outside the sphere and so affect the outside electric field of interest?

Now it is common in mathematical physics to simplify problems by using a single *scalar potential* instead of a three-component vector to describe a three-dimensional situation. In electronics, the *potential* is defined as the work required to move an electric charge some specified distance, and is measured in the familiar *volts*. The relation between potential and the electric field is given by Maxwell's third equation (Gauss' law),

$$\nabla \circ E = \frac{\rho}{\varepsilon}$$

where the force of the electric field in terms of potential is give by the gradient

$$E = -\nabla \varphi.$$

Combining the two equations gives Poisson's equation,

$$\nabla^2 \varphi = -\frac{\rho}{\varepsilon},$$

which is recognizable as the Laplacian of the potential function φ, and here is just the charge density divided by the dielectric constant. However, inside the sphere, there is no actual charge and thus no electric field, and outside the sphere, there is only the effective electric field E' to provide the potential. That is, the uniform charge distribution on the surface of the sphere producing the potential φ is the "idealized potential" (i.e., that potential that can reconcile the difference between the effective field and the average field).

The general definition of potential is the amount of work required by an electric field to push a point charge q that is subject to the force F of an oppositely charged point charge q', in accord with Coulomb's law

$$F = \frac{E}{q} \propto \frac{qq'}{r^2},$$

where the work is calculated by the line integral from point 1 to point 2 over the distance r_1 and r_2, so

$$Work = -\int_1^2 E \circ dS = -\int_{r_1}^{r_2} \frac{q}{r^2} dS = -q\left(\frac{1}{r_2} - \frac{1}{r_1}\right),$$

and it can be seen that electric potential decreases inversely with distance $(1/r)$.

Because

$$P = \eta \cdot E = \eta \cdot (-\nabla \varphi),$$

and within the sphere, the electric potential is obtained by integrating over the volume of the sphere,

$$\varphi = -P \circ \iiint \nabla(1/r) dV = -P \circ \nabla \iiint (1/r) dV.$$

Now, setting

$$\varphi_o = -\iiint (1/r) dV,$$

then

$$\varphi = P \circ \nabla \varphi_o,$$

where the function φ_o is the negative electric potential owing to a uniform charge density over the volume of the sphere; therefore, it satisfies Poisson's equation, and

$$\nabla^2 \varphi_o = 4\pi.$$

Because

$$-\frac{\partial \varphi}{\partial x} = \frac{\partial}{\partial x}\left(P_x \frac{\partial \varphi_o}{\partial x} + P_y \frac{\partial \varphi_o}{\partial y} + P_z \frac{\partial \varphi_o}{\partial z}\right) = P_x \frac{\partial^2 \varphi_o}{\partial x^2} + P_y \frac{\partial^2 \varphi_o}{\partial x \partial y} + P_z \frac{\partial^2 \varphi_o}{\partial x \partial z},$$

and $(\partial\varphi/\partial y)$ and $(\partial\varphi/\partial z)$ can be similarly expressed, due to the symmetry of a sphere relative to its center, that is,

$$\frac{\partial^2\varphi_o}{\partial x\partial y}=\frac{\partial^2\varphi_o}{\partial y\partial z}=\frac{\partial^2\varphi_o}{\partial z\partial x},$$

and

$$\frac{\partial^2\varphi_o}{\partial x^2}=\frac{\partial^2\varphi_o}{\partial y^2}=\frac{\partial^2\varphi_o}{\partial z^2},$$

then because of the above Poisson equation relation $\nabla^2\varphi_o=4\pi$, every term in the equation above is 1/3 of the whole, so

$$-\nabla\varphi=\frac{4\pi}{3}P.$$

Accordingly, the effect of the effective electric field E' inside the sphere on the molecules in terms of the electric field and charge polarization is

$$E'=E+\frac{4\pi}{3}P.$$

Every molecule's electric dipole moment p then has a directly proportional relation with the effective electric field as follows,

$$p=\alpha\cdot E',$$

where α is the "average polarizability" of the molecules. If there are N molecules, then the entire electric dipole moment is

$$P=Np=N\alpha E'.$$

Using the three equations above and $P=\eta\cdot E$, canceling E' and E, the dielectric susceptibility and the average polarizability relation is

$$\eta=\frac{N\alpha}{1-(4\pi/3)N\alpha}.$$

Again using Gauss' law on the spherical charge distribution in Gaussian units, $\nabla \circ D = 4\pi\rho$, and $\varepsilon = 1 + 4\pi\eta$ (the 4π factor comes from the Gaussian units), the relationship between the dielectric constant and the average polarizability is,

$$\varepsilon = \frac{1 + (8\pi/3)N\alpha}{1 - (4\pi/3)N\alpha}.$$

Or conversely, the relationship between the average polarizability and the dielectric constant is,

$$\alpha = \frac{3}{4\pi N} \cdot \frac{\varepsilon - 1}{\varepsilon + 2}.$$

The equations can also be written in terms of the index of refraction n in terms of the average polarizability,

$$\alpha = \frac{3}{4\pi N} \cdot \frac{n^2 - 1}{n^2 + 2}.$$

The above equations are known as the Lorentz–Lorenz equations; they form the connection between Maxwell's electromagnetic theory with the atomic level coefficients of a material medium, such as the average polarizability and dielectric constant, and the index of refraction of light. There will be occasion to use these relational equations in the development of the theory of liquid crystal displays.

Interestingly, the almost identically named H.A. Lorentz (Dutch) and L. Lorenz (Danish), published almost identical calculations in the same scientific journal at almost the same time. In respect of that parity, the equations came to be known by their peculiar double appellation. The first Lorentz incidentally received in 1902 the second Nobel Prize for Physics, and is more famous in the physics world than the second Lorenz [2].

References

[1] Born, M. and Wolf, E. 1999. *Principles of Optics*, 7th edition. Cambridge, pp. 89–93.

[2] Lorentz, H.A. 1880. *Wiedem. Ann.*, **9**; and Lorenz, L. 1880. *Wiedem. Ann.*, **11**.

4

The Polarization of an Electromagnetic Wave

In the previous chapters, the observation of double refraction of light and the *polarizing* effect of calcite were noted, and then light was shown to be an electromagnetic wave that when passed through a medium, was affected by the molecular structure of that medium. In the sinusoidal wave solutions to the electromagnetic wave equation, for anisotropic media, there was found a phase difference between the electric vector components of the light, and that caused a phase lag that *polarized* the light. The mathematical description of the polarization came from the Maxwell equations, but what does the *polarization of light* mean physically?

Unpolarized Light

Naturally occurring light is unpolarized, but according to solutions of the Maxwell equations given in Chapter 2, it was found that elliptical polarization is the state where there are no special correlations between the amplitudes and phases of the components of the electric vector; so what then is "unpolarized light"? The sum of the vibrations of the electric field components in natural light actually always form an ellipse (the most general form

Liquid Crystal Displays, First Edition. Robert H. Chen.

of polarization), but the polarization modes are constantly randomly changing through ellipses of different ellipticity (including lines and circles), and if the changes are more rapid than can be detected, then the light is considered unpolarized because all of the polarization effects are averaged out. In the words of Richard Feynman [1],

> light is unpolarized only if one is unable to determine whether it is polarized or not!

Then practically, a beam of light can be tested for polarization by passing it through a polarizer, and if turning the polarizer does not change the intensity of the light passing through, it is deemed *unpolarized* (like unpolarized sunglasses when rotated having no effect on light). There was no correlation between the orthogonal components of the electric vector of the light as it passed through the medium, so the direction and magnitude of the resultant electric vector varied rapidly and randomly in time, which is just the definition of "unpolarized."

As discussed in Chapter 2, light is an electromagnetic wave so light can be represented by a wave comprised of an electric vector and a magnetic vector. As depicted in Figure 4.1, the electric field vector of the **E** and the magnetic induction vector **B** (represented by the gray-shaded sine curve) in a sinusoidal electromagnetic wave are always mutually perpendicular ("orthogonal"), and at the same time, both vectors are perpendicular to the direction of propagation of the wave. The angle that the plane of vibration of the E and B vectors makes with the axis of the propagation direction is random in an unpolarized beam; that is, looking along that axis the E and B vectors will be radially emanating from that axis at many different angles (as shown schematically by the arrows in the figure below at right). By convention, the electric field vector **E** is chosen to represent the polarization state of an electromagnetic wave and thus of light; in fact, the magnetic B vector could just as well be used to represent the polarization state, but the point is that the mathematical description of polarization requires just one of them.*

Elliptical, Linear, and Circular Polarization

Polarization depends on the electric field of light being a vector; that is, the oscillating electric field vector **E** of the light must be resolvable into compo-

* The B vector is of course the "magnetic induction vector," but for simplicity, it will sometimes just be called the "magnetic field vector."

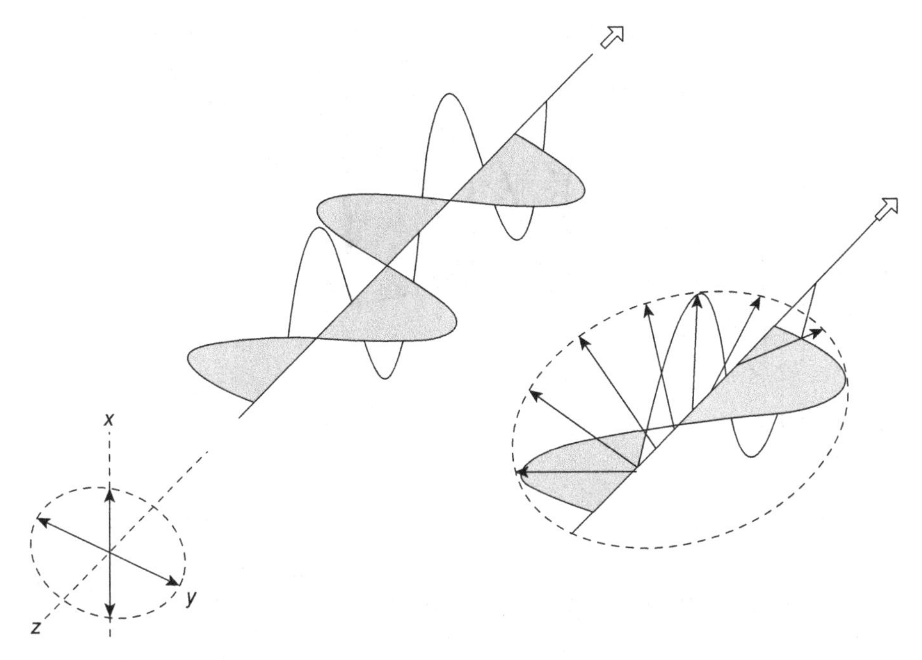

Figure 4.1 An electromagnetic wave has an electric vector and a magnetic vector.

nents in order to produce any polarizing effects. A *linear polarizer* transmits one component and blocks the other component; the component that is passed is said to be aligned with the polarizer's *transmission axis* (also called the *optical axis*). The earliest known polarizer material is tourmaline, which has an orderly parallel crystal structure, and passes the light's electric field vibrations perpendicular to its parallel structure direction and absorbs those vibrations parallel to the structure. Modern polarizers use a sheet of Polaroid™ which is a thin layer of small crystals of herapathite (iodine and quinine salt) all aligned with their crystal axes parallel. Polarizers commonly used in liquid crystal displays are stretched polymer films, like polyvinylacetate (PVA) doped with iodine. The light that passes through the linear polarizer is *linearly polarized*, and the plane defined by that single direction of oscillation moving along with the wave in the z-direction is the *plane of polarization* (so linearly polarized light is also called *plane-polarized* light). That is, all the E vibrations in planes that are in directions other than parallel to the axis of transmission of the polarizer are blocked, as shown schematically in Figure 4.2.

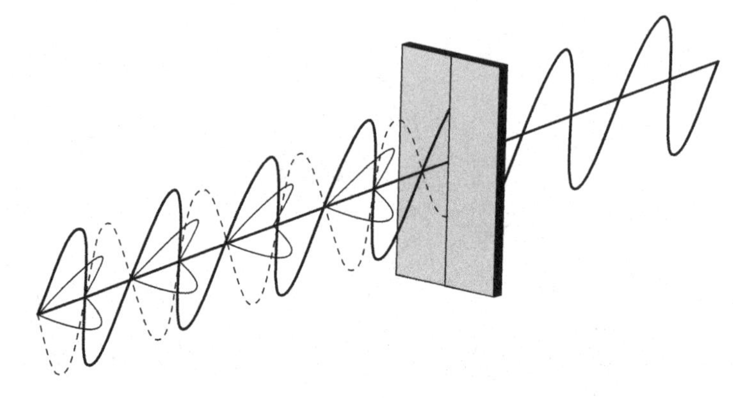

Figure 4.2 A linear polarizer polarizes light in one vector component direction.

The description and analysis of polarization by material bodies, including liquid crystals, all begin with a beam of linearly polarized light incident on the body in question; then the effects of the structure of the body on the linear polarization state of the incident light is analyzed in respect of the further polarization of the light by the material.

An anisotropic structure can polarize light. A mechanical analogy with tissue paper may be helpful; if one tears the tissue in a certain direction, the tissue will part easily and smoothly, but if torn in a direction perpendicular to that smooth direction, the tear will be jagged and difficult. Of course, the smooth tear is along the grain of the fibers of the tissue, and the jagged tear is against the grain. In the optical case, it is the electric vector of the electromagnetic wave that encounters different crystal structures in different directions in an anisotropic crystal, and the speed of the electromagnetic wave thereby changes, although the direction remains the same.

In order to describe the state of polarization quantitatively, a Cartesian coordinate system (x,y,z) can be superimposed on a polarizing crystal (such as calcite) and aligned with the crystal's structural order, then with respect to the light passing through, the index of refraction of the material can have different values, n_x, n_y, and n_z in different directions in the crystal. Placing a coordinate system on the crystal is for purposes of the delineation of direction in preparation for a mathematical description of polarization, so the coordinates can be lined up with the structure of the crystal in a fashion that will be most convenient for calculations. The *refractive index* describes the amenability of the crystal to the transmission of the electric

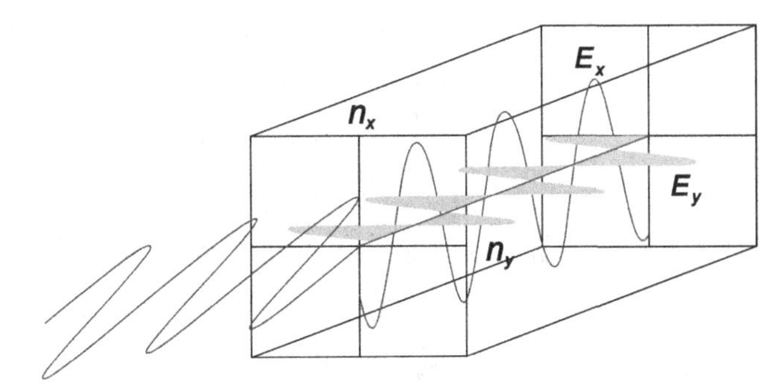

Figure 4.3 Electric vector of linearly polarized incident light is resolved into *x*- and *y*-components in the crystal.

wave vector of the incident light, and can be different for different directions in the crystal.

If the linearly polarized light beam is incident upon the crystal and the direction of its electric field vector is parallel to the x-component of the index of refraction n_x, then it will be influenced by the n_x index of refraction only, and since the electric vector cannot be decomposed into components, there is no change in the polarization state of the incident light (its speed may be reduced however, depending on the index of refraction of the crystal). The same can be said for a beam of linearly polarized light parallel to the y-axis where only the n_y index of refraction has any effect; the light will go through without any further polarization.

If the linearly polarized beam meets the crystal with its E vector vibrations at an angle as shown in Figure 4.3, then its electric field vector $\mathbf{E}$ can be seen as separated into components along the x-axis and the y-axis; that is, the E vector is resolved into its E_x and E_y (shaded sine curve) components. As the linearly polarized light beam traverses the crystal, the E_x and E_y components then may be dissimilarly influenced by the n_x and n_y indices of refraction.

The relative amplitudes of the E_x and E_y components will be determined by the angle of incidence of the linearly polarized light beam; for instance, if the plane of the linearly polarized light lies closer to the x-axis, the amplitude of the E_x component will be larger than the amplitude of the E_y component, and vice-versa for lying nearer the y-axis, thus the amplitudes of the electric vector components may be different as shown in Figure 4.3.

As the beam traverses the crystal, because of the different values of the indices of refraction n_x and n_y in an anisotropic crystal, the light

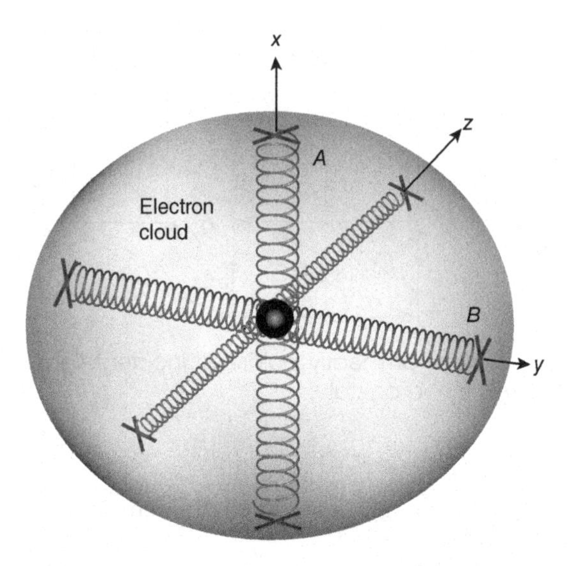

Figure 4.4 Springs of different spring constant illustrate the anisotropy of the response of electrons in a crystal.

beam's electric field components will be confronted with different crystal structures, and the electric field components may have different speeds through the crystal, and this will cause a *phase lag* between the electric field components.

But how is it that the E_x and E_y components' relative amplitudes and phase differences determine the state of polarization? Aside from the mathematical description given in Chapter 2, polarization can be illustrated by a mechanical construct of the mathematical situation. First, the spring model of the electron motion in a solid crystal used in atomic physics is depicted schematically in Figure 4.4, wherein springs of different spring constant in different directions represent the crystal structure's anisotropic effect on the motion of the electrons in the crystal in response to the electric forces from the electric vector of light.

The electrons in the crystal moving in response to the electric field vector of the light passing through will generate an electric field of their own, and alter the electric field of that light, and since the electrons are constrained by the crystal structure as represented by the springs, they transfer the crystal's structural character to the light by changing the speed of propagation through the crystal of the different electric components of the light. The mutually perpendicular motions of two electrons "attached to springs" affect the motion of the components of the light beam's electric field vector; the springs thus also can represent the motion of the electric field components of light as they pass altered through the crystal.

The simplest oscillatory motion of a body attached to a spring is *simple harmonic*; that is, the motion is linearly dependent on a restoring force in accordance with Hooke's law ($F = kx$, where k is the spring constant), and the motion can be described as sinusoidal waves in time.

Recalling that the solution to the Maxwell wave equation was a sinusoidal function, if the spring body A of Figure 4.4 is the E_x component of the electric field vector and B is the E_y component, then the E vector of the electromagnetic wave can be described by the simple harmonic motion model above. Polarization states thus can be represented by the electric field vector moving as springs described above, and the arrowhead of the resultant vector of the components E_x and E_y ($E^2 = E_x^2 + E_y^2$) will trace an ellipse, a straight line, or a circle depending on the components' respective amplitudes and phase difference, and that gives rise to the elliptical, linear (plane), and circular polarization states.

Elliptic Polarization

As shown in Figure 4.5, if the angle of incidence of the plane of polarization of the linearly polarized light is some arbitrary value, the amplitudes of the electric vector components generally will not be equal (but can be), and if the E_x wave has traveled through the crystal faster than the E_y wave (because of the anisotropy of the crystal), and at exit the E_y wave lags the E_x wave by some phase angle θ, the incident plane-polarized light will be *elliptically polarized* by the crystal, meaning that as the waves progress along the z-axis, the resultant vector of the E_x and E_y components rotate and trace an ellipse. The elliptical polarization is the most general state of polarization.

To illustrate the relative motion of the electric components of the electric vector that result in a turning vector that prescribes an ellipse, a vector phase diagram is shown at the bottom half of Figure 4.5. In the diagram's first quadrant, taking the figures going from right to left in turn, as the E_x and E_y waves progress along the z-axis, it can be seen that as the E_x component increases, the E_y component decreases, but not in a fashion that can exactly make up the E_x increase, so the resultant of E_x and E_y is a vector E with a vector arrowhead that traces an ellipse, and the arrow turns in a counterclockwise direction. This is called *right-hand elliptical polarization*, where "right hand" means using the right-hand rule with fingers curling in the direction of arrow rotation.

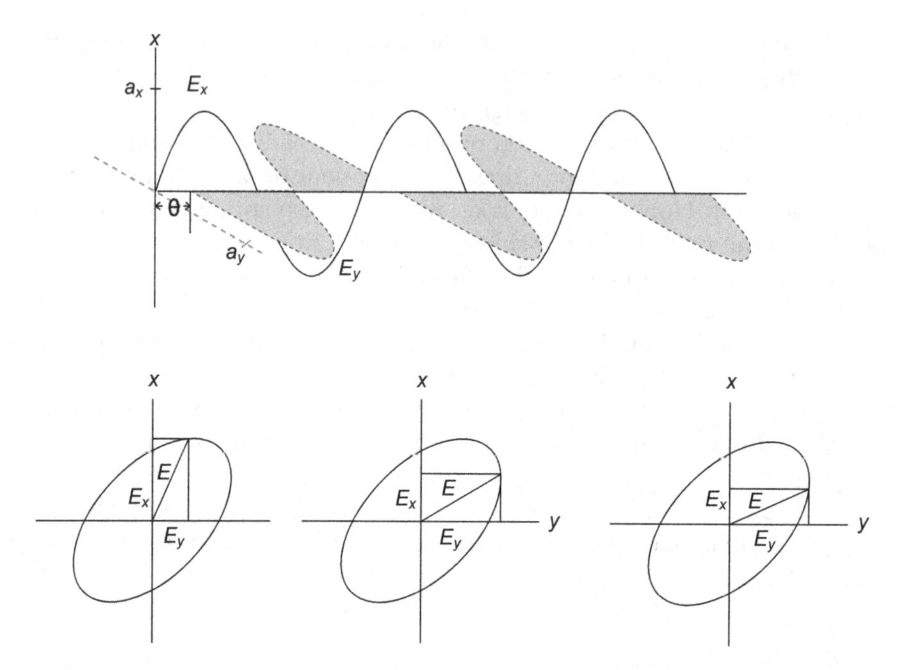

Figure 4.5 The resultant electric vector from in-phase electric vector components vibrates in a plane at an angle with the direction of propagation.

If in the first quadrant the component E_y is increasing and the E_x component is decreasing (taking the figures from left to right in turn), the E_x component decrease does not exactly offset the E_y increase, the resultant vector arrowhead prescribes an ellipse, and is turning in a clockwise direction; this is called *left-hand elliptical polarization.*

Linear Polarization

If plane-polarized incident light in traversing the crystal undergoes no phase lag (in-phase), or if the phase lag is exactly 180° (exactly out-of-phase) at exit from the crystal, the resultant electric vector will change magnitude (vibrate), but only in that one plane that is at the same angle as the angle of incidence as it travels through the crystal in the direction of propagation z. The resultant vector does not rotate in space as the wave progresses along the z-axis. The polarization state of the incident plane-polarized light has not been altered by the crystal if there is no phase difference. From Figure 4.6, it can be seen that if the E_x and E_y sine waves are in phase, that is, the spring con-

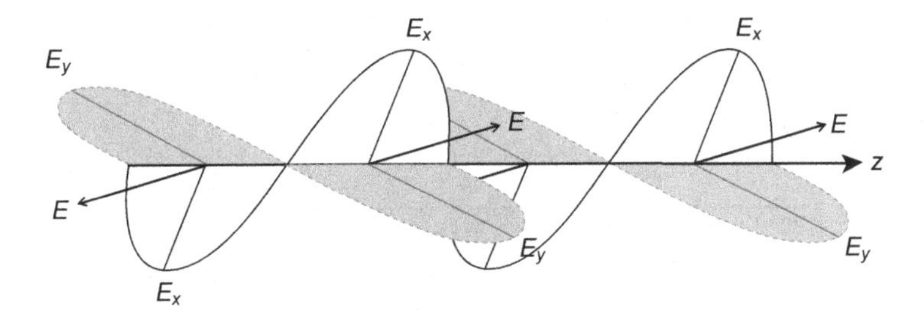

Figure 4.6 The resultant electric vector from electric vector components that are out-of-phase by an arbitrary angle θ will rotate and generally trace an ellipse.

stants in Figure 4.4 are the same, the resultant E vector will vibrate always in the same plane as the wave moves along in the z-direction, and it will not rotate.

Circular Polarization

As shown in Figure 4.7, if the linearly polarized incident beam of light meets the crystal's n_x, n_y structure exactly at an angle of 45°, then the amplitudes of the electric field vector components E_x and E_y will be the same, and as the components move through the crystal, if by virtue of the crystal structure's different index of refraction for different directions, one of the components is slower and thereby phase lags the other by exactly 90° at exit, then the light beam emerging from the crystal will be *circularly polarized*. Circular polarization then is a particular state of elliptical polarization wherein the amplitudes of the light's electric vector components must be equal, and the phase lag is 90° at exit.

As illustrated in the vector phase diagram at the bottom of Figure 4.7, when the phase difference between two sinusoidal motions having the same amplitudes is 90°, the vector sum of E_x and E_y is a vector E with a vector arrowhead that traces a circle. In the diagram's first quadrant, it can be seen that as the E_x component increases, the E_y component decreases just an amount to make up for the E_x component increase, and taking the figures from right to left in turn, the sum arrow will trace a circle and turn in a counter-clockwise direction—this is called *right-hand circular polarization*. If the component E_y is increasing in the first quadrant, and the E_x component is decreasing (taking the figures from left to right in turn), then the resultant

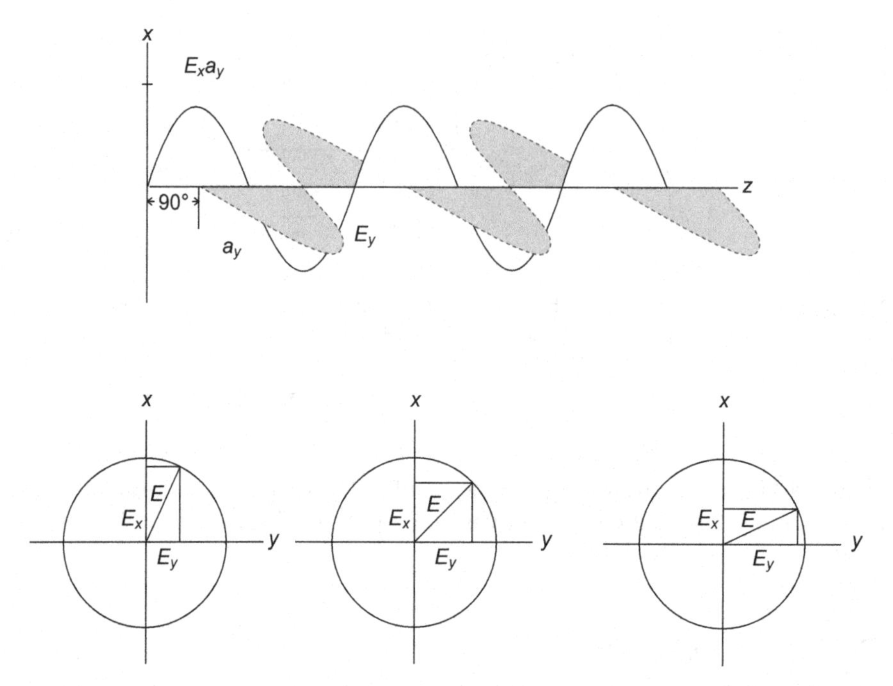

Figure 4.7 The resultant electric vector from electric vector components that are of equal amplitude and have a phase difference of exactly 90° will rotate and trace a circle.

arrowhead traces a circle that turns in a clockwise direction; this is called *left-hand circular polarization.*

That the resultant vector exactly prescribes a circle further can be shown by representing one wave E_x by a sine function; then if the waves are out-of-phase by exactly 90°, since $sin(\theta + 90°) = cos(\theta)$, the other wave E_y can be represented by a cosine function. As shown at the bottom half of Figure 4.7, if the waves are drawn in a plane as lines progressing forward with time, by the Pythagorean theorem, their vector sum is $E^2 = E_x^2 + E_y^2$, and since $sin^2 \theta + cos^2 \theta = 1$ is the equation of a circle, if E is "normalized" to equal one by setting the amplitudes of the E_x and E_y waves equal to one, then the vector sum is indeed a circle, both geometrically and mathematically.

In summary, as shown in Figure 4.8 for equal E_x and E_y component amplitudes, going from left to right as the phase difference δ (radians) increases, when there is negative phase difference (meaning right-hand polarization),

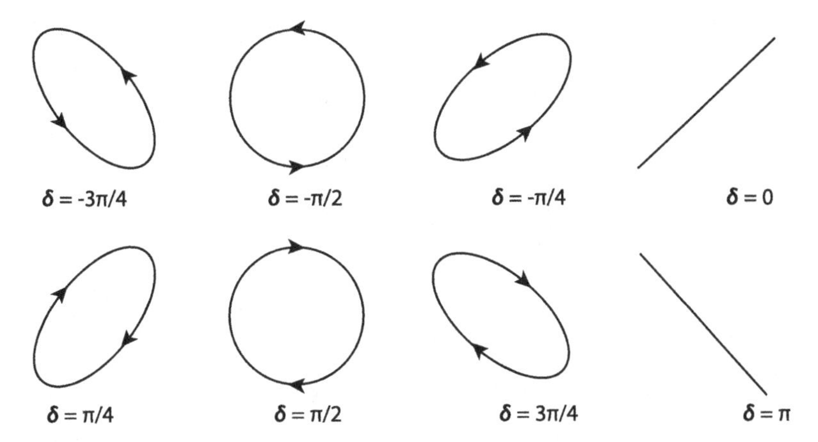

Figure 4.8 The polarization states for electric components of equal amplitude and phase difference given by δ in radians.

at $\delta = -3\pi/4$, the ellipse is tilted to the left; when $\delta = -\pi/2$ (90°), since the amplitudes are equal, the ellipse fattens up to form a circle; when $\delta = -\pi/4$, the ellipse tilts in the other direction, and when $\delta = 0$, the light is linearly polarized and the ellipse is flat (linear). Taking the second group of ellipses, as the phase difference is positive and increases from $\delta = \pi/4$ to $\delta = 3\pi/4$, the ellipses are left-hand circularly polarized, but otherwise the same as for the negative phase difference counterpart.

For unequal amplitudes ($a_x = 0.5\ a_y$) of the E_x and E_y components, as shown in Figure 4.9, the ellipses still change tilt and are right and left-handed polarized as in Figure 4.8 for the equal amplitude case, but they are slimmer, since the different amplitudes impart different ellipticities, and there is no circle at $\delta = -\pi/2$, since the E_x and E_y amplitudes are not equal.

The three principal polarization states of electromagnetic waves, linear, elliptical, and circular depend on the phase differences between the electric field vector components E_x and E_y, and the circular polarization further requires equal E_x and E_y amplitudes as well. As shown in Figure 4.10, if there is no phase difference, the resultant of the components will be no different, and there will be no change in the polarization state of an incident linearly polarized wave, and it will remain linearly polarized (Figure 4.10a). If, however, E_x and E_y have a random phase difference, the resultant vector from the components will result in an elliptical polarization, which is the general case (Figure 4.10b). If the E_x and E_y amplitudes are the same, and there is a phase difference of exactly 90°, then the resultant of the

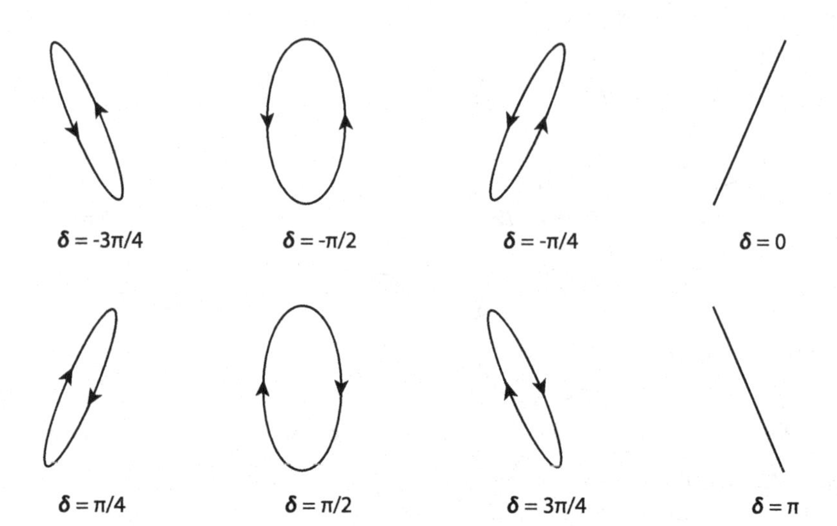

Figure 4.9 The polarization states for electric components of different (one-half) amplitude and phase difference given by δ in radians.

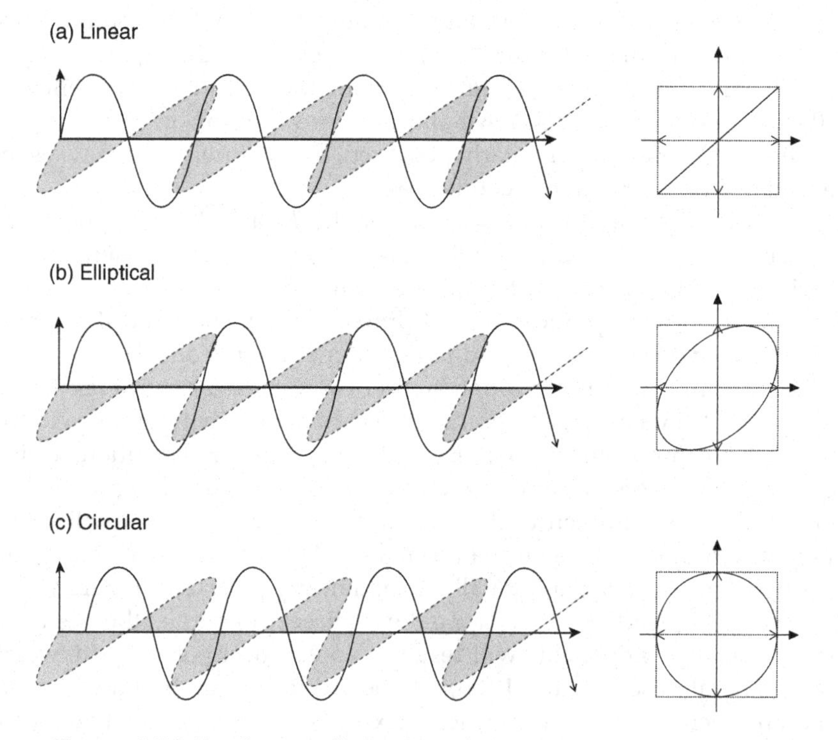

Figure 4.10 The linear, elliptical, and circular polarization states.

components will be a circular polarization state (Figure 4.10c). Again, the linear and circular polarization states are just special cases of the general elliptical polarization.

When a linearly polarized light beam meets a second linear polarizer (called the *analyzer**) that has its transmission axis exactly perpendicular to the transmission axis of the first linear polarizer (*crossed polarizers*), the linearly polarized light beam will be completely blocked by the crossed analyzer, ideally producing what is called in liquid crystal display technology the *black* (or *dark*) *state*; this state is critical to the image quality of displays, as will be seen in the chapters to follow.

To emphasize the concept of polarization of light by anisotropic crystal structures that is critical to the understanding of the operation of liquid crystal displays, if a linearly polarized beam of light is incident on the crystal at an angle 45° to the birefringent indices of refraction, then the electric field vector will be resolved into its E_x component along the x-axis and its E_y component along the y-axis, and the amplitudes of those components will be the same because of the exactly 45° angle between them. Further, if the different index of refraction encountered by the E_x and E_y components, one of the components will be slower, or lag, the other component, if that lag is exactly 90°, then the resultant vector from the components E_x and E_y will rotate as the wave progresses through the crystal, and trace a circle at exit from the crystal.

Birefringent materials having a thickness that produces exactly a 90°-phase retardation, because 90° is one-fourth of a full 360°, are called *quarter-wave plates*, which are commonly used in research in optics and in liquid crystal displays.

From the above analysis, it can be seen that different birefringent materials can be chosen according to their particular anisotropic features in order to produce desired polarization effects.

Birefringence

The above describes how a crystal polarizes light; the question now is what kinds of materials do the polarizing, and how can this polarizing be described physically. It turns out that except for the symmetric cubic crystals (such as common table salt), almost all other crystal structures are somewhat

* The second polarizer is called an "analyzer" because it was commonly used by experimentalists to "analyze" the light (i.e., examine its polarizations state) emerging from the liquid crystal under study.

different in structure in different directions, and thus are birefringent and can polarize light. How they polarize light will be analyzed using "extraordinary" means.

Bartholinus observed the double refraction of light passing through his piece of calcite because of the *dielectric anisotropy* of calcite that effectuates different values for the index of refraction in different directions. As described above, the electric vector of a linearly polarized beam of light incident on a birefringent crystal at an arbitrary angle will be resolved into its E_x and E_y components, and so the light will proceed through the crystal in two different modes because of the different dielectric effects on the light in different directions in the anisotropic crystal. In the terminology of the physics of crystals, one mode is through the so-called *fast axis* of the crystal, and the other through the *slow axis*. Because the orthogonal components of the light's electric vector vibrate in mutually perpendicular directions, they encounter different structures within the crystal that differentially affect their speed through the crystal. It is apparent then that what each light electric vector "sees" in the crystal will depend on the angle of the light electric vector incident on the birefringent crystal, and that is why Huygens saw the effect when turning the second piece of calcite in his hands.

From the naturally observed birefringence in calcite, in the following, an intuitive explanation of birefringence will be followed by an introduction to the physical and chemical theories of the interaction of light with matter that is at the heart of the operation of liquid crystal displays.

Almost everybody knows that the speed of light is *the* constant of Nature; the basis of Einstein's theory of relativity is that the speed of light is immutable, and that everything else, space, time, and mass, are all "relative." But of course, we have seen that the speed of light is constant only in vacuum, and it does change in passing through material media, and that the change is denoted by the index of refraction n which is equal to 1 in vacuum and for example, the speed of light in water is about 2/3 the speed in vacuum (i.e., $n = 1.33$ for water).

The scientific investigation of the interaction of light and liquid crystals is based on the physics of crystals and the theories of fluid dynamics. Crystals in the fluid state are the basic material of the liquid crystal display. The indicatrix of the effect on the speed of light by a material is the index of refraction n, and for complex material structures, the index of refraction may be different for different directions within the crystal; those indices of refraction were denoted as n_x, n_y, and n_z for the mutually orthogonal directions x, y, and z in the crystal.

A uniform, homogeneous structure has just one index of refraction, and therefore all of the indices of refraction will be equal ($n = n_x = n_y = n_z$), and the crystal material is *isotropic*, since there is no difference in n with respect to direction. For example, completely transparent glass is isotropic, so the components of the **E** vector of a beam of light, no matter what direction, will be the same, and indeed this is just the basic use of transparent glass, to transmit an undistorted image.

For materials that have different indices of refraction for different directions, for instance if any two are not the same, say $n_x \neq n_y$, the effect of that material on the electric vector components of light will be different for the x and y directions, and the material is called *anisotropic*. These materials are classified by crystallographers as *uniaxial* because there is only a single transmission axis through which light may pass and not be polarized. Those materials having a different index of refraction for each of the three base directions ($n_x \neq n_y \neq n_z$) are also anisotropic (actually super anisotropic), and are called *biaxial* because their structure allows for two transmission axes.

As noted in Chapter 3, the speed of light in a material having an index of refraction n is just

$$v = \frac{c}{n}.$$

From the above equation, it can be seen that if the index of refraction is not the same, then the speed of the light will not be the same. A larger index of refraction will cause the light to travel more slowly, and conversely a smaller index of refraction will travel faster through the anisotropic material. Taking the refractive index literally, if a material were found with refractive index less than 1, then the speed v could be greater than the speed of light in vacuum; or even more intriguing, if n were negative, then light would be "reversed" and the material would be rendered invisible.

Once again, the speed of light changes while passing through an anisotropic material, because that the light is an electromagnetic wave, and is itself composed of an oscillating electric field, represented by its electric field vector, and that vector will cause the electrons in the atoms of the material to vibrate. But because those vibrating electrons are constrained by the forces holding the crystal structure of the material together, they cannot freely move, but they do generate an electric field themselves. The superposition of the two electric fields will be different from the electric field of the light by itself, produced as it were by the effect of the constrained electrons in the material. So a new electric field is generated from the original light electric field, carrying within the effect of the molecular structure of the material. The new light wave thus will have a different speed in the material reflecting the molecular structure of that material through its refraction.

Ordinary and Extraordinary Waves

For a uniaxial crystal, if the indices of refraction are such that, for example, $n_y = n_z$, $n_x \neq n_y$, and $n_x \neq n_z$, then n_y and n_z are called the *ordinary index* (because they are the same), and can be designated by a single index n_o, and the unequal index of refraction n_x is called the *extraordinary index* (since it is different), and is designated by n_e. Taking the incident light propagating along the z-axis, the electric field vector component vibrating in the y-direction (E_y) is then called the *ordinary wave*, and the electric field component vibrating in the x-direction is called the *extraordinary wave*, the latter so designated because n_x is different from the two other indices of refraction, and so the extraordinary wave will travel at a different speed from that of the ordinary wave. It is important, however, to remember that the direction of the light beam is in the z-direction, and the electric field vector components always are vibrating in planes that are perpendicular to the direction of propagation of the light beam, but in planes that form different angles in the xy-plane.

The larger the index of refraction, the slower the speed of light through the material; for example, light passing through a uniaxial material with an ordinary index of refraction of $n_y = n_z = 1.3$ and an extraordinary index of refraction of $n_x = 1.6$, has its electric component E_y traveling the fast axis as the ordinary wave, and its electric component E_x traveling the slow axis as the extraordinary wave.

Since the speeds of the ordinary and extraordinary waves are different, the sinusoidal waves are not in step and there will be a *phase difference* between them. Owing to the larger index of refraction with which it must contend, one wave will be slower than the other, so this phase difference is also called a *phase retardation*. The different phase retardations (usually measured in radians) will produce different polarization states of the emerging light as described above. The appropriate generation of phase retardation in the incident light is the core of the engineering of polarizing materials, among which are the liquid crystals that form the heart of the liquid crystal display.*

The *birefringence* of a material can be defined by the difference between the extraordinary and the ordinary index of refraction,

$$\Delta n = n_e - n_o.$$

The anisotropic character of a uniaxial crystal is determined by the relative values of the ordinary and extraordinary indices of refraction; that is, if $\Delta n > 0$ meaning $n_e > n_o$, then the uniaxial crystal exhibits *positive birefringence* (also called *positive anisotropy*); its extraordinary wave is the *slow wave* (because n_e is larger), and its ordinary wave is the *fast wave* (because n_o is smaller). If conversely, $\Delta n < 0$ meaning $n_o > n_e$, then the material is *negatively birefringent* (also called *negatively anisotropic*); its extraordinary wave is the *fast wave* (because n_e is smaller), and its ordinary wave is the *slow wave* (because n_o is larger).

Materials that have three different indices of refraction for three different directions ($n_x \neq n_y \neq n_z$) are called *biaxial*, and although naturally occurring biaxial materials are difficult to find, man-made artificial biaxial compound materials are many.† A biaxial material has two optical axes (and thus it is called "biaxial"); light passing through an optic axis is by definition not phase retarded, but light incident at any other angles will undergo phase retardation. In liquid crystal displays, biaxial materials are primarily used as compensation films.

The asymmetric crystal structure of uniaxial and biaxial materials creates a dielectric anisotropy that can polarize light passing through. The structural

* The more general wave mechanics theory of light propagating through a uniform nonmagnetic uniaxially anisotropic material uses a dielectric tensor to describe the material, and directionally invariant (polarized) eigenmodes to describe the light propagation through the material. The eigenmode travels at a speed of c/n_i where n_i is index of refraction in the ith direction, which is just the eigenvalue corresponding to the ith eigenmode. A noneigenmode wave's electric vector can be resolved into two electric field components concordant with the eigenmodes; the corresponding light does not change direction, but propagates at a different speed, thus engendering phase retardation and causing subsequent changes in the polarization state of the light. From wave matrix calculations, it can be shown that the eigenmodes are orthogonal, the energy flux is given by the classical Poynting vector ($S = E \times H$), and two eigenmodes are just the ordinary and extraordinary waves. Yang, D.K, and Wu, S.T. 2006. *Fundamentals of Liquid Crystal Devices*, John Wiley & Sons, Ltd., p. 47ff.

† de Gennes says the smectic-C is biaxial; de Gennes, P.G. and Prost, J. 2007. *The Physics of Liquid Crystals*, 2nd edition, Oxford, p. 163ff.

order of the crystal, its translational, rotational, and mirror symmetries, and the maintenance or destruction thereof, thus can be gainfully described by the mathematical theory of symmetry groups and their permutations [2]. The deeply theoretical symmetry group methods of modern mathematical physics can be brought to bear on the study of liquid crystals, as well as solid crystals. Symmetry group theoretical physics is a major branch of solid state and particle physics theory, and is extremely well-developed, and has been used in liquid crystal research [3].

Quantum Mechanical Polarization

The retardation of light by a material can be conceptually understood as simple vibrating electrons interacting with the electric vector of light, but rigorous theories require relativistic quantum electrodynamics (QED) to explain the interaction of light photons with material electrons in terms of probability waves. But even in this esoteric theory, the electrons in a crystal can be still be described as little balls connected by springs (the attractive forces of the crystal structure) executing simple harmonic motions, and the oscillating electron electric dipole field, utilizing the field superposition principle can then be described by the Maxwell equations, altered by the requirements of quantum dynamics [4]. The quantum mechanical theory of photon–electron interactions requires a departure from the purely wave concept of light and a purely particle concept of electrons toward a wave-particle duality. For the light, discrete photons "particlerizes" the light wave, and for the electrons, the simple harmonic oscillator turns the particle-like electrons into electron waves (or at least electron wave generators). But the wave–particle duality, while successfully addressing the conundrum, will come at a steep price: the devolution from strict determinism in physics to the physics of probability.

To start at the beginning, for light, the simplest geometrical optics principle surely is Snell's law of reflection, which can be demonstrated by a minima calculation. In 300 BC, the eminent first geometer Euclid noted that the angle of reflection of light from a surface was equal to the angle of incidence. But it was not until 400 years later that Heron of Alexandria, in his chronicle *Mirrors*, established a critical natural principle, that the incident light somehow mysteriously chooses the shortest distance for its journey, and it is this fact that makes the angle of reflection to equal the angle of incidence.

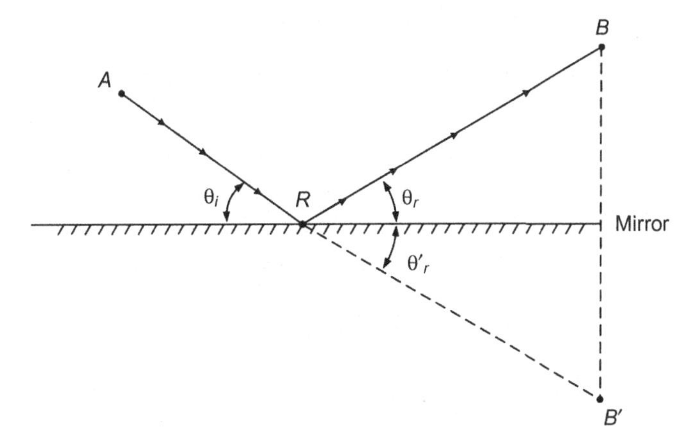

Figure 4.11 Snell's law of reflection from Heron's principle of least distance.

That light always travels by way of the shortest distance can be used to quite convincingly demonstrate Snell's law of reflection; that is, that the angle of reflection of a beam of light incident on a plane mirror will have be the same as the angle of incidence. From Figure 4.11, it can be seen that if the incident beam passes through the mirror, according to Euclid's geometry, the straight line *ARB′* is the shortest distance between two points.* So *RB′* is the shortest distance, and if the angle θ_r of the reflected beam (*RB*) is not equal to θ_r', *RB* will not be the shortest distance, so θ_r must equal θ_r', and from simple geometry $\theta_r = \theta_r' = \theta_i$, the angle of reflection equals the angle of incidence; Snell's law of reflection is so "proved."

From reflection, it is natural to next investigate refraction. Looking down when standing in water, it appears that one's feet have changed position, and a passing wave distorts their shape. This is caused by the refraction of light by the water. As noted previously, light passing through different media have different velocities, $v_1 \neq v_2$, and with θ_i, the angle of incidence from medium 1 to medium 2, and θ_r the angle of refraction, Snell's law of refraction is

$$\frac{\sin\theta_i}{\sin\theta_r} = \frac{v_1}{v_2}.$$

* Actually, even this simple statement requires the variational calculus to prove. Also, in Einstein's non-Euclidean general theory of relativity, where space–time is curved, it is not necessarily so that the shortest distance between two points is a straight line.

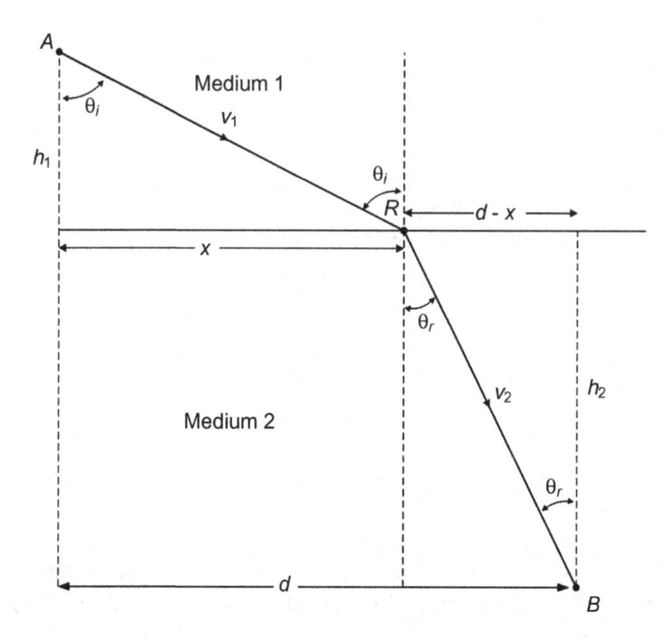

Figure 4.12 Snell's law of refraction from Fermat's principle of least time.

The incident light beam passing through a different medium will change direction, and in contrast to Snell's law of reflection, understanding the above equation requires more than just geometric intuition, it also requires the minimization calculus and Fermat's principle of least time (and not Heron's principle of least distance). From Figure 4.12, given the velocities v_1 and v_2, and simple geometrical considerations, the time required to travel from A to B is

$$T = \frac{\sqrt{h_1^2 + x^2}}{v_1} + \frac{\sqrt{h_2^2 + (d-x)^2}}{v_2}.$$

Further from Figure 4.12 and some simple trigonometry, it is clear that

$$\sin\theta_i = \frac{x}{\sqrt{h_1^2 + x^2}}$$

$$\sin\theta_r = \frac{d-x}{\sqrt{h_2^2 + (d-x)^2}}.$$

Because at the time (1658), the calculus had not yet been invented (discovered?), the French mathematician Fermat employed some algebraic skill and a type of limiting process close to calculus to successfully derive Snell's law of refraction [5]. It was not until 1682 that the great German mathematician Leibniz used his just-invented calculus to calculate the minimum by setting the derivative of the time (T) with distance traveled to zero, and he then easily derived Snell's law of refraction, as follows,

$$\frac{dT}{dx} = \frac{1}{2} \cdot \frac{1}{v_1} \cdot \frac{2x}{\sqrt{h_1^2 + x^2}} + \frac{1}{2} \cdot \frac{1}{v_2} \cdot \frac{-2(d-x)}{\sqrt{h_2^2 + (d-x)^2}} = 0,$$

and thus

$$\frac{x}{v_1\sqrt{h_1^2 + x^2}} - \frac{d-x}{v_2\sqrt{h_2^2 + (d-x)^2}} = 0.$$

Substituting in the $\sin\theta_i$ and $\sin\theta_r$ equations above immediately gives Snell's law

$$\frac{\sin\theta_i}{\sin\theta_r} = \frac{v_1}{v_2},$$

and as stated previously, the definition of the index of refraction n is just

$$v_j = c/n_j,$$

and so the speed of light in a medium will decrease by $1/n_j$, and

$$\frac{\sin\theta_i}{\sin\theta_r} = \frac{n_2}{n_1} \quad \text{and} \quad n_1 \sin\theta_i = n_2 \sin\theta_r.$$

Thus, the minimization procedure of the differential calculus as used in Fermat's principle of least time can be used to derive Snell's law of refraction.

One hundred years before Fermat, the legendary Leonardo da Vinci had at least conceptually understood the ubiquity of the concept of minimization, for he said

> Every action in Nature is governed by the principle of least action.

Indeed, the least action principle of physics neatly encompasses all the "least" principles described above and is indeed a basic tenet of modern physics.

Returning to the geometrical optics described above, the "ability" of light to know and choose the shortest route can also be described by a minima calculation of the variational calculus, that is, the position and momentum of a photon may be said to be determined by its tendency toward the path of least action. The minimization technique of the calculus of variations to be introduced in Chapter 7 used to calculate the Helmholtz free energy in liquid crystal systems is one in this succession of "leasts."*

Snell's laws of reflection and refraction today is learnt in middle school, but consideration apart from geometry and calculus requires an answer to a deeper question, "just how does the light know what is the shortest distance *before* embarking on its journey"; and in fact, this is an epistemological conundrum that requires the deepest modern theoretical physics to answer, namely relativistic quantum electrodynamics.

How light "knows" which path to take requires an explanation of the subatomic interaction between an electron and photon. The application of Maxwell's electromagnetic field theory in the subatomic world requires the incorporation of the particle-like photons of light in the theory in order to be able to address the subatomic level of interaction, and since the photons travel at light speed, the treatment must be relativistic.

Fortunately, a relativistic quantum dynamical photon can be modeled by the harmonic oscillators encountered in previous chapters, each oscillator having a *probability wave function*, which if visualized as a *wave packet* has a limited spatial extent, and this essentially quantizes the erstwhile continuous Faraday–Maxwell electromagnetic field. When the probability wave functions of the photons interact mutually and with the probability wave functions of the synonymous harmonic oscillator electrons, the square of the resultant probability amplitude reflects the superposition of the constituent probability amplitudes, and the shortest route is chosen through recognition of the highest probability. Again, this is not so difficult to state, but extremely difficult to *show*, and while the wave packet and simple harmonic oscillator

* As a gratuitous aside, the author has heard that the hopelessly optimistic main character in Voltaire's classic work *Candide* is satirically based on Leibniz as believing that "all is for the best in the best of all possible worlds"; if indeed Voltaire misconstrued the principle of least action as applying to human fortune, the esteem he has held as a philosopher should be greatly diminished.

provide conceptual help, the *state* of affairs lies in the essential indeterminacy of quantum mechanics.*

The deep understanding of the nature of the liquid crystal and the interaction with light and electric fields, at its essence must be based on a relativistic quantum electrodynamics that requires a probability wave function that is the devolution of the basic *wave–particle duality* conundrum of quantum mechanics; the photon wave packet mutual interaction and its interaction with the probability wave function of electrons can describe the basic phenomena of optics, including reflection, refraction, and interference, as well as the polarization central to the operation of liquid crystal displays [6]. Following is a basic description of quantum mechanical polarization by the eminent theoretical physicist P.A.M. Dirac that will hopefully also illuminate the concept of probability in quantum mechanics in the context of the polarization of light [7].

The quantum mechanical description of polarization is based on the principle of the superposition of *state functions*. To explain, as asserted by Dirac, "*every photon is in a certain state of polarization*" (linear [plane], circular, or elliptical). In an experiment with a polarizing piece of tourmaline crystal, a *beam* that is plane-polarized perpendicular to the optic axis of the tourmaline will go through, and a *beam* that is plane-polarized parallel to the optic axis will be absorbed (blocked); and if the beam is at an oblique angle α to the optic axis, a fraction $\sin^2 \alpha$ will pass through becoming circularly or generally elliptically polarized. For *beams*, this is all well and good, but for *photons*, the *perpendicularly plane-polarized photons* go through, and the *parallelly plane-polarized photons* are blocked, but what about the *obliquely plane-polarized photons*, do they split in two?

Taking a single incident photon, according to quantum mechanics, a detector behind the tourmaline sometimes will find a whole photon that is perpendicularly plane polarized, and sometimes it will find nothing, there will be no *parts* of photons detected. If this is repeated many times, there will be $\sin^2 \alpha$ of the incident photons found on the back side as perpendicularly plane-polarized photons. Thus, it can be said that a photon has a probability of $\sin^2 \alpha$ of passing through the tourmaline and a probability $\cos^2 \alpha$ of being blocked.

* Relativistic quantum mechanics is a compendium of the wave–particle duality, it is used to describe subatomic processes of any type of radiation, so it is also called the quantum theory of light and the quantum theory of radiation, but since quantum electrodynamics simplified is just QED, which is the Latin for *quod erat demonstrandum* ("so it is proven"), it is popularly used.

In this way, the individuality of the photon is preserved, but the *determinacy* of classical physics is sacrificed.*

> The most that can be predicted is a set of possible results, with a probability of occurrence for each . . . Questions about what decides whether the [obliquely plane-polarized] photon is to go through or not and how it changes its direction of polarization when it does go through cannot be investigated by experiment and should be regarded as outside the domain of science.

Dirac nonetheless continues by describing an obliquely plane-polarized photon as being partly in the state of polarization parallel to the axis and partly in the state of polarization perpendicular to the axis (analogous to the electric vector being resolved into parallel and perpendicular components).

> The state of oblique polarization may be considered as the result of some kind of superposition process applied to the two states of parallel and perpendicular polarization

This "special kind of relation between the various states of polarization" are now applied not to light *beams*, but to the *states of polarization of one particular photon.*

> This relationship allows any state of polarization to be resolved into, or expressed as a superposition of, any two mutually perpendicular states of polarization . . . When we make the photon meet a tourmaline crystal, we are subjecting it to an observation. We are observing whether it is polarized parallel or perpendicular to the optic axis. The effect of making this observation is to force the photon entirely into the state of parallel or entirely into the state of perpendicular polarization. It has to make a sudden jump from being partly in each of these two states to being entirely in one or other of them. Which of the two states it will jump into cannot be predicted, but is governed only by probability laws.

It should be emphasized that this is *not* a *statistical probability*, as Dirac points out in no uncertain terms,

> Some time before the discovery of quantum mechanics people realized that the connexion between light waves and photons must be of a statistical character. What they did not clearly realize, however, was that the wave function gives

* Dirac is quoted extensively here because the author believes that no one can express the principles of quantum mechanics better.

information about the probability of one photon being in a particular place and not the probable number of photons in that place.

So it can be said that every photon is in a certain state of polarization—linear, circular, or elliptical—and they interact among themselves and with electrons through the superposition of the various and sundry probability waves that can be described by state vectors. On the other side of the polarizer, the beam of photons is reconstituted in a state of polarization determined by the superposition of the polarization states.

In spite of Dirac's epistemological admonition, in quantum mechanics, conceptually one can say that the probability wave functions having no phase differences would have the largest amplitudes, so the shortest route through matter should be the wave combinations that result in the highest probability, and these are those that are in-phase; this then would coincide with the optical axis of the tourmaline crystal, and this is how the light "knows" what to do and where to go.

The detailed theories of the quantum mechanical polarization of light by electrons, atoms, and molecules, and vice-versa the charge polarization of those electrons, atoms, and molecules by light, is exceedingly complex and a subject of current research [8].

References

[1] Feynman, R.P., Leighton, R.B., and Sands, M. 1963. *The Feynman Lectures on Physics*. Addison-Wesley, Reading. Vol. I, 33–32.

[2] Weyl, H. 1950. *The Theory of Groups and Quantum Mechanics*. Dover, New York.

[3] Yang, D.K., and Wu, S.T. 2006. *Fundamentals of Liquid Crystal Devices*. Wiley, Chicester, Chapter 4.

[4] Feynman, R.P. 1985. *QED*. Penguin, London; Feynman, R.P., Leighton, R.B., and Sands, W. 1963. *Lectures on Physics*, Vol. II, 32–1, Addison-Wesley, Reading; and Zee, A. 2003. *Quantum Field Theory*, Princeton.

[5] Nahin, P. 2004. *When Least Is Best*. Princeton.

[6] Klauder, J.R., and Sudarshan, E.C.G. *Fundamentals of Quantum Optics*. Dover, New York; Feynman, R., Leighton, R.B., and Sands, W. 1963. *Lectures on Physics*, Vol. I, 26–7ff., Vol. II, 19–1. Addison-Wesley, Reading; and Baym, G. 1969. *Lectures on Quantum Mechanics*, Advanced Book Program Westfield, Perseus, New York.

[7] Dirac, P.A.M. 1981. *The Principles of Quantum Mechanics*, 4th edition. Oxford.

[8] Auzinsh, M.P., Budker, D., and Rochester, S.M. 2010. *Optically Polarized Atoms*. Oxford; and Auzinsh, M.P., and Ferber, R. 1995. *Optical Polarization of Molecules*. Cambridge.

5

Liquid Crystals

As described in Chapter 4, while traversing a birefringent material, the different speeds of the ordinary and extraordinary waves engender a phase retardation within the emerging light that alters its polarization state. If that phase retardation can be controlled, in conjunction with a second polarizer (the analyzer), the intensity of the light can be modulated by the altered polarization state. This is the essence of the operation of a liquid crystal display.

In preview, the long rod-shaped molecules of a liquid crystal provide sufficient anisotropic structure to have fast and slow axes, and linearly polarized light passing through, depending on the angle of incidence, can be decomposed into ordinary and extraordinary waves, engendering phase retardation that causes changes in the polarization state of that light just as in the solid crystal case described above. But a liquid crystal goes one better: The molecular orientation of the liquid crystal is easily changed, owing to its inherent fluidity, and thus its optical birefringence (derived from its dielectric anisotropy) is also easily changeable, and since an external electric field can control the birefringence, the polarization state of the light passing through can be precisely controlled, and in conjunction with a second polarizer, the intensity of light can be modulated.

Liquid Crystal Displays, First Edition. Robert H. Chen.

Carrots

The first recorded observation of a liquid crystal was in 1888 by the Austrian botanist Friedrich Reinitzer. In his investigation of the cholesterol in carrots, he noticed that there were what appeared to be phase changes at two different temperatures: at room temperature the compound was solid, but when the temperature was raised to 145°C, it appeared to be in a sort of *mesophase* between liquid and solid. When more heat was added until the temperature reached 179°C, the milky compound cleared to a transparent liquid. Unable to explain what he had observed, Rienitzer turned to his friend the German physicist Otto Lehmann. Lehmann not only was an expert crystallographer, in his laboratory sat the world's only temperature-controlled polarizing microscope. With it, Lehmann could observe the optical effects that structural changes wrought with increasing temperature. So Lehmann put Reinitzer's carrots compound between two glass slides, and, gradually increasing the temperature by adjusting the flame source, observed the phase changes in what was later to be identified as cholesteryl benzoate.

If light is passed through two polarizers having their axes of transmission perpendicular to each other (crossed), just as when the lens of two polarizing sunglasses are rotated so that their transmission (optic) axes are crossed, because of the complete canceling of the orthogonal components of the electric field vector of the light by the crossed polarizers, the light should be completely blocked. So light directed from the bottom of the polarizing microscope by the mirror (as shown in the schematic Fig. 5.1) should be blocked by the carrot sample. But to his surprise, Lehmann saw that the light not only passed through the carrot compound, even more surprisingly, it exhibited a double refraction effect like a birefringent solid crystal. It could be said that just at this instant, although he certainly could not have imagined it, the trillion dollar liquid crystal display industry was born [1].

Devices based on Lehmann's polarizing microscope are still used today to study liquid crystals. The liquid crystal shown in Figure 5.2 is similar to what Lehmann observed, but the image is from a modern-day polarizing microscope. The axes of the curly patterns of light and dark are perpendicular to the plane of the image (the so-called *Grandjean texture*). After several communications with Lehmann, Reinitzer in 1888 at the May meeting of the Vienna Chemical Society announced the discovery of a liquid crystal. In his report, Reinitzer noted that his carrot compound had two melting temperatures, could reflect circularly polarized light, and could change the polarization of incident light from left- to right-hand circular and vice-versa [2].

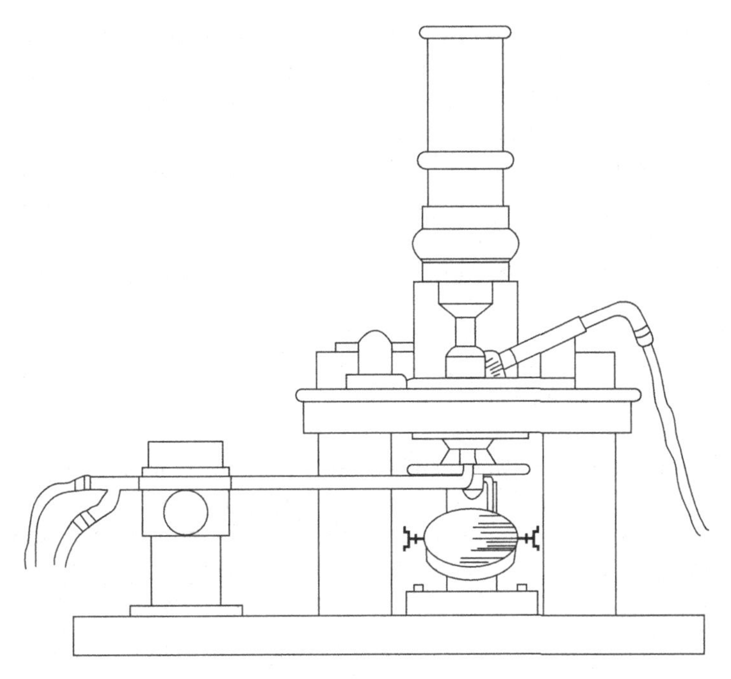

Figure 5.1 Lehmann's polarizing microscope used to study the first liquid crystals. From Collings, P.J. 2002. *Liquid Crystals,* 2nd edition. Princeton. p. 19.

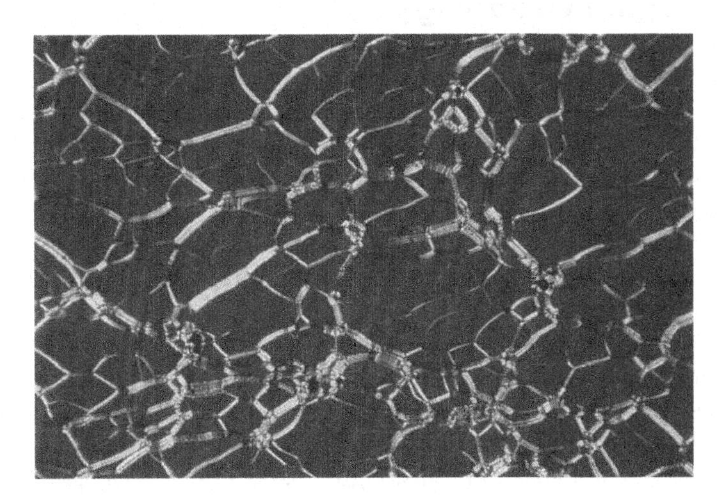

Figure 5.2 Polarizing micrograph of cholesteric liquid crystal. Courtesy of Professor H. Kitzerow.

Just as melting snow changes phase from a crystalline snowflake to liquid water, so the ordered orientational and positional anisotropic structure of the cholesteryl benzoate crystal, when heated, turns into an isotropic liquid. The significance of what Lehmann discovered was that the cholesteryl benzoate, while in its substantially solid state, would possess positional and orientational order, and in its liquid state would have little or no order, but in between, it would maintain an orientational order but lose some positional order. In this mesophase between the solid to liquid phases, the residual orientational order is just what allows the facile change of molecular orientation that makes a controllable birefringence possible. This kind of fluid crystal behavior in a thick liquid prompted Lehmann to call the cholesteryl benzoate a *"fliessender kristalle"* ("flowing crystal"), which later would become the English "liquid crystal." As early as 1907, the German chemical conglomerate Merck was producing and advertising *flussiger* ("liquid") and *fliessender kristalle*, as shown in Figure 5.3.

The science of liquid crystals was founded in botany, germinated in organic and physical chemistry, and came to fruition in physical optics. The advent of liquid crystal displays however required the electrical, mechanical, and material sciences, plus the invention of transistors and integrated circuits. As such, a liquid crystal display can be characterized as a device spanning the old and the new, and the low and the high technology.

Liquid Crystal Genealogy

Otto Lehmann and the French crystallographer Georges Friedel in 1922 took Reinitzer's discovery of liquid crystals, and in accord with its change of phase with the addition of heat, called them *thermotropic*. At the first phase transition temperature, the crystal structure melts to a murky anisotropic fluid; at the second phase transition temperature, the fluid clears to an isotropic liquid. The first and second phase transition temperatures are now commonly called the melting temperature (T_m) and the clearing temperature (T_c).

Thermotropic liquid crystals generally are constituted from long ellipsoidal (cigar-shaped) molecules, generally called *calamitic*, and disc-shaped molecules, appropriately called *discotic*. Today's liquid crystal displays almost all use calamitic liquid crystals, and discotic liquid crystals are primarily used in compensation films. Figure 5.4 is Friedel's categorization of the five types of orientational and positional order of calamitic liquid crystals.

Figure 5.3 Advertisement for liquid crystals by Merck in Germany in 1907. From Kawamoto, H. 2002. *Proc. IEEE*, **90**, 4, 462, April, courtesy of Ludwig Pohl.

It can be seen from Figure 5.4 that "solid crystalline" has both positional and orientational order; that is, each layer's calamitic molecules are neatly ordered and are oriented in substantially the same direction. At the other extreme on the far right-hand side, the "liquid isotropic" exhibits no ordering, either positional or orientational whatsoever. However, in the middle,

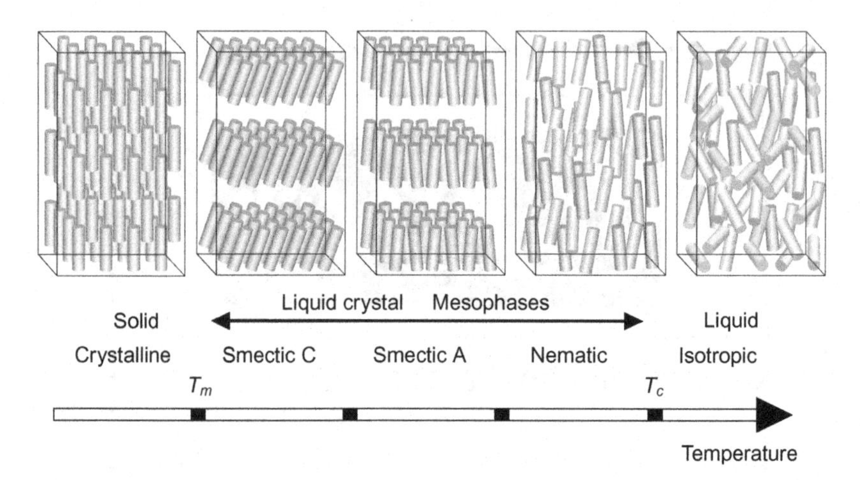

Figure 5.4 Orientational and positional order in the mesophases of thermotropic liquid crystals as categorized by Friedel. Adapted from Lueder, E. 2001. *Liquid Crystal Displays*. Wiley, p. 5.

the "liquid crystal mesophases," following the increase of temperature, the "smectic C" (Greek σμηγμα meaning "soapy") liquid crystals have positional and orientational order, with clear positional layer separation. Further increasing the temperature results in "smectic A," similar to smectic C, but apparently with the molecules having a more vertical orientation, but still clearly positionally layered.

As the temperature is further increased, the "nematic" (Greek υημα meaning "threadlike") domain is reached. The nematic liquid crystal molecules retain the vertical orientation of the smectic, but with molecules positionally interlayered in a construct that early observers quaintly termed *imbricated* [3].*

The greater positional and orientational order in the smectic liquid crystals results in a greater viscosity as compared with the less-ordered nematic liquid crystal. But it is just this lower viscosity that allows the nematics to respond more sensitively to external applied electric fields, which will be seen later as the reason why the nematic is the liquid crystal of choice for displays. The higher viscosity renders the smectics less suitable for use in fast-motion displays, but because smectic displays can retain an image in the absence of an external electric field, they consume less energy and thus may

* "Imbricated" means "overlapping in sequence," *Webster's Encyclopedic Unabridged Dictionary of the English Language* (Portland House, 1989).

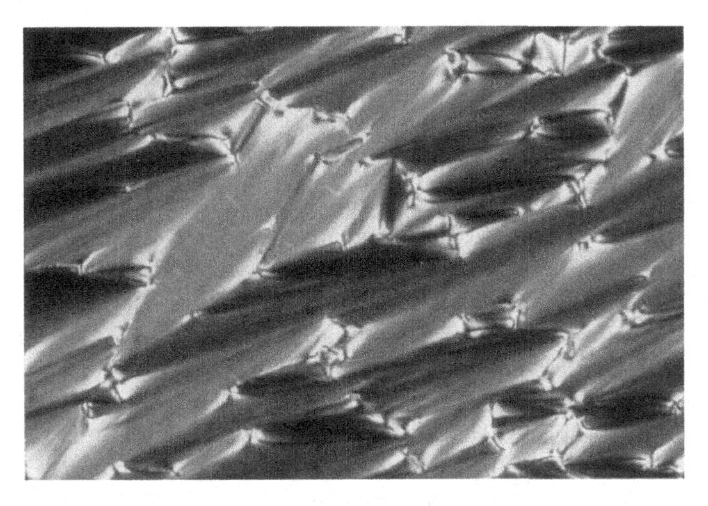

Figure 5.5 Polarizing micrograph of the layered smectic-A liquid crystal. Courtesy of Dr. Mary Neubert.

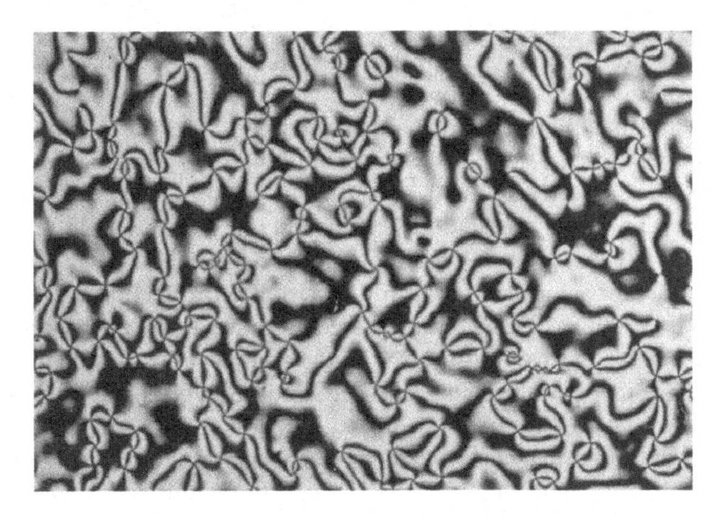

Figure 5.6 Polarizing micrograph of the stringy nematic liquid crystal. Courtesy of Professor H. Kitzerow.

be used efficiently in static or slower-moving displays, such as advertising panels, light-blocking and imaging windows, and electronic books.

Using a polarizing microscope, Figure 5.5 shows that the crystal structures of smectic-A and the nematic of Figure 5.6 are clearly different. The distinct layering in smectic-A indicates that the layers may slide in mutual translation, and in fact it is just this interlayer sliding that makes smectic liquid crystals suitable for use as soap through its ability to clean by removing dirt by sliding it away. Somewhat mundanely then, the exotic smectic liquid crystals can be found in the residual gunk at the bottom of soap dishes; it, however, can also be more attractively found in bottles of commercial liquid soap.

The stringy nematic as seen in Figure 5.6 has discernible order, but both the positional and orientational order are not neatly arranged, but rather appear to be separated into different domains having different angular orientations. The black patches are regions of crystal lattice *disclinations* (analogous to dislocations in solid crystals) that, depending on the orientation of the domain, engender polarization states in light passing through that is extinguished by the analyzer. That is, from the disclinations, the black spots extend into brush-like curlicues that are the result of the liquid crystal molecules being alternately perpendicular and parallel to the plane of polarization so that the light is alternately blocked by the crossed polarizer of the microscope. The nematic black brushes are called *Schlieren textures*. These liquid crystal disclinations were once objects of considerable scientific interest called "disinclinations," but they have been relegated to the realm of "understood" in physics, and now no longer subjects of much study [4].

In summary, a solid has a both positional and orientational permanent order, a liquid has neither, and the mesophases liquid crystals in-between has a little of both. It is just this benign combination of order and freedom that makes liquid crystals ideal as a display medium. Within the liquid crystal mesophase, the orientational and particularly positional order in the smectic is clearly much greater than that in the nematic, the latter possessing only a modicum of orientational and positional order. But it is just this modicum that allows the nematic molecules to act as a crystal and at the same time be able to turn fluidly, thus being capable of quickly changing the polarization state of light passing through. In other words, the fluid nematic is more sensitive to externally applied electric fields, but it maintains an integrity of order sufficient to produce birefringence, which can change the polarization state of light. Perhaps this is something that one can learn from the liquid crystal personally; that is, to be useful, one must possess both integrity and sensitivity!

The Chiral Nematic

Within the nematic liquid crystal category, there is another rather special convoluted genus. It was a cholesteric liquid crystal phase that was observed by Reinitzer in his discovery of the liquid crystal, and it exhibited the three optical effects reported in his original paper. Friedel recognized one as the Bragg effect familiar to crystallographers; that is, when white light passes through the nematic cholesteric, constructive interference could be observed only at certain angles of reflection, and those angles in turn depended on the angle of incidence and the wavelength of the light. The second effect was that the plane of polarization of linearly polarized incident light would be rotated by the liquid crystal as the light passed through; this is just the definition of *optical activity*. The third effect was that the liquid crystal would selectively strongly reflect left- or right-hand circularly polarized incident light; this is just the definition of *optical chirality* (mirror anti-symmetry). It will be seen that these observed effects were direct manifestations of the special structure of the cholesteric liquid crystal.

Because these effects were manifest originally in the cholesteric compounds studied by Reinitzer and Lehmann, Friedel named the liquid crystal "cholesteric" in his categorization. Within the structural layers of the cholesteric, the molecules are aligned in parallel with the long axis in a plane, similarly to the nematic, but the layers are very thin with a slightly different orientation of the molecules in each layer so that the successive layers of the cholesteric molecules cumulatively form a helix. Figure 5.7 schematically shows the cholesteric as viewed from the side of the nematic (as shown in

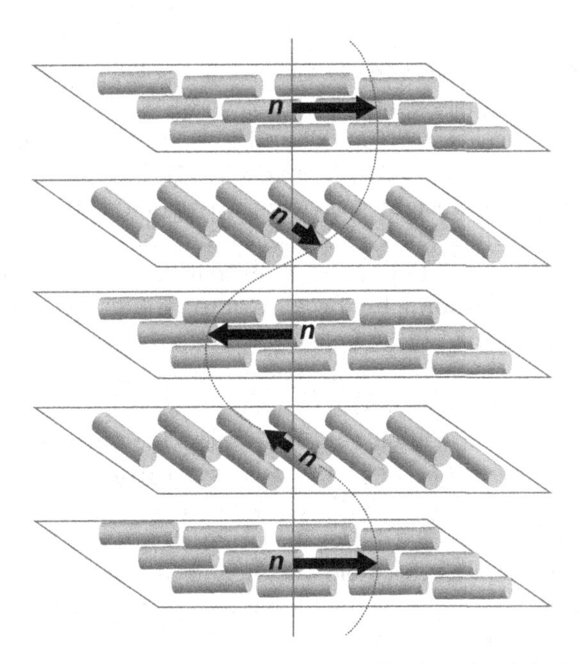

Figure 5.7 Cholesteric liquid crystal schematic showing interlayer twisting.

Figure 5.4), where the interlayer twisting is manifested by the different orientations of the molecules in successive layers. The ***n*** in Figure 5.7 is the *director*, which indicates the local average orientation and clearly delineates the spiral corkscrew molecular orientational structure.

The cholesteric helix has directionality in analogy with a screw on which a rotating nut will rotate with a forward progression. When the rotation reaches 360°, then the distance that the nut has traveled is called the *pitch* of the cholesteric liquid crystal, which then is a measure of the severity of the helix (the smaller the pitch, the more concentrated the twisting). The rotation can also be characterized as left or right handed, depending on the direction of rotation that induces forward progression.

Because the cholesteric natural helix rotates the plane of linearly polarized incident light (the optical activity), the light intensity does not change as its E vector spirals through the liquid crystal and can be aligned with the transmission axis of the analyzer to produce a high intensity (because there is no change in the state of polarization) bright state, called the *normally white* (NW) mode. Because of the natural optical activity, cholesteric liquid crystal displays are bright, and since contrast is defined as the ratio of the bright to the dark, contrast is high. Further, since the optical activity is natural to the structure, the cholesteric display does not need an electric field to maintain the normally white state and thus can conserve power. Since only when changing a state is power required, the cholesteric liquid crystal can be expeditiously used in, for example, electronic books, advertising displays, and privacy windows (see Chapter 27).

A full mathematical explanation of optical activity in cholesteric nematic liquid crystals is quite involved, but a very rough physical explanation can perhaps provide some insight. When linearly polarized light falls on the spiral molecule structure, the electric field vector of the light will drive the electrons in the molecules, but in directions that will be constrained to move not only horizontally around, but because of the spiral also vertically up. So the electric field radiated by the electron motion on opposite sides of the spiral will arrive at another point in space with a slight time delay, which is a phase retardation that generally will not be exactly π, but more by a factor of $\omega A/c$, where A is the diameter of the spiral, ω the angular velocity, and c the speed of light. Now that the phases do not exactly cancel out, in addition to the electric field vector of the linearly polarized light, there is a small transverse component added on to the original vector, which produces a slightly tilted resultant electric field vector, and as the light moves through the cholesteric liquid crystal, owing to that tilt, the plane of the electric field vector will rotate about the beam axis [5].

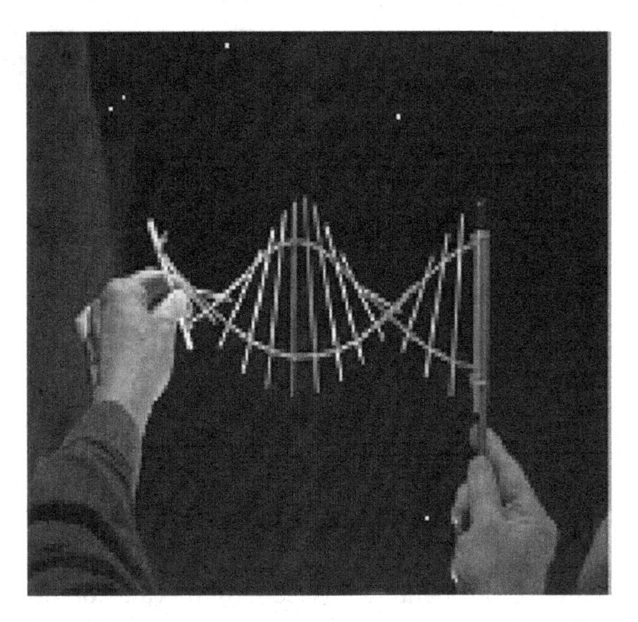

Figure 5.8 Stick-and-band model of a chiral structure. Courtesy of Fergason Intellectual Properties.

For historical reasons, "cholesteric" has continued to be used as the generic name for liquid crystals having this structure and exhibiting this type of behavior, but not all such spiral structures contain cholesterol, so a more apt term would be *chiral nematic*, wherein "chirality" refers to the mirror anti-symmetry of the right- and left-hand turning. Figure 5.8 is a stick-and-band model showing the twist of a chiral structure, and Figure 5.9 is a polarizing microscopic image of a chiral nematic compound with the spiral axis in the plane of the image (called *fingerprint texture* for obvious reasons); the spiral structure is clearly manifested.

If a nematic liquid crystal is doped with a chiral nematic, the chiral nematic *guest* in this so-called *guest-host compound* provides a directional twist impetus to the *host* nematic liquid crystal, effectively forcing it to twist in a predetermined direction (the twist direction of the guest chiral nematic) in a sort of charitable push. And because the pitch of the chiral nematic guest depends intimately on the concentration of the guest, the amount of twist provided can be precisely controlled; this is useful for pretilting molecules, in ferroelectric liquid crystals, and in super-twisted nematic (STN) displays, as will be seen.

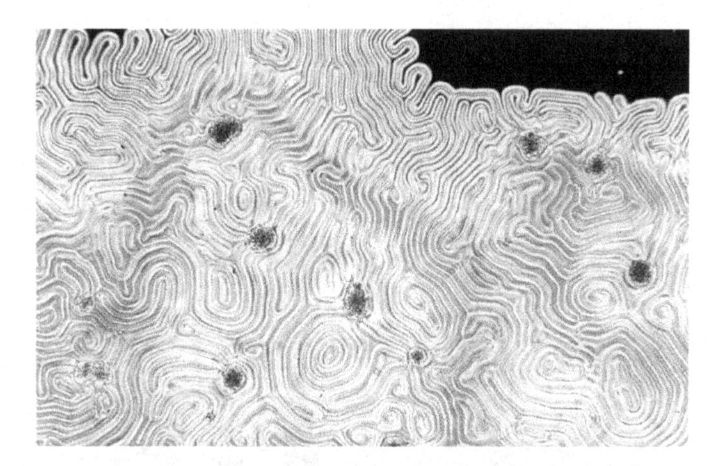

Figure 5.9 Polarizing micrograph of the fingerprint texture of a chiral nematic with the spiral axis in the plane of the image. Courtesy of Professor H. Kitzerow.

If the wavelength of incident light is close to the value of the pitch of the chiral nematic, that wavelength of light will undergo positive reinforcement and be reflected, and since the pitch of the chiral nematic varies with temperature (the absorption of heat will alter the molecular structure through thermal agitation), different temperatures then will produce different colors (wavelengths) of reflected light. Accordingly, chiral nematic liquid crystals can operate as very sensitive temperature gauges. Based on this, the American inventor James Fergason developed a chiral nematic liquid crystal film that could be coated on the skin to act as an all-body thermograph that could show the temperatures of the skin and underlying bone structure in a thermographic pattern for medical use as shown in Figure 5.10.

Soon thereafter, the irrepressible fashion toy industry produced a chiral nematic "mood ring," purportedly able to display the moods of the wearer by sensing the changes in body temperature incited by various emotions. The mood ring (as shown on the box in Fig. 5.11) was popular among teenagers for a few years, and then as all teen fads do, slowly faded away. The chiral nematic used in liquid crystal displays met a better fate, being commonly used in STN displays, but its use has also decreased as other types of liquid crystal displays have become popular.

The Ferroelectric Chiral Smectic-C

Returning to the Friedel's categorization of the thermotropic liquid crystals, the smectic-C is the first mesophase encountered with increasing tempera-

Figure 5.10 All-body thermograph made of chiral nematic liquid crystal film. Courtesy of Fergason Intellectual Properties.

ture after the solid phase. From the categorization in Figure 5.4, it can be seen that there are still distinct positional layers, and although the molecules *within* each layer are oriented in a substantially uniform direction, they are tilted at an angle. Now if a chiral dopant is added to promote a twist in the layers, rendering the molecules uniformly oriented within each layer, but with the twist progressing through the liquid crystal as a chiral helix, as shown schematically in Figure 5.12, this *chiral smectic-C* (also called the *smectic C**) will have an inter-layer twist so that the molecular long axis director (n) orientation is no longer unidirectional, and because of the intrinsic charge separation, the chiral smectic-C is polar. This is because the dipole moment of the chiral smectic-C is neither rotationally symmetric about the molecular axis nor reflectionally symmetric about the plane formed by the molecular axis and the director because of the twisting tilt; this different total configuration results in a different but discernible direction for

Figure 5.11 Chiral nematic mood ring.

the intrinsic dipole moment, producing what is evocatively called *spontaneous polarization*, meaning that the chiral smectic-C is naturally polar.

It is interesting to note that the chiral nematic itself is not polar because the structure is invariant for 180° rotations about the director, and also for rotations about the helical plane, as can be discerned from the chiral nematic molecular structure in Figure 5.12 (meaning that the tilt away from the vertical is the symmetry breaker). For smectic-A, the substantially vertical orientation with no tilt rules out rotational symmetry breaking, so smectic-A is nonpolar and would be nonpolar even with the addition of a chiral dopant [6].

This kind of intrinsic polarity is called *ferroelectric*, because the molecules have a definite natural directional orientation, much like the molecules in iron and other magnetic materials; of course there is no iron in the chiral smectic-C, and it is called "ferroelectric" because of its polarity. Ferroelectric liquid crystal displays presently are still in the research stage of development.

The Blue Flash

In his initial observations of liquid crystals, Reinitzer noticed a distinct blue flash of light during his temperature variation studies, but after duly record-

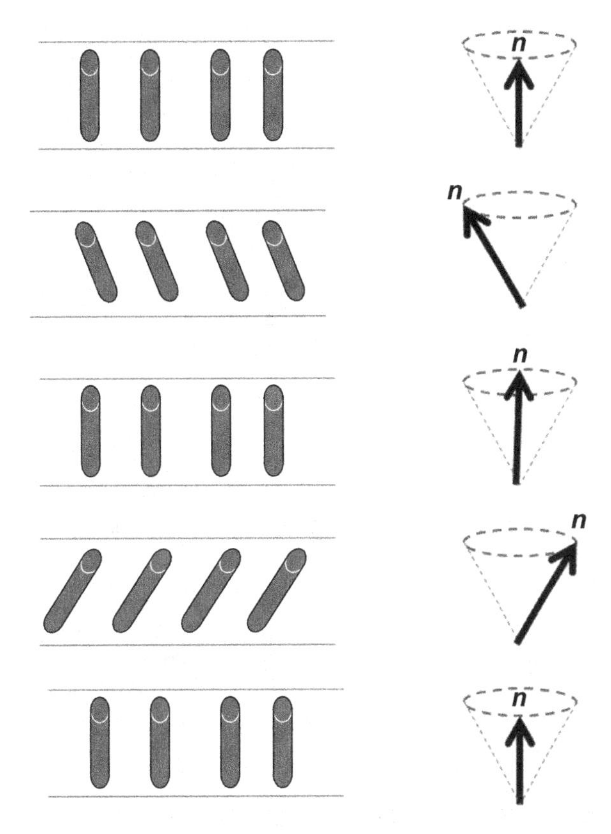

Figure 5.12 Chiral smectic-C schematic showing the polar chiral helix.

ing it, he paid no further heed to this *blue phase* of liquid crystals. It was to be 80 years later that this "double chirality" structure was recognized as an entirely new liquid crystal phase. For over 100 years, scientists had believed that the single helical structure was the most stable chiral nematic structure, but because of the different ways that molecules might fill in the spaces in the unique chiral structure, actually more complex multiple helical structures are possible that are just as stable and thus should be observable [7].

There are actually three different blue phases, each representing a first-order phase change, like that from liquid to cholesteric liquid crystal. Figure 5.13 is a schematic rendering of the molecular structure showing the blue phase liquid crystal's remarkable resemblance to the DNA double helix. It turns out that cholesteric nematic liquid crystals can exhibit different colors of the *blue flash*, which will appear as three-dimensional constructs, such as

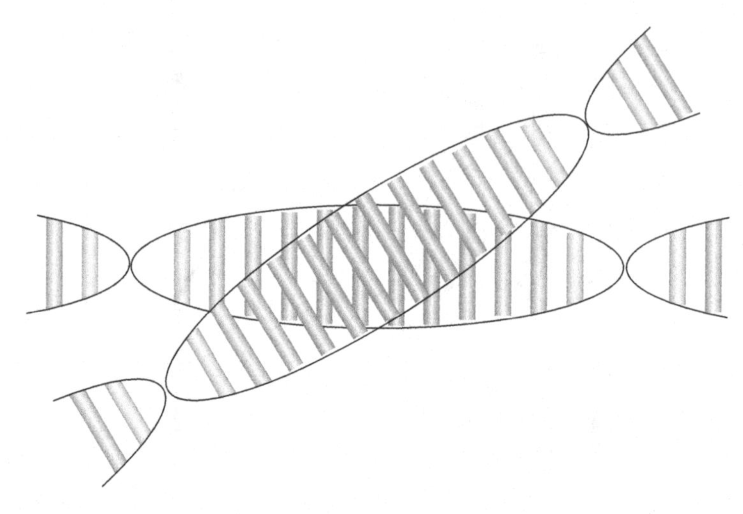

Figure 5.13 Blue phase liquid crystal schematic molecular structure showing the resemblance to the DNA double helix.

the blue, green, and red and their different planar structures as shown in Figure 5.14. Blue phase liquid crystal displays have been developed and are notable for their super-fast response times (see Chapter 27).

Lyotropic Liquid Crystals

The thermotropic liquid crystals introduced above change phase with temperature; there is another class of liquid crystals whose phase state is determined by the particular solvent present. The *lyotropic* liquid crystals are abundantly present in human body cells, brain, nerves, muscles, blood, and organs, and they play a critical role in biologic functions.* Life, after all, is water based, and bodily functions require an aqueous environment, for example, the proteins and enzymes that are transferred through organ membranes, to be effective in organic processes, must be fluid, but at the same time, they must maintain structure in order to perform their functions. It turns out that the lyotropic liquid crystal has the fluidity required for mem-

* "Lyotropic" means "any series of ions, salts, or radicals arranged in descending order relative to the magnitude of their effect on a given solvent," *Webster's Encyclopedic Unabridged Dictionary of the English Language* (Portland House, 1989).

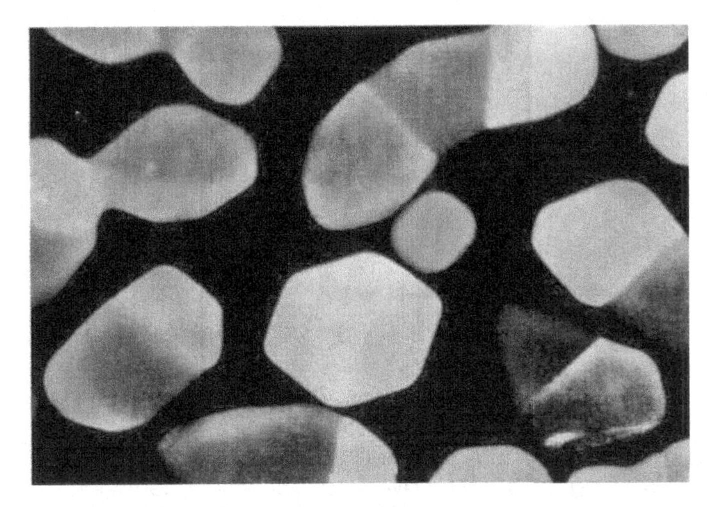

Figure 5.14 Blue phase planar structures photograph. From Hekimoglu, S. and Conn, J. 2003. *Liq. Cryst. Today*, **12**, 3, 1, September.

brane diffusion and the crystal molecular structure for biologic function. Diffusion depends on the structural order of the molecules and that order depends on the phase, which in turn depends on the solvent, temperature, pressure, concentration, viscosity, and the like, all the common chemical and physical parameters. The lyotropic liquid crystals are critical to understanding life mechanisms and disease, so they are subjects of study in biology, biochemistry, and medicine. But so far lyotropic liquid crystals have not been used in liquid crystal displays, the subject of this book, so there will be no further discussion of them hereinafter [8].

The Director and the Order Parameter

From the above discussions, it can be seen that the different properties exhibited by the different categories of liquid crystals, as well as by the solids and pure liquids, in essence are due to their varying degrees of orientational and positional order. Thus the deeper scientific analysis and understanding of liquid crystals requires a suitable metric to describe that all-important ordering.

As mentioned previously, if the long axes of the liquid crystal molecules in aggregate are grossly ordered in a specific direction, an *axis of symmetry*

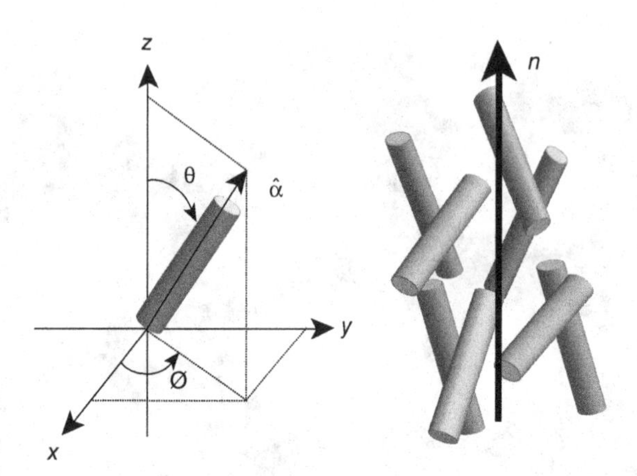

Figure 5.15 Polar coordinate system and schematic of molecules with the director ***n*** describing the sectoral average molecular orientation.

of the liquid crystal as a whole can be defined, and this axis, also known as the *crystal axis* or the *preferred axis* from solid state crystallography, describes the orientation of the liquid crystal as a whole. However, the orientation of the molecules inside the liquid crystal can change without the whole liquid crystal changing orientation, and indeed that is the principal reason why liquid crystals can be used in displays. So within the liquid crystal, an orientational vector called the *director*, is defined as the average of the molecules' long axis vectors in a local sector of the liquid crystal; that is, the orientation of the molecules in a given sector of the liquid crystal can be described by a local vector, the director ***n***. Referring to Figure 5.15, the polar angle θ in the polar coordinate system shown is just the angle between a given molecule's long axis vector $\hat{a}$ and the vector director ***n***, and with φ as the azimuthal angle of that molecule, the coordinates thereby fix the molecule's orientation in space. At the right in Figure 5.15 is a schematic drawing of some molecules in a local sector of the liquid crystal, with the director ***n*** describing the sectoral average molecular orientation.

If now some form of a statistical average of the polar angles θ described above is taken, then the variable internal orientation of the liquid crystal can be described mathematically. This is done by defining an *orientational order parameter* that can describe in aggregate the orientation of the molecules' directors in accord with observations, and thus serve as a parameter to indicate the changes in the molecular order within the liquid crystal for theoretical modeling. The problem then is to find some mathematical function to perform that task. It will be seen that the symmetry properties of the liquid crystal itself and the methods of mathematical physics will determine the selection of the most suitable function.

The simplest orientational order parameter would appear to be just the average of the polar angles of the directors. Now because of the three spatial dimensions, and the longitudinal symmetry of the long rod-shaped molecules, and the fact that there is no distinction between head and tail (a polar angle of 180° orientation is the same as 0°), and so there are many possible angles between 0° and 90° and greater, but there is only one direction that is exactly 0°, so it turns out that the average value is not 45° as one might expect, but rather 57°. This average value, while perhaps interesting, is only a number describing the whole liquid crystal and thus cannot describe any variations in the director orientation within the liquid crystal, so the information is not much more use than an axis of symmetry.

A finer order parameter was proposed by the Russian physicist Tsvetkov as the average of the second Legendre polynomial [9]. Why such a seemingly peculiar proposal should be made begs some knowledge of the particular features required of an orientational order parameter for describing liquid crystals; it also sheds some light on mathematical physics methodology [10].

Any orientational order parameter at the very least must obey two boundary conditions; namely in the high temperature unordered liquid state, it must be identically zero (absolutely no order), and in the low-temperature ordered solid state, it must be exactly one (completely ordered). In between, the parameter must be able to at least describe the distribution of polar angles of the molecules, call it $f(\theta)$, in the liquid crystal.

As the cosine function is often used in mathematics to delineate a projection, a good first choice for the orientational order parameter that obeys the boundary conditions might be the cosine of the angle between the long axis $\hat{a}$ of the molecule and its projection on the director $\boldsymbol{n}$. When θ equals 90°, the molecules' long axis is perpendicular to the director, and the molecules are thereby in complete disorder insofar as the director is concerned, and indeed $\cos(90°) = 0$, which satisfies the first boundary condition. When θ equals 0°, the molecules' long axis is aligned parallel to the director, and is thereby in complete order insofar as the director is concerned, and indeed $\cos(0°) = 1$, which satisfies the second boundary condition. So far so good, as it appears that the cosine function just might work.

Now it just so happens that the simple projective $\cos\theta$ is also the first Legendre polynomial solution of the Legendre equation, and the spatial average value (denoted by $\langle\cos\theta\rangle$) is by definition:

$$\frac{\int_0^{\pi} \cos\theta \cdot f(\theta)\sin\theta d\theta}{\int_0^{\pi} f(\theta)\sin\theta d\theta} = \langle P_1(\theta)\rangle = \langle\cos\theta\rangle,$$

where the integral over the angles of interest (0° to π = 180°) includes in the integrand the polar angle distribution function in the liquid crystal; when that is divided by the integral over the same angles of interest without the polar angle distribution, that is the definition of the average over all the polar angles θ.

Now, since in a liquid crystal, the projections angles of the vertically standing up and completely upside-down molecules have the same probability distribution, that is, for θ or $\pi - \theta$ (the negative projection of θ),

$$f(\theta) = f(\pi - \theta),$$

the integral calculation of the average value of the first Legendre polynomial $\langle P_1(\theta)\rangle$ above unfortunately turns out to be zero, $\langle\cos\theta\rangle = 0$, so like the simple average value of angles above, the cosine of the polar angle provides no useful additional information for variations of the orientation parameter behavior within the liquid crystal other than the boundary condition extremes, and indeed a look at the graph of the Legendre polynomial curves in Figure 5.16 shows that $P_1(\theta)$ is a simple straight line, symmetric with the origin, with a zero average value.

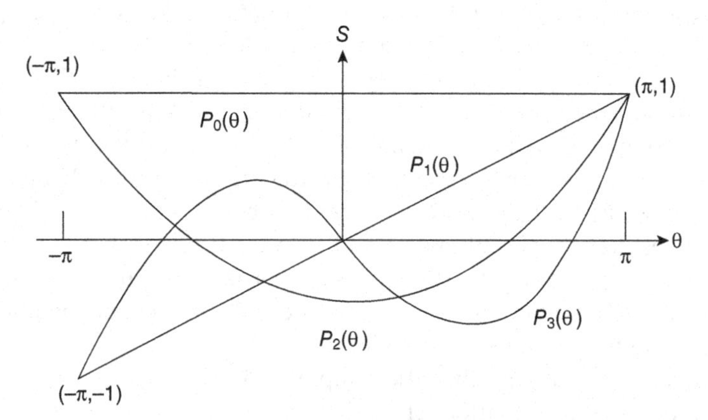

Figure 5.16 Graph of the Legendre polynomials.

But further looking at Figure 5.16, the second Legendre polynomial curve $P_2(\theta)$ appears to offer better prospects: its starting and ending points are the same as $P_1(\theta)$, but its average value is not zero, unlike $P_1(\theta)$ the curve is not symmetric about the origin of the graph.*

The second Legendre polynomial $P_2(\theta)$, for example, shows that when θ equals 90° ($\pi/2$ radians, between π and the origin), the average orientational order parameter is zero; that is, the molecules' long axes are perpendicular to the director, so they essentially have no order (which is the first boundary condition), and when the molecules are standing (straight up or upside down, $\theta = \pm\pi$), the orientational order parameter is just unity, meaning all the molecules are aligned with the director (which is the other necessary boundary condition). The curve of the second Legendre polynomial for θ in-between $\pi/2$ and π, where there should be a certain degree of order, also appear to be promisingly indicative of an orientational angle order. But is it certain that the second Legendre polynomial can fulfill the task of describing the orientational order parameter, and is really the best choice?

If a higher-order Legendre polynomials such as $P_3(\theta)$ were tried in the hope of gaining more detail in the order parameter, from the graph of the Legendre polynomials above, it can be seen that again there is symmetry about the origin and the average value again is zero, which disqualifies it just as the averages of the $\cos\theta$ and $P_1(\theta)$ were disqualified. Other functions can be tried as well, but the proof of the pudding is in the use of the second Legendre polynomial, as the parameter describing molecular order in the mathematical theory of liquid crystals, and it will be shown in the Chapters 9–12 that it performs this role quite satisfactorily. So although higher-order Legendre polynomials could be used, Tsvetkov's choice of the second Legendre polynomial is presently the liquid crystal order parameter of the day; with S denoting the sought-after order parameter, it is

$$S = \frac{\int_0^\pi \frac{1}{2}\langle 3\cos^2\theta - 1\rangle \cdot f(\theta)\sin\theta d\theta}{\int_0^\pi f(\theta)\sin\theta d\theta} = \langle P_2(\theta)\rangle = \frac{1}{2}\langle 3\cos^2\theta - 1\rangle.$$

* The Legendre polynomial is the solution to the differential equation $(1 - x^2)y'' - 2xy' + k(k+1)y = 0$, with $k = 2$ and where $y = (3x^2 - 1)/2$; $x = \cos\theta$ is used for the orientational order parameter.

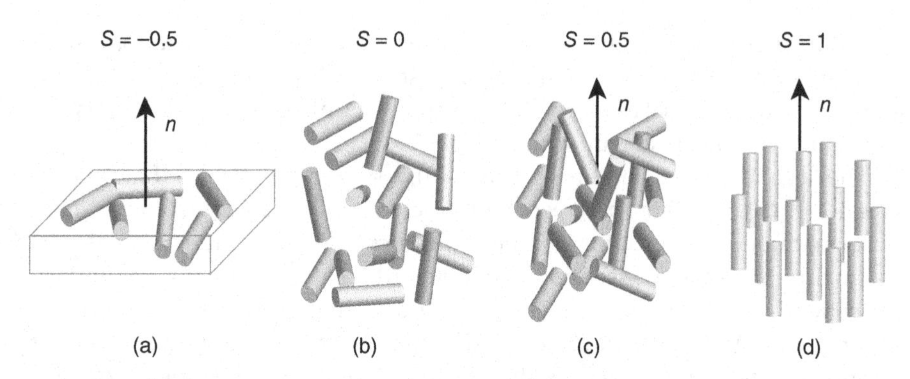

Figure 5.17 Orientational parameter *S* schematic of a few molecular orientational order states of a nematic liquid crystal.

Schematic drawings of nematic liquid crystal molecules in a few states of orientational order with the orientational parameter S are shown in Figure 5.17 to give an idea of its value *vis-à-vis* the molecular orientational order. In Figure 5.17a, the negative parameter $S = -0.5$ means that all the molecules are distributed at angles within a horizontal and not a vertical plane (where the polar angle is manifest).

The orientational order parameter describes a mean polar angle distribution of the director in a liquid crystal, and thus is an indicator of orientational order. The values of S range from $S = 0$ for absolutely no order (isotropic liquid) to $S = 1$ for complete order (anisotropic solid); in-between, liquid crystals have typical values of $S = 0.3$–0.9 [11].

If the director of a nematic liquid crystal is in the x-axis direction, then the y-axis will be perpendicular to the director, and the indices of refraction will be the nonequal n_x and n_y. In accord with the discussion of the preceding chapters, the E_x and E_y electric field vector components of light passing through the anisotropic uniaxial material will encounter different orientations of molecular structure and propagate at different speeds through the liquid crystal, and the phase lag between the E_x and E_y waves will produce a polarization of the light. Then an applied electric field, by changing the orientation of the liquid crystal molecules, can alter those polarization states, resulting in controllable changes of the light intensity when passed through a second polarizer. So the thermotropic calamitic nematic liquid crystals just described theoretically can be used as the basic modulating medium for liquid crystal displays.

The order parameter can be seen as associating the microscopic behavior of the molecules with the macroscopic behavior of the liquid crystal. The methods of x-ray diffraction can be used to investigate the microscopic molecular structure, and observations of the refractive effects on incident light of the liquid crystal can describe its macroscopic behavior. Since the orientation order parameter links the microscopic and macroscopic views, S is the basic metric for studying the molecular orientation and the integrated behavior of liquid crystals.

Stiff But Flexible

As for what kind of general liquid crystal structures have the appropriate dielectric anisotropy and dynamic response to electric fields necessary to provide controllable birefringence, experiments with many different chemical compounds have found that only the long rod-like molecules having a stiff spine and flexible ends can meet the requirements. Completely pliant molecular structures will bend too easily in response to external forces and thus be unable to maintain any of the positional or orientational order necessary for structural anisotropy. Rigid rods that are free to turn can predispose a smooth ordering along the molecular long axes. The flexible ends provide the interaction at the heads and tails of the molecules that further promote longitudinal ordering, and also facilitate anchoring with the glass substrate (Chapter 11). In addition, the Coulomb attractive force of intrinsically polarized long molecular structures will reduce intermolecular collisions and further promote parallel ordering.

A rough analogy perhaps can provide some insight. A helicopter precipitously dumps a jumble of freshly cut logs into a river; the logs will as a matter of course arrange themselves in a roughly parallel fashion, and more so as they proceed downstream, demonstrating that long rod-like structures can and will naturally manifest a longitudinal orientational order in a fluid medium. In contrast, leaves and twigs with no rigid core will just become entangled and drift in a clumpy mess. However, if the logs have some remaining root twigs and leaves at their ends, then in addition to the orientational ordering, the twigs and leaves on one log will have an entangling attraction to twigs and leaves on another log's end in a roughage analogy to a Coulomb attraction, so that the mutual entangling of ends will further promote parallel positional ordering. Although difficult to demonstrate experimentally, it is likely that completely rigid molecules without flexible ends might go directly from the solid crystal state to the liquid state, bypassing the liquid crystal mesophases.

Liquid crystal compounds presently used in displays are all derivatives of compounds having this long rod-like structure, such as, for example, the

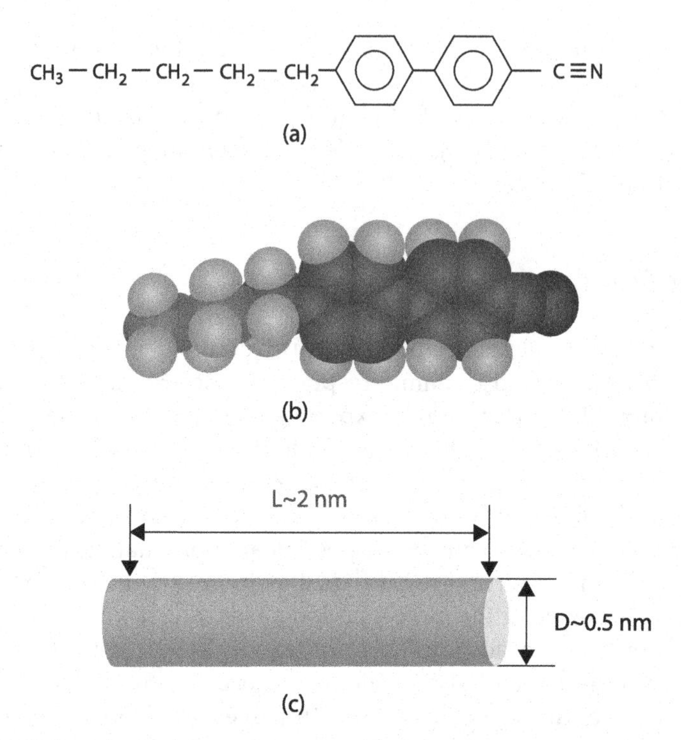

Figure 5.18 Structure of the 5CB liquid crystal. From Yang, D.K., and Wu, S.T. 2006. *Fundamentals of Liquid Crystal Devices*. Wiley, p. 2.

commonly used 4′-n-pentyl-4-cyanobiphenyl (5CB), shown as chemical schematics in Figure 5.18. The chemical forumula at top in Figure 5.18a shows that 5CB has an intrinsic electric dipole polarization triple bond (C≡N) moment. As shown in the molecular structure model of Figure 5.18b, the mid-portion of the molecule is a relatively rigid cyanobiphenyl, and the ends are relatively flexible pentyl hydrocarbon rings. The 5CB and the development of synthetic liquid crystals will be described in Chapter 14.

From Figure 5.18b, it can be seen that the 5CB molecule obviously is not a perfect cylinder, but owing to random thermal motions and the fact that there are no obstacles to rotation around its cylindrical axis, 5CB molecules can and do rapidly and freely rotate about the long axis, and since standing upright or upside-down is equivalent, 5CB physically can be represented by a hard cylindrical long rod as shown in Figure 5.18c.The long rods are used to model uniaxial liquid crystal molecules.

If the molecular structure is such that rotation about the long axis is inhibited, then the liquid crystal belongs to the more structurally complicated biaxial liquid crystal group. So far, no naturally occurring biaxial liquid crystal has been found, but some physical chemists believe that if the middle portion of the structure is curved instead of straight, then the axial asymmetry will produce a two optical axes structure, and this would manifest a natural biaxial-type double refraction [12]. Biaxial liquid crystals have been produced in the laboratory and have been used commercially as compensation films in liquid crystal displays.

The liquid crystal now, in addition to the requirements of physical integrity and sensitivity mentioned previously, further must display backbone, and also possess flexibility in extrema. Whether it is up to the task will depend on its character.

Liquid Crystal Character

The nematic state is a stable, commonly occurring liquid crystal phase, and because of its ordered molecular structure, the nematic phase can effect double refraction similar to solid crystals. But recall that Huygens had to turn the calcite pieces in his hand in order to observe the effect of polarization on the intensity of light, and in a solid, this can be done easily enough, but how can one control the orientation of the liquid crystal molecules? It turns out that because the individual liquid crystal molecules are intrinsic dipoles, an external electric field can induce dipole moments and change their orientation. That is, the natural separation of the positive and negative charges at the ends of the liquid crystal molecule has a mechanical equivalent called an intrinsic dipole moment. When an external electric field is applied, the intrinsic dipoles will turn to align with the field direction because of the Coulomb attractive force between opposite charges. A well-controlled external electric field thus can act as an extremely facile hand that can precisely turn the liquid crystal molecules with far greater precision than the relatively clumsy motions of the human hand. But attaining that precision will require an understanding of a liquid crystal's material character.

The basic features of a liquid crystal display, such as its response time, contrast, viewing angle, and operating voltage, plus the electric resistivity of the liquid crystal itself, are determined by the material character of the liquid crystal. For instance, the phase retardation of the light passing through depends on the anisotropic nature and thickness of the liquid crystal; the threshold voltage and resistivity depend on the dielectric constant and elasticity of the liquid crystal; and the response time depends on the viscosity, the thickness of the liquid crystal layer, and the temperature. The relation between the microscopic dynamics of the molecules and their dependent macroscopic features can be expressed in equations, the coefficients of which can be determined empirically. Following are introductions to the material coefficients that will be used to describe the liquid crystal.

Viscosity

A fluid's viscosity is a measure of its resistance to fluidic motion; that is, viscosity in a liquid crystal is the frictional force between neighboring layers of fluid. The viscosity at room temperature of a nematic liquid crystal, for example, is about 10 times that of water. Although there is no complete theoretical explanation of the viscosity of liquid crystal materials in relation to their other properties and structure, in an anisotropic fluid, viscosity is best described by a viscosity coefficient tensor which can handle different viscosity values in different directions (Chapter 12); and generally, a positive anisotropic liquid crystal having its intrinsic dipole moment aligned with the molecular long axis has a stronger anisotropy and the consequently easier turning makes for a lower viscosity. Viscosity is dependent on temperature exponentially as [13]

$$\xi = \xi_o e^{(E/k_B T)},$$

where in the equation above, ξ_o is an experimentally derived constant, k_B is the well-known Boltzmann constant of statistical physics, and E is the activation energy (the minimum potential energy required for molecular diffusion). It is clear that as the temperature rises, the exponential function $e^{(E/k_B T)}$ decreases rapidly and so the viscosity decreases as temperature rises. As an example, the viscosity of the engine oil in an automobile motor varies with temperature, so in order to achieve optimum performance, engine oils with different viscosity ranges should be used as the seasons change. If the temperature is so low as to cause the fluid to freeze, then the viscosity is infinite, just like in a solid, and the molecules cannot move at all relative to each other (this is clearly bad for the engine).

Since the principal operating motion of liquid crystal molecules in displays are rotational, when an external field is turned on, the viscosity is reflected in the *coefficient of dynamic rotational viscosity* commonly denoted by γ. Of course, the higher the dynamic rotational viscosity, the slower the response of the liquid crystal to the electric field. As will be shown in Chapter 12, the relationship between the response time t_s and the dynamic rotational viscosity γ, and the thickness of the liquid crystal layer d is directly proportional as follows,

$$t_s \sim \gamma \cdot d^2.$$

From the above relation, it can be seen that the larger the dynamic rotational viscosity, the slower the response time, but the thickness of the liquid crystal layer is even more important in determining response time, as it is squared in the relation to response time. Thus typical thicknesses of fast liquid crystal displays are in the micrometer range. Rapid switching on of the electric field would favor a low rotational viscosity, but in particular, turning off the electric field requires consideration of another material coefficient, the elasticity of the liquid crystal.

Elasticity

Elasticity generally refers to the distortion that a body undergoes in response to an external force and its ability to restore itself to its original shape. Under a moderate applied force, a solid body typically will change its exterior shape, but its interior positional and orientational molecular order will not substantially change. One difference between a liquid and a solid then is that the molecules of the former are not rigidly fixed as in the latter, so that when a liquid receives a force, the interior molecules will slide relative to each other, while the exterior shape, if held in a vessel, does not permanently change. This interior sliding and exterior restoration can be likened to a spring under compression; it will undergo twisting, curving, and bending inside, but the coils will only slide relative to each other while retaining the same overall spring shape.

Theoretically, the response of a liquid crystal undergoing an external force can be divided into three different basic types. As shown in Figure 5.19a, when there is no force, the equilibrium configuration maintains positional and orientational order. The distortion shown in Figure 5.19b is called *splay*, indicating an outward widening (also quaintly but accurately described as the response to "pulling a cow by a ring through the nose"), and is denoted by the K_{11} elastic constant. Figure 5.19c shows the *twist* distortion represented by the K_{22} elastic coefficient, and Figure 5.19d is the *bend* distortion represented by K_{33}. Any and all distortions of a body can be described by a combination of these three elastic coefficients. For example, the combination can be described by the *curvature strain*, which will be addressed in Chapter 12.

In preview, a curvature strain will arise from the anchoring of the molecules' tails by the rubbing compound and the distortion caused by the

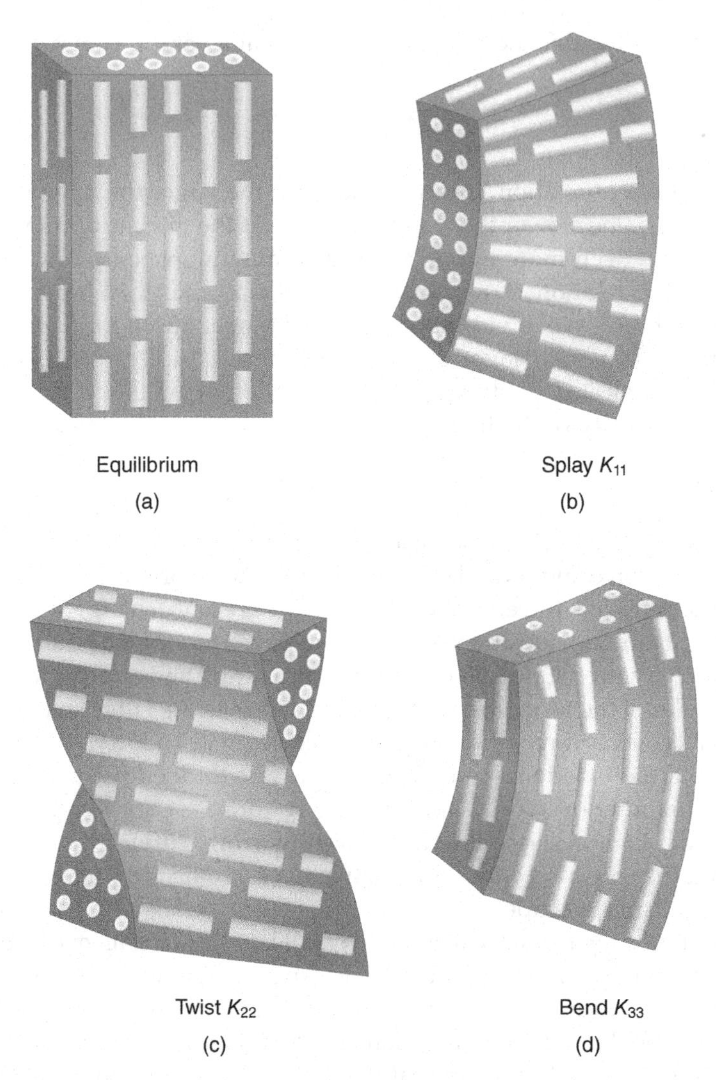

Figure 5.19 Elastic deformations categorization. Adapted from Lueder, E. 2001. *Liquid Crystal Displays*. Wiley, p. 10.

applied electric field; then potential energy will be pent up due to the distortion of the liquid crystal just as in the twisting of a spring. The physical analysis employs the Helmholtz free energy (F) which comprises the change in energy as a function of the orientation of the molecules' director $\hat{n}$ when undergoing the distortions described above. The simplified equation below gives an indication of how the contributions of each of the elastic forces plays out in the energy change of the liquid crystal,

$$\delta F = \frac{1}{2}\left[K_{11}\left(\frac{\partial n_x}{\partial x}\right)^2 + K_{22}\left(\frac{\partial n_y}{\partial y}\right)^2 + K_{33}\left(\frac{\partial n_z}{\partial z}\right)^2\right].$$

In the equation above, the incremental change in the energy (δF) of the liquid crystal is calculated by the splay, twist, and bend components of a distortion times the square of the spatial derivatives of the components of the director. A more detailed description of this type of energy analysis will be given in Chapters 11 and 12.

Upon suffering an external force such as an aligning electric field, the internal structure of the liquid crystal will change, and the elastic coefficients will determine how the molecules will react through the restoring torque; in other words, just how much pressure can a liquid crystal withstand, and how it responds to such pressure is a test of its material character as indicated by its elastic coefficients. With their directors describing the internal molecular orientational order, when an electric field is applied, the molecules' intrinsic dipole moments will turn to align with the field, but the liquid crystal's viscosity and the restoring torque will resist the aligning turn. The tension between the forces can be described by an energy equilibrium analysis that will be introduced in Chapter 12. Using those energy equations, values of the K_{ii} coefficients for different liquid crystal compounds may be determined, and optimum choices of liquid crystal material can be based on the various calculated K_{ii} coefficient values.

The Induced Dipole Moment

In the absence of any external electric field, the aggregate of a uniform uniaxial nematic liquid crystal, even with intrinsic dipoles, is structurally electrically nonpolar. For an intrinsic dipole along the molecular long axis (positive anisotropy), since the molecules in the aggregate have the same probability of being upright or upside-down, there is head–tail symmetry, and thus no preferred charge direction (the definition of *polarity*) in the liquid crystal in spite of the intrinsic dipole moment along the molecular long axis. For intrinsic dipoles perpendicular to the molecular long axis (negative anisotropy), there is rotational symmetry about the director, and reflectional symmetry about the plane perpendicular to the director, resulting in cylindrical symmetry, thus there is equal probability of any directional orientation perpendicular to the director, and therefore again no preferred charge direction, and thus the negative anisotropy liquid crystal is also non-polar.

An electric field acting on the liquid crystal aggregate will induce a dipole moment. For the molecules having an intrinsic dipole moment, there will be three contributions to the induced dipole moment (P_{induced}), in order of magnitude of effect: (1) reorientation of the intrinsic molecular dipole, (2) relative displacement of the atoms of the molecule, and (3) deformation of the electron clouds in the atoms of the molecule; that is,

$$P_{\text{induced}} = P_{\text{molecules (dipole)}} + P_{\text{atoms}} + P_{\text{electrons}}.$$

The magnitude of these relative contributions to the charge polarization depend on the frequency of the electric field acting on the liquid crystal. The turning of the large molecules is sluggish (relatively high inertia), and contributes to induced polarization only up to the relatively slow megahertz frequencies (and of course zero frequency dc electric fields); the applied electric field used to change the birefringence of the liquid crystal by turning the molecules operate in this frequency range.

The atoms of the liquid crystal molecules are relatively well-fixed in the molecule, and although lighter than the molecules, are still heavy, and so the atoms respond only up to the relatively low frequencies of infrared light. Thus atomic-induced dipole moments are not very significant in liquid crystal-induced dipole moment analysis.

The motion of the electrons, being the smallest and lightest, is of course the fastest and contributes to induced polarizations up to the frequencies of ultraviolet light. So for visible light incident on the liquid crystal, the induced charge polarizations will be almost totally due to the electron vibrations in the molecules. For liquid crystal displays then, the induced dipole moment of interest insofar as light is concerned is that arising from electron vibrations, which will be analyzed following.

For light incident on the liquid crystal, using the classical forced oscillator with damping model, a simple analysis of the response of the liquid crystal takes into consideration three forces: (1) the elastic restoring force (Hooke's law),*

$$F_{\text{elastic}} = -kx,$$

* Of course Hooke's law is just an approximation for relatively small distortions that can be described by a simple linear relationship; the intermolecular attractive forces comprising electrons, atomic, and molecular structure are very complicated.

where k is the elastic constant and x the linear displacement from the equilibrium position; (2) the damping force from viscosity,

$$F_{viscosity} = -\gamma \frac{\partial x}{\partial t},$$

where γ is the coefficient of viscosity; and (3) the electric field force from the light, using the convenient complex exponential to denote the force from the light's electric vector,

$$F_{\text{electric}} = -eE_o e^{i\omega t},$$

where e is the electron charge, E_o is the amplitude of the light electric vector, and ω is the frequency of the light. The equation of motion is the familiar second order differential equation of motion,

$$m\frac{d^2 x}{\partial t^2} = -kx - eE_o e^{i\omega t} - \gamma \frac{\partial x}{\partial t},$$

where m is the mass of the electron. The well-known solution is

$$x = x_o e^{i\omega t},$$

where

$$x_o = -\frac{eE_o}{m(\omega^2 - \omega_o^2 + i\gamma\omega)},$$

is the amplitude of the resulting electron vibration, and the induced dipole moment is

$$p = -ex_o = \frac{e^2 E_o}{m(\omega^2 - \omega_o^2 + i\gamma\omega)},$$

where $\omega_o = \sqrt{k/m}$ is just the oscillator frequency. The expression above is called the *transition dipole moment* in quantum mechanics, and this term will be used or understood as referring to the dipole moment of the liquid crystal molecules in response to electric vector of incident light; that is, to good approximation, $p = P_{electrons}$ [14].

When a dc (or relatively low frequency ac) external electric field is applied, as in liquid crystal displays through thin film transistors acting on the ITO electrodes, there is little or no displacement of the atoms in the molecule, and the electrons are more affected by the fast vibrations of the incident light, so the principal reaction of the liquid crystal is that the intrinsic molecular dipole will turn to align with the applied field. An energy analysis of the dipole moment orientation (refer to Chapter 10) shows that the dielectric anisotropy is indeed proportional to the order parameter S, and the birefringence as denoted by the indices of refraction is also approximately linearly proportional to S. This means that the molecular orientational order of the liquid crystal can indeed be changed by an applied electric field to produce a controllable birefringence; this is the essence of the operation of a liquid crystal display.

Upon the application of the external electric field, although the liquid crystal's viscosity will resist the change in molecular orientation, the molecules' turning to align their dipole moments with the field will not be slow and muddling, but rather swift and decisive. This is because the molecules in the middle layers of the liquid crystal are far from the anchoring effects of the rubbing compound on the glass substrates, and thus are relatively free to move, and any voltage above the threshold will result in a sudden change in orientation, making the liquid crystal capable of operating as an *optical switch* or *valve*.

The threshold voltage (V_{th}) is dependent on the liquid crystal's dielectric anisotropy and its elastic character. High dielectric anisotropy makes for low threshold voltages, and high elastic coefficients makes for high threshold voltages. These relationships will be derived and explained in later chapters; for now, an example of an equation relating the threshold voltage with the dielectric anisotropy ($\Delta\varepsilon = \varepsilon_{\Updownarrow} - \varepsilon_{\perp}$), where the subscripts indicate dielectric constants parallel* and perpendicular to the molecular long axis respectively, ε_o is the electric permittivity of free space, and the bend elastic coefficient K_{33} is

$$V_{th} = \pi\sqrt{\frac{K_{33}}{\varepsilon_o \Delta\varepsilon}}.$$

* The double-headed arrow subscript is used to denote the up-down symmetry of the nematic liquid crystal molecules. Auzinsh, M., and Ferber, R. 1995. *Optical Polarizaion of Molecules*, Cambridge, p. 5.

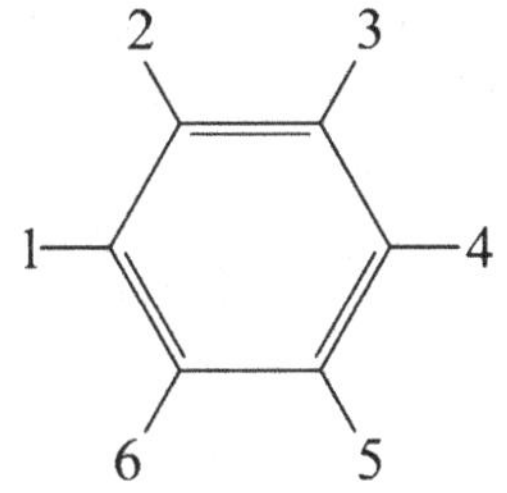

Figure 5.20 Phenyl ring model schematically showing dipole moment positions.

The dielectric constants determining the anisotropy ($\Delta\varepsilon$) of a liquid crystal are dependent on the average charge polarization α, the orientational order parameter S, and the angle β between the dipole moment of the molecule and the liquid crystal director. The dielectric anisotropy of liquid crystals having an intrinsic (permanent) dipole moment is dependent on that angle β and the temperature. For example, the dipole moment in the organic chemistry phenyl ring model is at positions 1–6 as shown in Figure 5.20. If the dipole moment is at an angle β less than 55°, then the dielectric anisotropy $\Delta\varepsilon$ is positive; for β values greater than 55°, the dielectric anisotropy $\Delta\varepsilon$ is negative (the relation will be shown in the phenomenological equation for the dielectric anisotropy developed in Chapter 10).

From this, the positive or negative dielectric anisotropic nature of the liquid crystal and the orientation of the intrinsic dipole moment can be determined. Liquid crystals having a positive $\Delta\varepsilon$ are typically used in twisted nematic and in-plane switching displays, and negative $\Delta\varepsilon$ liquid crystal compounds are generally used in vertically aligned displays. The significance of the positive and negative anisotropies of the liquid crystal will become clear in the chapters to follow.

To reiterate, using an external electric field to control the orientation of the liquid crystal molecules can change the dielectric anisotropy of the liquid crystal, and this will change the polarization state of linearly polarized light passing through. Just as Huygens turned the calcite crystals in his hand so that their relative orientation would change the intensity of the light passing through them, so the modern-day liquid crystal display utilizes an electric field to control the liquid crystal molecular orientation so as to change the polarization state of light coming from an internal light source, and, in conjunction with a second polarizer, thereby control its intensity.

Both solid and liquid materials indeed can double refract light, but solids require gross mechanical means to turn the whole body and thus turn its molecules (one might say the molecules of a solid are too constrained); liquid molecules on the other hand have literally no positional or orientational order to speak of, and their turning and moving is almost uncontrollable (they are too free). Only the mesophase liquid crystal molecules enjoy a liquid's freedom to turn and thus are extremely sensitive to applied external fields, but at the same time possess to some extent a solid's internal positional and orientational order, allowing the liquid crystal to exhibit birefringence (it can be said to possess just the right amounts of freedom and order). It is precisely this mesophase character that makes the liquid crystal the soul of the flat panel display, as shall be seen.

References

[1] Kawamoto, H. 2002. The history of liquid-crystal displays. *Proc. IEEE*, **90**, 4, 461; Castellano, J.A.. 2005. *Liquid Gold*. World Scientific, Singapore; Lehmann, O. 1904; and Collings, P.J. 2002. *Liquid Crystals*, 19, 2nd edition, Princeton.

[2] Reinitzer, F. 1888. Beitrage zur Kenntniss des Cholesterins. *Wiener Monatschr, Fur Chem.*, **9**, 421.

[3] Bernal, M.D., and Crowfoot, D. 1933. *Trans. Faraday Soc.*, **29**, 1032.

[4] Frank, F.C. 1958. *Disc. Faraday Soc.*, **25**, 19; Chandrasekhar, S. 1992. *Liquid Crystals*, 2nd edition., Cambridge, p. 117; and de Gennes, P.G., and Prost, J. 1993. *The Physics of Liquid Crystals*, 2nd edition. Oxford, p. 163ff.

[5] Feynman, R., Leighton, R.B., and Sands, M. 1962. *Lectures on Physics*. Addison-Wesley, Reading, Vol. I, pp. 33–36; and Yang, D.K., and Wu, S.T. 2006. *Fundamentals of Liquid Crystal Devices*. Wiley, Chicester, p. 57ff.

[6] Yang, D.K., and Wu, S.T. 2006. *Fundamentals of Liquid Crystal Devices*. Wiley, Chicester, p. 117ff.; and Chigrinov, V.G. 1999. *Liquid Crystal Devices: Physics and Applications*. Artech House, Boston, p. 165ff.

[7] Hekimoglu, S., and Conn, J. 2003. *Liq. Cryst. Today*, **12**, 3, September; Collings, P.J. 2002. *Liquid Crystals*, 2nd edition. Princeton, p. 181ff.; and Collings, P.J. and Hird, M. 2004. *Introduction to Liquid Crystals*. Taylor & Francis, London, p. 125ff.

[8] Collings, P.J. 2002. *Liquid Crystals*, 2nd edition. Princeton, p. 186ff.

[9] Tsvetkov, V. 1942. *Actya Physicochim (USSR)*, **16**, 132.

[10] Tanenbaum, M., and Pollard, H. 1985. *Ordinary Differential Equations*. Dover, New York, p. 591ff.; and Farlow, S. 1980. *Partial Differential Equations for Scientists and Engineers*. Dover, New York, p. 284.

[11] Saupe, A. 1972. *Mol. Cryst.*, **16**, 87; and de Gennes, P.G., and Prost, J. 1993. *The Physics of Liquid Crystals*, 2nd edition. Oxford, p. 41ff.

[12] Madsen, L.A., et al. 2004. *Phys. Rev. Lett.*, **92**, 145; and Yang, D.K., and Wu, S.T. 2006. *Fundamentals of Liquid Crystal Devices.* Wiley, Chicester, p. 10ff.
[13] Chigrinov, V.G. 1999. *Liquid Crystal Devices: Physics and Applications.* Artech House, Boston, p. 46ff.
[14] Yang, D.K., and Wu, S.T. 2006. *Fundamentals of Liquid Crystal Devices.* Wiley, Chicester, p. 24ff.

6

Thermodynamics for Liquid Crystals

In the preceding chapters, the polarization of light as an electromagnetic wave resulting from the interaction with crystal structures was described, with the light electric vector in an anisotropic medium analyzed as ordinary and extraordinary waves that travel at different speeds through the crystal, resulting in a phase lag that produces crystal birefringence. The mesophases of liquid crystals then were introduced as having a crystal-like fluid structure characterized by the different degrees of positional and orientational order of the different mesophases. That order created structural anisotropies that would produce different optical effects in light passing through much as crystals do, but owing to the fluid nature of liquid crystals, the birefringence was alterable by applying an electric field. An orientational order parameter was then devised to parameterize that fluid molecular order for use in theoretical studies. Finally, the liquid crystal as a material medium responsive to stresses was characterized by its viscosity and elasticity.

A theory of the behavior of liquid crystals at the very least must be able to physically and quantitatively describe the dielectric anisotropy that produces birefringence and the subsequent controllable interaction with light that forms the basis of liquid crystal displays. The theory also must address the solid, mesophase, and liquid phase transitions that are

Liquid Crystal Displays, First Edition. Robert H. Chen.

the essence of the liquid crystals themselves. Models are formulated from constructs based on physics, such as an electric dipole moment field, and results are produced from the models by solving the model equations using mathematical techniques, for example, finding the states of lowest orientational potential energy through minimization of the Helmholtz free energy using the calculus of variations. Since physical systems naturally seek the lowest potential energy states, such minima calculations are used in all the theories.

Although armed with the basic understanding of polarization as a probability state of an individual photon, while some of the other fundamental ideas of the quantum mechanics of photon–electron interactions will be touched on in chapters to follow, upon delving deeper than the interaction between a single photon and electron—or atom or molecule—the theories do not have analytical solutions, and the approximations employed are so dauntingly complex as to force a step back and take a coarser view.

Retreating all the way back to Newtonian mechanics, the motion of an individual dipole molecule under an applied force can be adequately described, but the sheer number of molecules in even a small sample of liquid crystal make mechanistic descriptions based on the individual molecules impossible.

However, taking the averages of the effect of many molecules to produce a mean field is a well-developed statistical mechanics theory for gases, and this was applied to liquid crystals successfully by Maier and Saupe, among others, by using a mean molecular dipole moment field model and calculating the phase transition energies as functions of the orientational order parameter by minimizing the Helmholtz free energy, and deriving an equation for the liquid crystal anisotropy as a function of the orientational order parameter.

In another approach that has the allure of a greater functional simplicity, by taking a power series expansion of the orientational order parameter to model the Helmholtz free energy, and fitting the coefficients to experimental observations, the de Gennes phenomenological model also can produce the phase changes from minimization of the liquid crystal free energy as functions of the orientational order parameter. Then, based on the induced dipole moment, an explicit expression for the anisotropy as a function of orientational parameter can be derived.

For the static and dynamic behavior of the liquid crystal medium as a whole, particularly under the influence of anchoring boundary conditions and electric and magnetic fields, the classical field theory of continuum physics was used by Ericksen and Leslie, among others, to model the anisotropic liquid crystal using an asymmetrical stress tensor in a complex theory grounded in fluid dynamics.

Of course, any good theory requires physical and mathematical acumen, but even elegant equations painstakingly derived must pass the test of experimental verification. Only then will the theory have any value to predict results for display design, or to point out further avenues of research and development. Further, useful theoretical models must not be overly cumbersome mathematically or fraught with coefficients that render the theory expedient but offer little physical insight. So the question to ponder during the discussion of these theories is: Do the theories pass the tests of the observations of the real world, and do they provide guidance and promote understanding, but even if so, at what cost?

Before beginning discussions of the various theories, the physics needed to describe the system—thermodynamics—and the mathematics required for calculations of the physics of the system—the variational calculus—must first be introduced; these will be used repeatedly in the analysis of liquid crystal systems using the theoretical models to be described in Chapters 8–12.

The Three Laws of Thermodynamics

The macroscopic thermodynamic parameters are likely the most familiar physical phenomena because they are encountered in everyday life. For instance, adding heat to boil water to make tea requires a certain amount of heat to raise the temperature so many degrees, which is just the specific heat. When the water boils, it begins a phase change, and the volume of water vapor thermally expands as a gas producing the familiar whistle through the tea kettle spout. Conversely, if the gas is compressed to point of a phase change back to a liquid, it becomes incompressible. But in winter when the ice on the lake freezes in a phase change from liquid to solid, at the freezing temperature the volume of the ice expands with no further change in temperature, which is called the isothermal compressibility. Lastly, the energy required for a phase change, say from ice to water and water to vapor, is what schoolchildren learn as the latent heats of melting and evaporation. The latent heat then is a measure of just how different the phases are, for example, it requires about 80 calories/g to melt ice to water, and 540 calories/g to evaporate water to steam.

The discipline of thermodynamics can be used to investigate these everyday encounters with heat and its effects, and thermodynamics is also indispensable to the study of liquid crystal displays.

The first law of thermodynamics is just the principle of the conservation of energy. For example, in a liquid crystal system, the energy may change form from the elastic potential energy of molecules being rotated to other forms of energy, such as the kinetic energy of motion and heat energy of friction, and although changing form, the total quantity of energy is always the same. But the first law does not impart knowledge of the direction of the changes; for instance, it is known that a falling stone changes its gravitational potential energy to kinetic energy, and upon hitting the ground, the kinetic energy changes to thermal energy of the stone and the ground. However, the reverse will not happen, that is, the thermal energy of the ground and stone will not gather to cause the stone to shoot up into the air, spontaneously gaining kinetic and gravitational potential energy to return to its original point. There is a directionality in the processes of energy transformation.

The second law of thermodynamics in its earliest form states succinctly that the direction of energy transformation is from a hotter place toward a colder place, so for instance, putting the tea kettle on the burner will not result in the water in the kettle becoming colder and the burner flame hotter. In a later form, the second law states that spontaneous changes in a system will go towards states of greater disorder as quantified by its *entropy*.* But the second law gives no baseline for the measure of disorder.

The third law of thermodynamics in one form states that the disorder, as measured by its entropy, of a solid crystal in thermal equilibrium (meaning that there is no net flow of heat, which implies no change in temperature within the system and with the surroundings) approaches zero as the temperature approaches absolute zero (Kelvin), thereby setting the baseline for order and its subsequent journey to disorder.

The three laws of thermodynamics are used to investigate the nature of the liquid crystal phases themselves, and to calculate the orientational distribution of the liquid crystal molecules using the technique of minimizing the liquid crystal's orientational potential energy. Understanding of the phases helps to select appropriate liquid crystals for display use, and calculations of the minimum potential energy (stable) states provides guidance for the design of liquid crystal displays.

Phase Transitions

To form the mesophases, the latent heat required for the solid to liquid crystal transition is about 60 calories/g, and the liquid crystal to isotropic

* To show the equivalence of the early form of hot to cold and the later spontaneous directionality toward disorder of the second law is not a simple matter.

liquid transition requires only about 2 to 7 calories/g (depending on the type of liquid crystal), showing at the outset that a liquid crystal is more liquid than crystal. The transitions from solid to liquid crystal mesophase to liquid equilibrium phases of liquid crystals are achieved through the addition of heat, which causes the molecules to move away from orientational and/or positional order. In the phase transition from solid to fluid, the positional order will begin to dissolve, and the molecules will commence an interlayer sliding, but the molecules may retain a parallel-like orientational order, becoming in the process a "liquid-like solid" or liquid crystal. This phase transition can be parameterized by the macroscopic thermodynamic parameters of volume and entropy.

Continuing to slowly add heat will cause the ordered fluid crystal to gradually lose any positional and orientational order it may have had and phase into a disordered anisotropic pure liquid. This phase transition is *discontinuous* in that it jumps between subphases at two different temperatures (as will be seen). This so-called weak first-order (discontinuous) second-phase transition is a manifestation of the molecules' long-range orientational disorder as described by the orientational order parameter S of the liquid crystal, and can be further parameterized by relative increases in the thermodynamic specific heat and thermal expansion, while the volume and entropy remain only slightly changed.

Historically, the solid to fluid phase transition has been called the "first," since it is the first phase transition from adding heat; the fluid (or mesophase) to isotropic liquid phase transition is called the "second" phase transition, since it is the next phase transition with added heat, but since the latter is also deemed "first-order" (because the orientational order parameter undergoes a discontinuous change), and since a "second-order" or continuous phase transition is when the orientational order parameter S changes continuously as the temperature decreases through the phase transition, calling the first-order phase transition a "second phase transition" may lead to confusion. Therefore, in the descriptions of the liquid crystal phase transitions, where appropriate, the solid to mesophase phase transition will be called just that, and the mesophase to isotropic liquid phase transition important to the theories of liquid crystals will be called the "mesophase-to-liquid," or more specifically the "nematic-to-isotropic" (NI) phase transition.

Entropy

The disorder of a system is quantified in the second law as *entropy*, a measure of disorder with a baseline in the crystal solid state at absolute zero (third law). Entropy then is a measure of the disorder of a system of liquid crystal molecules, and for spontaneous changes (what happens when a constraint on the system is removed), the total energy transformed will be the same (first law), and the disorder of the system will always naturally increase (second law).

This mysterious entropy is encountered in everyday life, but probably not recognized for what it is (which is admittedly not so clear-cut). A simple example is that while walking, if your shoelaces becomes loose and start to unravel, they have been released from the constraint of being tied, and as you continue walking, your shoelaces will spontaneously further unravel as your walking system continues; and so it is that the laces will not miraculously retie themselves as you walk, but rather the longer you walk, the looser your shoelaces become. That is, in your walking system, the shoelaces have become more and more randomly *disordered,* and the entropy of your walking system has spontaneously increased.

The spontaneous increase of entropy can be explained by the probability of occurrence of states in the physical theory of statistical mechanics. To illustrate, if five coins are tossed, the probability of all heads or all tails is very low, but the probability of four heads and one tail (or vice-versa) is five times larger because five different arrangements of heads and one tail satisfy the criterion (since there are five different coins which can be the odd-out). For three heads (tails) and two tails (heads), there are 10 cases each that satisfy the criterion (3-2 either way), and thus each is ten times as likely to occur. So the highly ordered state of all heads or all tails has only one state, and the more disordered state of three heads and two tails has 10 possible states. Therefore, the high-disorder state is more probable because there are more available states.

In addition to the heads/tails state, which could be called the "orientational state" of the coin, its position after being thrown and landed is also a "state of position," but of many more possibilities. The total energy of a system in thermal equilibrium (uniform temperature) will be equally partitioned among all the possible different states (the degrees of freedom, such as position and orientation) of the constituents according to the principle of *equipartition of energy.*

If 100 coins are tossed, the number of ways for all combinations of heads and tails is about 10^{30}, and the probability of all heads is practically negligible at $1/10^{30}$. Of course, the number of possible positions of the coins is even more formidable. For typical molecular systems of say 10^{23} molecules, the possible states of where the molecule may be (position) and its pointing direction (angular orientation) are correspondingly enormous, and the probability that systems of many molecules will spontaneously go toward states where the possible positions and orientations are the most is almost a certainty; that is why one never sees, for example, the molecules of a gas in a box spontaneously all line up and congregate in a corner. And that is also why your untied shoelaces will not spontaneously re-tie themselves as you continue walking. In other words, the tied shoelace is highly ordered, having few states, while the untied shoelace has many possible different unordered states, and is thus overwhelmingly more probable.

The direction priority of the natural spontaneous transfer of heat (energy) is determined by that probability and expressed by the second law of thermodynamics. In the dance of order and disorder irremeably performed in the classical physics view of Nature, with every spontaneous event, the arrow of time flies only forward, and as natural events are irreversible, they always proceed towards the greater disorder. Actually all events, natural or otherwise, are irreversible; some are closer to being reversible (like turning the five coins back to all heads), but strictly are still irreversible (the positions of the coins cannot be exactly the same as before); the mathematical construct of reversibility is really just to minimize entropy gain for thermodynamic calculations.

Entropy can be quantified as defined by the father of statistical mechanics Ludwig Boltzmann, and if N is the number of possible states:

$$\Omega = k_B \ln N,$$

where k_B is the Boltzmann constant, and here Ω is used to denote the entropy (usually it is S, but since S has been and will be used to denote the orientational order parameter, Ω is used instead to avoid confusion). The use of the natural logarithm (*ln*) in Boltzmann's entropy equation is not only to make the expression of huge numbers involved in the entropy of ordinary systems of molecules more convenient, it also shows that for interdependent systems, the entropy is *additive* (it can be negative); that is, for two systems interacting, the entropies of each are added according to the Boltzmann entropy equation.

For example, in describing the entropy of the spontaneous formation of life forms on Earth, since living matter clearly is in a more ordered state than the random distribution of its constituent chemicals, it is natural to ask: How can life spontaneously occur with an increase of *order* rather than disorder? Is the Second Law incorrect as regards the formation of life?

The answer is that a decrease in entropy can occur only in respect to the entropy increase of a coupled system, with the result that the total entropy is increased by addition. In the example of life formation, green plants use photosynthesis to convert carbon dioxide and water to produce chemically more orderly glucose, so the coupled system is the Sun, and the exothermic releases of solar photons through fusion reactions in the Sun results in the increase of entropy there, which is greater than the decrease in entropy in the production of glucose.

At any rate, because the total *energy* must be the same coming in and going out, and the energy from the Sun is composed of higher energy ultraviolet photons as opposed to the lower energy Earth reradiated infrared photons, the Sun photons must have lower entropy by way of fewer photons than those reradiated by Earth. Thus the glucose has high free energy available to drive life-forming chemical reactions [1].

This kind of argument, however, seems to allow a way out of the conundrum of ever-increasing entropy simply by expanding the "system" to include something where the entropy increase will offset any entropy decrease in the system in question, which may be somewhat unsatisfying, for it leads inevitably to the "system" always having to be the whole Universe [2].

For the immense system of the Universe, it can be said that entropy started at a zero value at the creation of the Universe (whence it was apparently highly ordered because of a uniform, and therefore low entropy, hydrogen gas that is the origin of all the stars, including our Sun) and has increased until it will reach its maximum value at the ultimate destruction of the Universe, probably at the edge of the last giant black hole. But why should a black hole have maximum entropy?

A black hole can be conceptualized as a situation where the Newtonian escape velocity from a body $(2GM/R)^{1/2}$, which is square root directly proportional to the mass and inversely proportional to the radius of the body, may exceed the speed of light when the mass is large enough and the radius is small enough. Thus not even light can escape, making the body a "dark star." Then considering quantum mechanical complications when the body decreases to atomic size; that is, the Pauli principle forbids two identical electrons, protons, or neutrons (*fermions*) from being in the same state (*degeneracy*), but if the mass of the body is greater than about 1.6 times the Sun's mass, relativity overcomes the degeneracy and the body collapses into a *white dwarf* (electron degeneracy overcome) or a *neutron star* (neutron degeneracy overcome). If sufficiently massive, say 10 times the Sun's mass, the frictional heat of the imploding matter may trigger an explosive *supernova* (which have been observed, and they are so gigantic that they are not difficult to see), and if sufficient mass is retained and the collapsing radius becomes small enough, then the escape velocity may indeed increase beyond the speed of light, and everything is immediately and irremeably absorbed into that body—the dark star is now a black hole.*

The black hole may be seen as having maximum entropy because all that mass is concentrated in one point in space; a point which must be very disordered because of so many particles in one point. In between those cataclysmic times, entropy is a parameter in the second law of thermodynamics accompanying the evolution (or devolution) of time.

For anyone still unconvinced, the primacy of the second law was asserted by the eminent astrophysicist Arthur Eddington as follows [3]:

> If someone points out to you that your pet theory of the universe is in disagreement with Maxwell's equations—then so much the worse for Maxwell's equations. If it is found to be contradicted by observation—well, these experimentalists do bungle things sometimes. But if your theory is found to be against the second law of thermodynamics I can offer you no hope; there is nothing for it but to collapse in deepest humiliation.

Fortunately, the operational theories of liquid crystal displays do obey the Second Law of Thermodynamics and Maxwell's equations to boot, and the experimentalists have verified the basic theories fairly well.

The Boltzmann Distribution

So the three laws of thermodynamics, based on observation and experimentation, can provide macroscopic parameters, such as temperature, pressure, volume, and entropy, to study matter, including liquid crystals. The statistical mechanics, based on the atomistic theory of matter, then describes how atoms and molecules are distributed in energy (or velocity) space.

The so-called *Boltzmann distribution* from statistical mechanics can describe the energy distribution of an ideal gas, but as mentioned previously, liquid crystal molecules do interact chemically and electrically, and thus a rigorous description of the orientational energy distribution of liquid crystal molecules would require consideration of the so-called *short-range molecular interaction*. Fortunately, if the interaction between molecules is sufficiently weak, such that adding to the energy of individual molecules will result only in

* The theoretical estimate of the entropy at the end is 10^{123}, Penrose, R. 2004. *The Road to Reality*. Vintage, p. 707. A neutron star supernova explosion was recorded by Chinese and Japanese astronomers in July 1054; it is visible today as the Crab Nebula; other supernovas were recorded in 1572 by Tycho Brahe and 1604 by Johannes Kepler; refer to Maffei, P. 1978. *Beyond the Moon*. Avon and MIT Press, p. 153, and Moore, P. 1980. *Astronomy*. Simon & Schuster, p. 7.

the energy of the whole system increasing (thus as far as the energy is concerned, approximating an ideal gas), then the Boltzmann distribution can be used to describe the liquid crystal's orientational energy distribution as well, with the caveat that using the Boltzmann distribution to describe the energy distribution of the entire fluid cannot explicitly consider any short-range intermolecular attractive forces.

In speaking of a *distribution*, the concept hopefully can be illuminated by a simple example. The test grades of a class of students can be measured by the average score of all the students in the class, for instance if it is 65, then although we don't know how each individual student did on the test, we have an idea of the situation as a whole (the class), much like a macroscopic thermodynamic parameter such as the temperature. At the other extreme, in more competitive schools, the grades of each individual student are posted next to their names, analogous to knowing the kinetic energy of each individual molecule in a gas. Here the student (the molecule) has no place to hide, everyone knows his or her grade. More commonly, the test grades may be categorized on a grade scale such as the commonly used 100–90 = A, 89–80 = B, 79–70 = C, 69–60 = D, 59–0 = F system. In so doing the individual students are assigned to a level, but their precise individual score is not revealed, although a good idea of their test-taking ability is known.

If the results form a so-called normal, or *Gaussian* distribution (the familiar "bell curve"), then most students will be in the "C" range. For an ideal gas in thermal equilibrium, Maxwell showed that the distribution of velocities will be in accord with half of the Gaussian distribution, which is just an exponential decay with energy (the *Maxwell distribution*).

The kinetic theory of gases describes the gross behavior of matter from collisions of the constituent molecules. Statistical mechanics describes the distribution in energy space of the molecules resulting from an applied force on the matter that is in thermal equilibrium.

An *ideal gas* is composed of inert atoms that obey the *ideal gas law* where pressure and volume are determined by temperature alone

$$PV = Nk_BT,$$

where P is the pressure, V is the volume, N is the total number of molecules, k_B is the Boltzmann constant, and T is the temperature. If n is the number of molecules per unit volume ($n = N/V$), then the ideal gas law can be written as

$$P = nk_BT.$$

In the earth's atmosphere, due to gravity, the most air is at the bottom of a column of air, where the pressure is the highest (since the air near the surface has to hold up all the air above it). The difference in pressure at an altitude h is just the weight of the gas in the disk of air between h and $h + dh$, where dh is just the altitude differential. The force of gravity is given by mg, where m is mass of the disk of gas, and g is the gravitational acceleration, so the pressure difference is

$$dP = -mgndh.$$

Since $P = nk_BT$, and T is assumed constant in the disk, eliminate P to get a differential equation for the density of the air molecules,

$$\frac{dn}{dh} = -\frac{mg}{k_BT}n.$$

In the above equation, the density n has a derivative with respect to altitude that is proportional to itself, and a function that has such a property is just the exponential function, so a solution to the differential equation is just

$$n = n_o e^{(-mgh/k_BT)},$$

where n_o is the molecular density at point $h = 0$. So the distribution of the air molecules is exponentially decreasing with altitude depending on the gravitational potential energy mgh and temperature T.

In physics, probabilities (including distributions) are multiplied for resultant probabilities, and energies are added for total energy; only an exponential function will turn a multiplication into an addition, so probabilities must depend on energy exponentially. The gravitational potential energy in the above equation then can be generalized to any potential energy, and so *Boltzmann's law* gives the spatial distribution of molecules under a potential energy as a function of the exponential of the negative potential energy divided by the temperature

$$n = n_o e^{(-E/k_BT)},$$

where E is the potential energy. Boltzmann's law thus describes the journey of the molecules to the Maxwell distribution thermal equilibrium state through the impetus of the second law of thermodynamics by way of an exponential function.

Since any potential energy can be used in the equation, the energy density distribution of liquid crystal molecules responding to an orientational potential energy in a liquid crystal display can be described by Boltzmann's law. The resulting so-called *Boltzmann distribution* will be used in all the liquid crystal theoretical models to be introduced below.*

The Minimization of Free Energy

Applying an electric field to, or using boundary alignment chemicals to constrain the orientation of, a liquid crystal will cause the dipolar molecules to rotate, and the liquid crystal is forced into a more ordered orientational state of alignment with the field or boundary condition, and gains elastic energy in the process. The energy disposition of the molecules then will seek the lowest potential energy state in that constrained orientational situation. If the electric field is turned off, the molecules are released from a constraint, and in accordance with the first law of thermodynamics, the elastic energy is converted to kinetic energy, and through inter-molecular collisions, following the second law of thermodynamics, the molecules spontaneously revert to a more random disordered orientational distribution, in the process giving off heat.

Any physical system disturbed by an external force will have its energy increased and then seek the most stable state of minimum potential energy under the constraint in an "unstable equilibrium" configuration. Upon release from the constraint, that system will then spontaneously seek its lowest potential energy state, which is a stable equilibrium state. Both equilibrium states are attained by the minimization of the system's so-called *free energy*.

This free energy is potential energy that is naturally minimized by the system. This can be illustrated for example by throwing a ball into a bowl, which results in the kinetic energy transforming into friction heating both the ball and the bowl, but the ball will eventually come to rest with minimum gravitational potential energy at the bottom of the bowl. The mean field and continuum theories of liquid crystals both utilize the pursuit of lowest poten-

* The Boltzmann constant is defined by the ideal gas law as $k_B = PV/TN$. For more on statistical mechanics, refer to *Feynman* I: 40–22 and Planck, M., *Treatise on Thermodynamics* (1897) (Dover, 1945), and Genault, T., *Statistical Physics* (Routledge, 1988). Boltzmann committed suicide, it has been surmised, because of depression wrought by the aversion to the atomic theory of matter in Austria in the 19th century.

tial energy of the system to calculate the stable equilibrium states of the system; the free energy is usually called the *Helmholtz free energy*, which is defined as

$$F = U - T\Omega,$$

where U is the molar potential energy, T is the temperature, and Ω is the entropy. It can be seen that the free energy is a potential energy, and that as the entropy increases, the free energy decreases.

In textbooks and the literature, this "free energy" has various names, all of which will be used in this book as appropriate, and in the theoretical models to be introduced here, they usually refer to the same thing but sometimes slightly restricted by the issue at hand. Thus the *free energy* is also called the *Helmholtz free energy*, *orientational energy*, *distortion energy*, *elastic energy*, and *nematic energy*, all meaning the potential energy of the system that is "free" and can be calculated under the particular circumstances and minimized to obtain the equilibrium states of the system.

The basic physics that will be used in the theories to be described below have been set forth above. Now the mathematical analysis used in the calculations of the stable potential energy states that determine the orientational order of the liquid crystal molecules and the subsequent effect on incident light will be introduced: it is the *Calculus of Variations.*

References

[1] Oxtoby, D.W., Nachtrieb, N.H., and Freeman, W.A. 1994. *Chemistry*, 2nd edition. Harcourt Brace, Philadelphia, p. 462.
[2] Penrose, R. 2004. *The Road to Reality*. Vintage, New York, p. 686ff.
[3] Eddington, A.S. 1929. *The Nature of the Physical World*. Cambridge.

7

The Calculus of Variations

The physical basis for the calculation of energy states in the various liquid crystal theories is that a physical system will always tend to its lowest potential energy state; this is manifested by changes to the system's free energy, which will seek a minimum value. The states of the minimum free energy are calculated by the methods of the variational calculus, an extremely useful mathematical discipline, introduced below with some of its rather interesting history.

Maxima and minima (together *extrema*) calculations are commonly used in the natural and the engineering sciences. For example, in using Newton's fluid dynamics to determine the very practical problem of the best shape of a ship's hull to offer the lowest resistance to the water, it turns out unsurprisingly that it is just the minimum surface between two points, which result is easy to accept, if not so easy to prove. The extremum problem was also encountered in everyday life, as for instance in feudal Europe, land was ceded from father to sons according to how much land each son could mark off in one day given ropes of equal length, the objective of course being to encompass the maximum area possible.

Historically, this type of maximization has roots going back three thousand years (900 B.C.) to the era of the Phoenicians and their Princess Dido. In escaping from her tyrannical brother, the Princess sought asylum in what

Liquid Crystal Displays, First Edition. Robert H. Chen.

is today Tunisia on the northwestern shore of Africa. The king there granted asylum but dismissively bequeathed her "all the land that could be contained in a bull's skin" as her kingdom. The analytical Princess then proceeded to cut the bull's skin into thin strips, tying them together to form a very long rope, and placing one end at a point on the shoreline, played out the line in semicircle of considerable radius to form a substantial area, which years later was to become the mythical great city of Carthage, over which the Princess reigned to become Queen Dido [1].

This appealing story of the maximization of the area encompassed by a curve with arbitrary endpoints became known affectionately as *Dido's Problem*, and later generically as the *isoperimetric problem* in mathematics. Everyone should know the answer to Dido's Problem; the largest area is of course that delineated by a circle, an answer that is apparently eminently obvious. But mathematicians are strange ducks and are more interested in the proof that a circle does indeed encompass the greatest area rather than in the actual answer, and the mathematical demonstration of that proof is not so obvious. In fact, many attempts at proof before the invention of the calculus in the 17th century were made, among them the inscription of a polygon inside a circle, and as the sides of the polygon are increased to form an *n-gon*, as n increases, the area will increase, and as $n \rightarrow \infty$, the *n-gon* will approximate a circle, which will be the maximum area *n-gon*; this intuitive method approaches the idea of limits in calculus. But this is only a thought experiment resulting in an approximation, not a mathematical proof. It was not until the invention of the variational calculus that the *Lagrange multiplier* could rigorously prove the hypothesis, and it was the great mathematician Weierstrass who produced the first formal proof of the isoperimetric problem [2].

The Brachistochrone Problem

Every schoolchild knows the story about Galileo dropping stones from the leaning tower of Pisa to find that freely falling bodies will hit the ground at the same time regardless of weight, and his later study of the motion of the planets around the Sun.* In line with these observations, a simple problem strangely became popular in mathematical circles:

* Alas, the story is apocryphal, as it has been "proved" that Galileo could not have performed such a demonstration precisely because of air resistance and the limitations of time measurement at the time; no record of such an experiment has been found in Galileo's papers; the story comes from a biography by Galileo's assistant Vincenzo Viviani, who may have embellished the investigation. Feinberg, G. 1965. *Am. J. Phys.*, **33**, 501, and recounted in Giudice, G.F. *A Zeptospace Odyssey* (Oxford, 2010).

> For a wire fixed at two different points of different height, what is the path for a body to fall the fastest from one point to the other.

The problem came to be known as the *Brachistochrone problem*, from the Greek *brachisto*, meaning "the shortest path," and *chronos*, meaning "time." At first, the problem appears amenable to the simple minimization technique of the differential calculus (as performed in Chapter 4 in proving Snell's law of refraction), but that is only good for finding the minimum *point* (value) of a function, whereas the Brachistochrone problem requires finding the minimum *function* (curve) of all possible functions. After many years and many attempts by the preeminent mathematicians of the time, the problem finally was solved by the Swiss mathematician Johann Bernoulli, and his techniques gave rise to a completely new branch of mathematics known today as *the calculus of variations*.*

Interestingly enough, Bernoulli's first attempt at solving the brachistochrone problem involved the use of Snell's law of refraction and Fermat's principle of least time described in Chapter 4. In a remarkable display of insight, Bernoulli first turned Chapter 4's Snell's law of refraction (Figure 4.12) upside-down, and, as shown in Figure 7.1, the medium of lower refractive index (1) is placed below the medium of higher refractive index (2).

Snell's law of refraction for this case can be written,

$$\frac{\sin\theta_1}{v_1} = \frac{\sin\theta_2}{v_2} = const.$$

According to the above and the relation between index of refraction and the speed of the light beam,

$$n_j = c / v_j,$$

and from this it can be seen that the smaller n_j is, the larger (faster) v_j will be, so the downward-traveling light beam will be going faster and faster. If more and more monotonically lower index of refraction layers are added below the bottom-most layer, then the light beam should be traveling faster and faster, and as the layers are made thinner, the light beam path gradually turns, as shown in the lines of Figure 7.2.

---

* The mathematicians with the family name Bernoulli include the father *Nikolaus*, the younger son *Johann*, the elder *Jakob*, and his son *Daniel*, as well as a nephew also named *Nikolaus*. They all were outstanding mathematicians, and, unfortunately, in different languages, the different spellings of their given names also caused confusion, for example German, English, and Belgian spellings *Johannes*, *John*, and *Johann*.

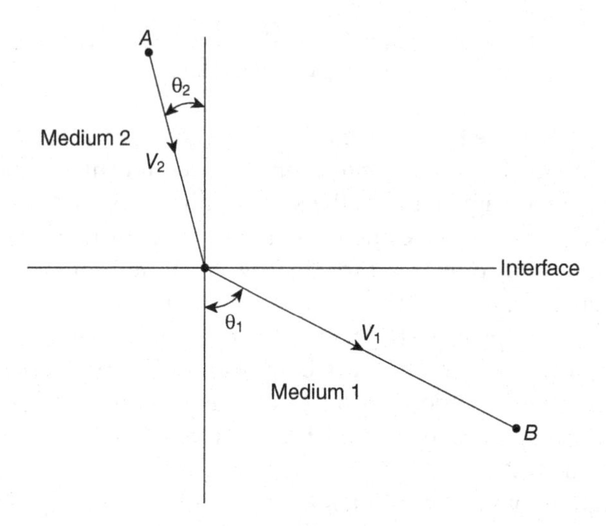

Figure 7.1 The medium of lower refractive index (1) is placed below the medium of higher refractive index (2).

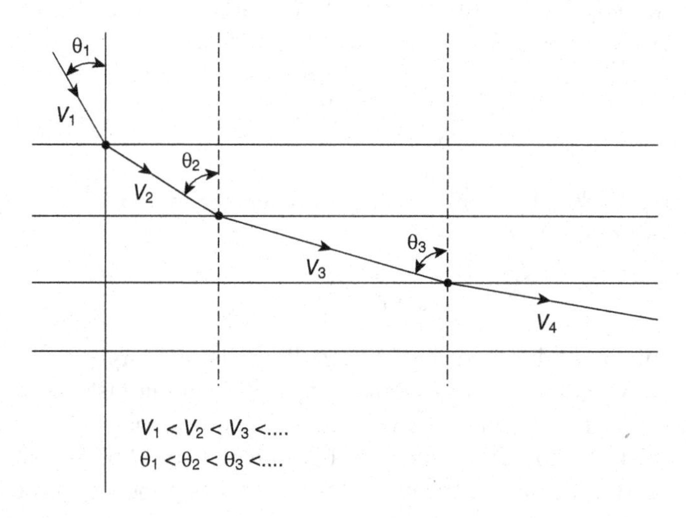

Figure 7.2 Monotonically lower index of refraction layers added causes the light beam to travel faster and the light beam path gradually turns.

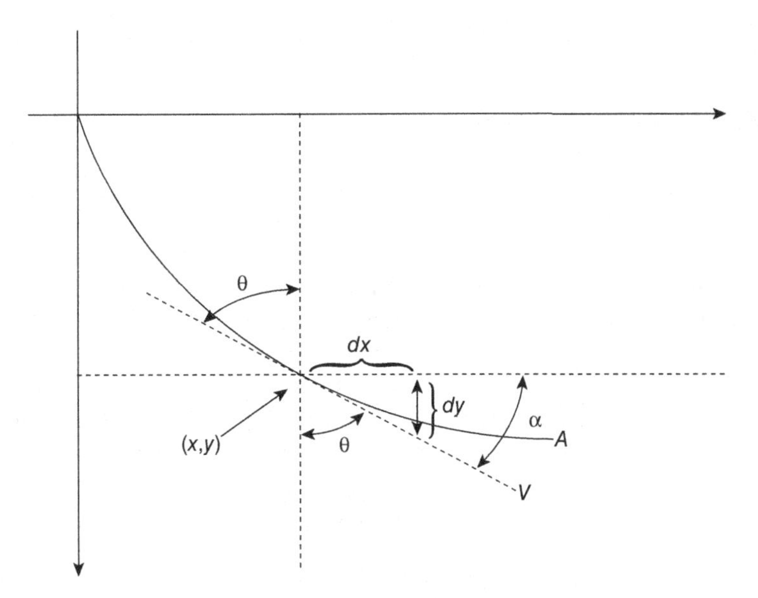

Figure 7.3 Taking infinitesimally thin slabs of lower indices of refraction, the body on the curve of fastest descent from B to A has a speed of $v(dy/dx)$.

If now the number of layers of decreasing index of refraction media is increased infinitely, the thickness of each layer will approach zero in the limit, and the straight light beam paths become a smooth curve as depicted in Figure 7.3, and at every point on the curve,

$$\frac{\sin\theta}{v} = const.$$

Bernoulli's insight is that the above procedure will minimize the time that the light beam will travel, and that by using this result, the path of least time is revealed, thus solving the brachistochrone problem [3]. It should be noted that Fermat's principle of least time for a body to fall in a path from one point to another is inherent in the construct through the speed of light. However, although the curve can be drawn and shown in Figure 7.3, the equation of the curve was still unknown, and it would be found by Bernoulli, again using considerable physical insight.

From Figure 7.3, at any point (x,y) on the curve from B to A, the speed of the body is v, which is just the tangent at that point, and according to the law of conservation of energy, the mass m under the acceleration of gravity g transforms its potential energy to kinetic energy as follows,

$$\tfrac{1}{2}mv^2 = mgy,$$

so the speed of the body is thus

$$v = \sqrt{2gy}.$$

Ingeniously, Bernoulli now analogized the speed of the light beam through the infinitesimal layers to the speed of a body falling through the layers v, and because the light beam traverses the layers according to Fermat's principle of least time, the body is falling at the fastest speed. From the analytical geometry of Figure 7.3, a simple change of variables and trigonometric identities, the integration of the derived differential equation yields the equation of the shortest path, which is given by the parametric equations

$$x = a(\beta - \sin\beta)$$
$$y = a(1 - \cos\beta),$$

where $\beta = 2\varphi = 2\cdot\tan^{-1}(\sqrt{y}/C - y)$, and thus $C = y[1 + (dy/dx)^2]$, and these equations describe a cycloid.

This was surprising, even to Bernoulli, that the curve of the fastest descent is a cycloid, the locus of a point on a circle rolling in a straight line on a plane surface, as shown in Figure 7.4.

Turning the cycloid upside-down will then just be the curve of fastest descent of the body, as shown in Figure 7.5. The reader might think a bit, why should the curve of fastest descent between two different points have anything to do with a point on the circumference of a rolling circle, a circle

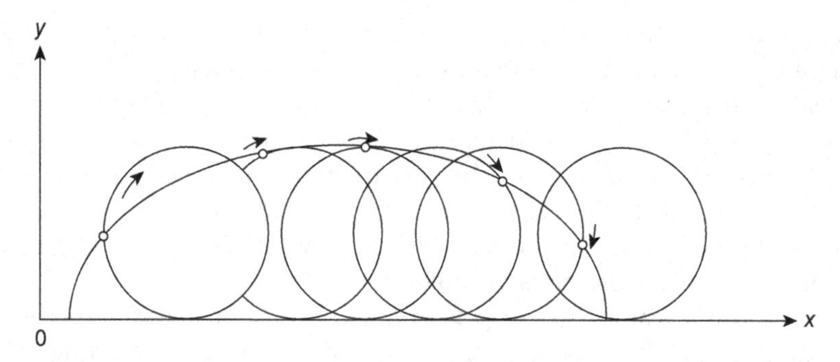

Figure 7.4 The locus of a point on a circle rolling in a straight line on a plane surface is a cycloid.

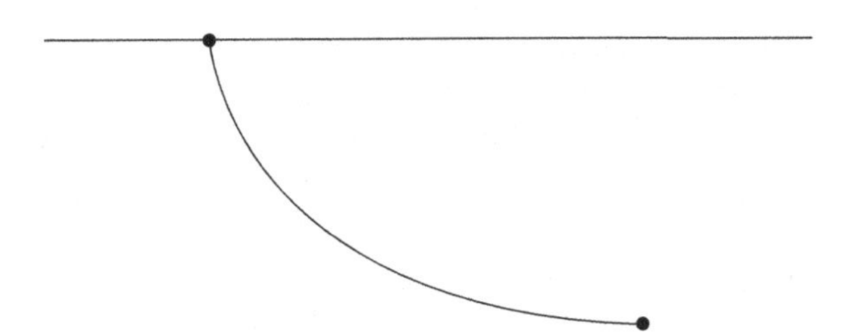

Figure 7.5 The cycloid turned upside-down is the curve of fastest descent.

that encompasses the maximum area possible? Is the point on the circumference on its way down really the fastest route down it can follow?

Today, the brachistochrone manifests itself in a good roller coaster's initial hair-raising descent from the precipice down towards its nadir. The details of Bernoulli's derivation of the cycloid from Snell's law of refraction and Fermat's principle of least time are not presented here because after the invention of the calculus of variations, the cycloid could be very easily derived from the Euler–Lagrange equation to be introduced below.

Catenary and Suspension

Another analytical geometry problem in engineering solvable by the variational calculus is to find the mathematical curve of a hanging chain (like a heavy electric cable), the *catenary*. Intriguingly, the tension imparted by an electric field exerted on a liquid crystal, which will be dealt with in great detail in Chapter 12, is similar to the weight exerted by gravity on the cable. Galileo, in thinking about the catenary, at first incorrectly believed that the hanging cable should be parabolic, but it turns out that it is a suspension cable holding a bridge deck that forms a parabola ($y = x^2$), whereas the catenary turns out to be the hyperbolic cosine ($y = \cosh x$). Again it was Bernoulli who solved both problems using the variational calculus. The difference between the catenary and the parabola of course comes from the different distribution of weight on the cables; in a catenary, the weight is that of the cable itself and is evenly distributed along the length of the cable, whereas for the suspension bridge cable, the weight is distributed on the cable coming from the weight of a deck hanging horizontally below; showing, logically enough, that different weight distributions result in different curves.

Interestingly, in the same intuitive manner that the brachistochrone problem was solved by Bernoulli, the force exerted on a catenary arises from the earth's gravity resulting in a pure tension; and as well known to structural engineers, if an architectural structure is to be stable, it must avoid tension and rely on a pure compression. In the 17th century, Robert Hooke conceptually froze the catenary and turned it upside-down to form an arch, and since the catenary depended entirely on gravity for its shape, the arch conversely thereby relied wholly on gravitational compression from the weight of the material and thus was entirely free of tension. Even today, building and bridge arches are constructed on this principle. It is evident that the variational calculus could very well handle problems that arose from utilitarian engineering concerns, but its development proceeded into the realm of pure mathematics as well [4].

Although the great mathematical physicists Descartes, Fermat, Newton, Leibniz, and Bernoulli all contributed to the early use of mathematical extrema in physics (dare one say "physical extremism"?), it was not until 1744 that the variational calculus was developed by Bernoulli's protégé the great genius Leonhard Euler, and together with its subsequent applications in mechanics by Joseph Lagrange, the physical application of the variational calculus has become known as the *Euler–Lagrange Equation.*

The calculus of variations began from the study of very practical extrema problems such as the minimum surface, isoperimetric, brachistochrone, catenary, and suspension curves; it then evolved to the rather esoteric calculations of minimum surfaces of soap bubbles, but then returned to its practical roots when used in the Hamiltonian energy analysis of physics and today's modern optimization and control theories. From this, it can be seen that the calculus of variations spans the ancient to the modern, and delves deeply into pure mathematics.

Returning to liquid crystal displays, it is the thermodynamics of liquid crystals as manifested in the minimization of the Helmholtz free energy that forms the theoretical physics basis for the study of liquid crystal systems, and the Euler–Lagrange equation is its indispensable mathematical instrument.

The Euler-Lagrange Equation

Generally, a line integral gives the value of a curve that possesses a minimum value $y(x)$, where $f(x,y)$ is a continuous function describing the physical system. This is different from the simple minimization calculation in the differential calculus because the dependent function $y(x)$ is itself a curve, and not just a point on a curve. In its most general form, $f(x,y)$ is a function of the independent variable x, the dependent variable y, and their derivative dy/dx (i.e., the change of y with x), as

$$f\left(x,y,\frac{dy}{dx}\right),$$

and the value of $f(x,y,dy/dx)$ from the arbitrary points x_A to x_B is given by line integral,

$$I=\int_{x_A}^{x_B} f\left(x,y,\frac{dy}{dx}\right)dx.$$

Because one does not know beforehand what the minimum curve $y(x)$ is, a *test function* is presumed $\bar{y}(x)$, which is just the difference from the desired minimum function $y(x)$ as expressed by an additional term $\varepsilon\mu(x)$, that is, ε is a parameter, and $\mu(x)$ is a function of the independent variable x, and because $y(x)$ is continuous, $\mu(x)$ also must be continuous. The trick of the variational calculus lies in the definition of $\bar{y}(x)$ as being equal to $y(x)$ when ε equals zero,

$$\bar{y}(x)=y(x)+\varepsilon\mu(x).$$

Since the objective is to find the minimum value of a curve, the end points of course must be set, so $\mu(x)$ can be set to satisfy two boundary conditions, namely at the integration limits $\mu(x)$ must be zero, so $\mu(x_A)=0$ and $\mu(x_B)=0$. When $\varepsilon=0$ then, $\bar{y}(x)$ equals $y(x)$, and the test function then becomes the desired minimum function. The line integral of the curve with the test function as independent variable is then just the minimum value that is sought,

$$\bar{I}=\int_{x_A}^{x_B} f\left(x,\bar{y},\frac{d\bar{y}}{dx}\right)dx.$$

Now the objective is to minimize the above line integral $\bar{I}$, keeping in mind that $\bar{I}$ is a function of the parameter ε; and in this case, calculation of the minimum value with respect to a single variable can invoke the well-known minimization technique of the differential calculus. Minimization of the function occurs when $\varepsilon=0$, so the necessary condition for the minimization of $\bar{I}$ is found by setting the derivative with respect to the parameter ε at $\varepsilon=0$ equal to zero,

$$\left.\frac{d\bar{I}}{d\varepsilon}\right|_{\varepsilon=0} = 0.$$

Substituting the line integral of $\bar{I}$ from above into the derivative and then differentiating gives

$$\frac{d\bar{I}}{d\varepsilon} = \frac{d}{d\varepsilon}\int_{x_A}^{x_B} f\left(x,\bar{y},\frac{d\bar{y}}{dx}\right)dx.$$

For convenience, it is customary to write $(d\bar{y}/dx) = \bar{y}'$, where the prime denotes the derivative with respect to x, so the above equation may be written more simply as

$$\frac{d\bar{I}}{d\varepsilon} = \frac{d}{d\varepsilon}\int_{x_A}^{x_B} f(x,\bar{y},\bar{y}')dx.$$

From Leibniz' rule, when the limits of integration (x_A and x_B) are not functions of the variable of differentiation ε, then the derivative of an integral is just the integral of a derivative,

$$\frac{d\bar{I}}{d\varepsilon} = \int_{x_A}^{x_B} \frac{d}{d\varepsilon}[f(x,\bar{y},\bar{y}')]dx = \int_{x_A}^{x_B} \frac{\partial f}{\partial \varepsilon}dx.$$

According to the chain rule of differentiation,

$$\frac{\partial f}{\partial \varepsilon} = \frac{\partial f}{\partial \bar{y}}\cdot\frac{\partial \bar{y}}{\partial \varepsilon} + \frac{\partial f}{\partial \bar{y}'}\cdot\frac{\partial \bar{y}'}{\partial \varepsilon} + \frac{\partial f}{\partial x}\cdot\frac{\partial x}{\partial \varepsilon},$$

then

$$\frac{d\bar{I}}{d\varepsilon} = \int_{x_A}^{x_B} \frac{\partial f}{\partial \varepsilon}dx = \int_{x_A}^{x_B}\left[\frac{\partial f}{\partial \bar{y}}\cdot\frac{\partial \bar{y}}{\partial \varepsilon} + \frac{\partial f}{\partial \bar{y}'}\cdot\frac{\partial \bar{y}'}{\partial \varepsilon} + \frac{\partial f}{\partial x}\cdot\frac{\partial x}{\partial \varepsilon}\right]dx.$$

Because $(\partial\bar{y}/\partial\varepsilon) = \mu(x)$, $(\partial\bar{y}'/\partial\varepsilon) = \mu'(x)$, and $(\partial x/\partial\varepsilon) = 0$, then

$$\frac{d\bar{I}}{d\varepsilon} = \int_{x_A}^{x_B}\left[\frac{\partial f}{\partial \bar{y}}\mu(x) + \frac{\partial f}{\partial \bar{y}'}\mu'(x)\right]dx.$$

When $\varepsilon = 0$, $\bar{y} = y$ and $\bar{y}' = y'$, so

$$\left.\frac{dI}{d\varepsilon}\right|_{\varepsilon=0} = 0 = \int_{x_A}^{x_B}\left[\frac{\partial f}{\partial y}\mu(x) + \frac{\partial f}{\partial y'}\mu'(x)\right]dx.$$

According to the rule for integration by parts,

$$\int_{x_A}^{x_B} u\,dv = uv\Big|_{x_A}^{x_B} - \int_{x_A}^{x_B} v\,du,$$

so for the second term on the right-hand side of the above minimization equation, set

$$u = \frac{\partial f}{\partial y'} \quad \text{and} \quad dv = \mu'(x)dx,$$

and because

$$dv = \frac{d\mu(x)}{dx}dx, \text{ and thus } v = \mu(x),$$

then

$$\frac{du}{dx} = \frac{d}{dx}\left(\frac{\partial f}{\partial y'}\right) \Rightarrow du = \frac{d}{dx}\left(\frac{\partial f}{\partial y'}\right)dx,$$

so

$$\int_{x_A}^{x_B}\left[\frac{\partial f}{\partial y'}\mu'(x)\right]dx = \frac{\partial f}{\partial y'}\mu(x)\Big|_{x_A}^{x_B} - \int_{x_A}^{x_B}\mu(x)\frac{d}{dx}\left(\frac{\partial f}{\partial y'}\right)dx.$$

Because of the boundary conditions on $\mu(x)$ stated above, $\mu(x_A) = 0$ and $\mu(x_B) = 0$, the first term on the right hand side of the equation above equals zero,

$$\frac{\partial f}{\partial y'}\mu(x)\Big|_{x_A}^{x_B} = 0,$$

and

$$\left.\frac{dI}{d\varepsilon}\right|_{\varepsilon=0} = 0 = \int_{x_A}^{x_B} \left[\frac{\partial f}{\partial y} \mu(x) - \mu(x) \frac{d}{dx}\left(\frac{\partial f}{\partial y'} \right) \right] dx = \int_{x_A}^{x_B} \mu(x) \left[\frac{\partial f}{\partial y} - \frac{d}{dx}\left(\frac{\partial f}{\partial y'} \right) \right] dx.$$

Now the *fundamental lemma of the variational calculus* states that between (x_A, x_B), since

$$\int_{x_A}^{x_B} \mu(x) \left[\frac{\partial f}{\partial y} - \frac{d}{dx}\left(\frac{\partial f}{\partial y'} \right) \right] dx = 0,$$

and because $\mu(x)$ is an arbitrarily chosen function, then it is the expression in the square brackets that must vanish, and that is just the Euler–Lagrange equation,

$$\frac{\partial f}{\partial y} - \frac{d}{dx}\left(\frac{\partial f}{\partial y'} \right) = 0.$$

The Euler–Lagrange equation will be extensively employed in the theory of liquid crystal displays to be described in the following chapters, to calculate the minimization of the Helmholtz free energy, from which the stable equilibrium states of the system can be determined. From these states the orientation of the liquid crystal molecules can be ascertained.

Deeper Meanings of the Euler–Lagrange Equation

The role that the Euler–Lagrange equation plays in calculating the Helmholtz free energy, critical as it will come to be seen, is not all that the calculus of variations can achieve; for even the Maxwell equations can be derived from the Euler–Lagrange equation.

Even deeper, the calculus of variations can be used to deduce the fundamental conservation laws of physics. The general form of the Lagrange equation (the "Lagrangian") is just $L = K - V$, meaning kinetic energy minus potential energy; the minimization of the action line integral then can be expressed as the principle of least action (where "Δ" means "a change in")

$$\Delta \int_{x_A}^{x_B} L dx = 0.$$

If the Lagrangian possesses some form of continuously smooth symmetry, then there will be associated with that a conservation law; that is,*

If the Lagrangian is invariant with time, then energy is conserved.
If the Lagrangian is invariant in space, then momentum is conserved.
If the Lagrangian is invariant with angular rotation, then angular momentum is conserved

And if the Lagrangian possesses the invariance of a complex number phase change (the gauge, a rotation in the complex plane) for time and space in a quantum mechanical electromagnetic interaction (the so-called *gauge invariance*), then electric charge is conserved [5].

But unfortunately, a general coordinate transformation invariance does *not* imply gravity conservation; meaning that Einstein's general theory of relativity has not yet been successfully incorporated into a unified field theory of relativistic quantum mechanics. Such a theory would span the edges of the universe and delve deep into the quantum world of subatomic particles; the Grand Unified Theory, the long-sought Theory of Everything.

Well, that's life. For the purposes of calculating changes in the Helmholtz free energy in this book, the Euler–Lagrange equation is more than adequate. By substituting the Lagrangian L into the Euler–Lagrange equation above, the calculus of variations can be used very expeditiously in the theoretical analysis of liquid crystals, the equation of use is [6]

$$\frac{\partial L}{\partial y} - \frac{d}{dx}\left(\frac{\partial L}{\partial y'}\right) = 0.$$

* The almost mystical deduction of the conservation laws of physics from the variational calculus was developed by the mathematical genius Emmy Noether in 1920, and is known as "Noether's theorem." Noether was then at Gottingen, the world's preeminent institute of mathematics, under the direction of the famed mathematicians Felix Klein and David Hilbert. But because of the blatant discrimination against women in German academia in those days, although recognized by all as a once-in-a-lifetime prodigy, Noether was continually denigrated by the administration and not allowed to teach courses at Gottingen. The mathematics department would only hire Noether as an assistant to Hilbert, and she could only teach in Hilbert's absence as a substitute. In the 1930s, when Hitler's fascists came to power and began their purge of non-Aryans, the Jewish Noether and many other outstanding scientists emigrated to Britain and North America. Noether accepted a position at Princeton University to continue her research and be free to teach the abstract algebra that she pioneered. For interesting accounts of life in German academia, refer to Reid, C. *Hilbert* (Springer-Verlag, 1996) and *Courant* (Springer-Verlag, 1996).

This very brief introduction to the variational calculus will help the understanding of calculations in the theories of the liquid crystal to be outlined in the following chapters.

References

[1] Nahin, P. 2004. *When Least Is Best*. Princeton, p. 35ff.
[2] Courant, R., and John, F. 1989. *Introduction to Calculus and Analysis II/1*. Springer-Verlag, New York, p. 365ff.
[3] Nahin, P. 2004. *When Least Is Best*. Princeton, p. 214ff.; Courant, R., and John, F. 1989. *Introduction to Calculus and Analysis II/1*, p. 329ff., Springer-Verlag, New York; and Gelfand, I.M., and Fomin, S.V. 1991. *Calculus of Variations*. Dover, New York, p. 26ff.
[4] Thomas, G.B. 1965. *Calculus and Analytic Geometry*, 3rd edition. Addison-Wesley, Reading; Courant, R., and John, F. 1989. *Introduction to Calculus and Analysis II/1*. Springer-Verlag, New York, p. 329ff.; and Gelfand, I.M., and Fomin, S.V. 1991. *Calculus of Variations*. Dover, New York, p. 21ff.
[5] Weyl, H. 1950. *The Theory of Groups and Quantum Mechanics*. Dover, New York.
[6] Nahin, P. 2004. *When Least Is Best*. Princeton, p. 214ff.; Lanczos, C. 1970. *The Variational Principles of Mechanics*, 4th edition. Dover, New York, p. 381ff.; p. 384ff., and p. 352ff.; and Messiah, A. 1958. *Méchanique Quantique*. North Holland & Wiley, New York.

8

The Mean Field

In Chapter 5, the liquid crystal was viewed as a solid crystal analyzed at its atomic structure level using dielectric constants, polarizabilities, induced dipoles, and other microscopic parameters. In this chapter, the mean field theory of the liquid crystal as a dipole moment field will be introduced, and the macroscopic thermodynamic coefficients of temperature, pressure, volume, specific heat, thermal expansion, isothermal compressibility, latent heat, and entropy will serve as the bases of investigation of the liquid crystals themselves, with the orientational order parameter S playing a critical role.

Ideal Gas in Crystal Lattice

In accord with the peculiar fluid/solid character of liquid crystals, the simplest theoretical liquid crystal model would be to order an ensemble of spherical ideal gas atoms in a crystal lattice structure. In the lowest energy (and entropy) state, the billiard-ball atoms are highly ordered just as in a solid. When heat is added, the lattice will melt, and the chemically non-reactive atoms will start to lose their positional order and stray from their

Liquid Crystal Displays, First Edition. Robert H. Chen.

lattice positions, and when as many atoms move as are in their solid lattice position, the matter is in a state of gas-like disorder.

Based on this model, using the Helmholtz free energy analysis of a gas, various macroscopic parameters of the first strong phase transition from solid to fluid may be calculated that roughly agree with experiment. However, the so-called *LJD model* by the British physicists Lennard-Jones and Devonshire obviously can only describe the positional order changes for phase transitions and cannot handle the critical orientational order changes for anisotropy changes in a liquid crystal, the reason being of course that immutably uniform hard spheres can have no "orientation" to speak of.

In an effort to accommodate the molecular orientation, Pople and Karasz suggested the addition of two places in the crystal lattice atoms that represent different "orientation states" in the liquid crystal that are responsive to the "pressure" of the first strong phase transition. Although an improvement, unsurprisingly, the addition of only two direction states was clearly insufficient to describe the second weak phase transition and its many directional changes in the intra-mesophase phase change [1].

Long Rod Models

Because the double refraction exhibited by liquid crystals is intimately related to its molecular structure, a better model naturally would be one that considered the rod-like molecular shape of the nematic liquid crystal molecules. Again, to avoid the complexity that chemical reactions would bring, the rod-like molecules should be nonreactive; that is, "hard" in the gas dynamics terminology. So in 1968, the Norwegian chemist Lars Onsager proposed a fluid ensemble of hard-rod molecules that relied on their long axial structure to manifest the changes in orientational order of the molecules under phase transition pressure.

Intuitively, just as a bundle of sticks will naturally align along their long axes under external pressure (as in the analogy with logs in a stream of Chapter 5), so the Onsager hard-rod model of a liquid crystal will tend to manifest an axial long-line orientational order. Calculations using this longitudinal anisotropic uniaxial symmetry do indeed produce some thermodynamic parameters that agree with experiment, but further calculations based on the mechanics of the individual rods using the classical Hamiltonian energy equations* unfortunately resulted in rather complex nonlinear equation systems. Of course, nowadays, such nonlinear equations can be solved numerically on a computer, but since Onsager's hard-rod model has since been superseded by more refined models, such computations are of little use. Furthermore, it turns out that the ratio of length to diameter of the Onsager rods must be greater than 100 for the orientational effect to comport with experiment, so although useful to model the very long molecules of macromolecular polymers, the Onsager model is not currently used for the micromolecular liquid crystals [2].

Derivative models have been proposed, such as rectangular parallelopipeds, hard rod plus cap, spherical cylinders, prolate and oblate spheroids, but all the hard-rod models suffer in the same way from their lack of consideration of the electric charge in molecules, which obviates the electrical interactions among molecules that are necessary for more complete explanations of the phase transitions. So it became clear that a model that allowed charge-attractive interactions among the molecular rods was needed; that is, a model that encompasses the charge polarization intrinsic to the liquid crystal molecules.

The Composite Electric Field and Average Index of Refraction

In accord with the discussion of the induced dipole moment in Chapter 5, the electric fields within the liquid crystal include the field from the intrinsic polarization of the molecules ($\boldsymbol{E}_p$), the electric field of the impinging light beam ($\boldsymbol{E}_l$), and the electric field induced by the light electric field primarily through the motion of the electrons of the molecules ($\boldsymbol{E}_{li}$). If an external electric field ($\boldsymbol{E}_E$) is applied, it will also induce electric dipole fields in the liquid crystal ($\boldsymbol{E}_{Ei}$). According to the principle of superposition, these fields can be vector added to produce the composite electric field,

* Regarding the Hamiltonian, whereas the Lagrangian introduced in Chapter 6 is a function of the generalized coordinates of space and their time derivative velocities ("generalized" meaning any coordinate system) and is the *difference* between the kinetic and potential energies ($L = K - V$), and the Euler–Lagrange equation (Chapter 7) gives all the Newtonian equations of motion; the Hamiltonian generalized coordinates are position and momentum, and the Hamiltonian gives the *total* energy of the system, and the equations of motion are from the relations $(dp_r/dt) = -(\partial H/\partial q^r)$ and $(dq^r/dt) = (\partial H/\partial p^r)$ where p_r is the momentum, q^r is the position, H is the Hamiltonian and the subscript and superscript r runs through all the coordinates.

$$E = E_p + E_l + E_{li} + E_E + E_{Ei}.$$

Because the fields are mutually influential, the complexity of the interactions can be imagined, and thus the efficacy of Faraday's superposition principle in intuitive field theory truly can be appreciated. However, on the microscopic photon-electron interaction level, as described previously, the problems still are formidable, and the consideration of average, or mean, electromechanical effects were a natural route to take to simplify the calculations in the theoretical models.

The descriptions of the liquid crystal models to follow will utilize average indices of refraction to describe the effects of the liquid crystal on light. The dielectric constant discussed previously can represent the anisotropy of a material, and its relation with the index of refraction can be derived from Maxwell's electromagnetic equations, as shown in Chapter 3. In a nonmagnetic liquid crystal material,

$$n = \sqrt{\varepsilon},$$

where, as before, n is the index of refraction and ε is the dielectric constant of the material. Recall that the index of refraction relates light to a material, and the dielectric constant is a property of the material itself. In the Lorentz–Lorenz equation derived in Chapter 3, it was shown that the relation of the charge polarization parameter α (polarizability) of a material is related to the index of refraction as

$$\alpha = \frac{3}{4\pi N} \cdot \frac{n^2 - 1}{n^2 + 2}.$$

With the Vuks assumption that the internal electric field in a crystal can be taken as the *average field* acting on the molecules, and it is the same in all directions, then [3]

$$E_{\text{int}} = \frac{\langle n^2 \rangle + 2}{3} E.$$

Vuks then derived the Lorentz–Lorenz equations in terms of the average index of refraction $\langle n^2 \rangle$ and the indices of refraction of the extraordinary and ordinary waves (n_e and n_o). The Vuks equation takes into consideration the extraordinary and ordinary charge polarization (α_e and α_o) in expressions for the average index of refraction in terms of the extraordinary and ordinary indices of refraction to provide values closer to experimental results (N is number of molecules per unit volume) [4]:

$$\frac{n_e^2 - 1}{\langle n^2 \rangle + 2} = \frac{4\pi}{3} N\alpha_e,$$

and

$$\frac{n_o^2 - 1}{\langle n^2 \rangle + 2} = \frac{4\pi}{3} N\alpha_o,$$

where

$$\langle n^2 \rangle = \frac{n_e^2 + 2n_o^2}{3},$$

is the relation of the average index of refraction to the extraordinary and ordinary indices of refraction. Thus, if the molecules of a liquid crystal are charge polarized in terms of extraordinary and ordinary polarizabilities (α_e, α_o), then it can exhibit double refraction as manifested by the extraordinary and ordinary indices of refraction (n_e, n_o), which, as has been shown in Chapter 4, will polarize light passing through. Vuks' average index of refraction thus takes into consideration the anisotropy of the medium; it is used in the mean field and continuum theories to be discussed in the following chapters.

Starting with the dipole moments produced by the composite electric field just described, the total mechanical effect of all the electric fields may be considered in terms of a mean dipole moment that engenders a mean dipole moment field. The liquid crystal mean field theories to be introduced below all use the concept of electric field forces being transformed to mechanical forces through the electric dipole moment as manifested by a mean dipole moment field.

In the theories to be introduced following, the total energy per unit volume of liquid crystal is the total of the applied electric field energy and the orientational potential energy of the liquid crystal as manifested in the potential energy of the mean dipole moment field. Certain static molecular configurations resulting from anchoring boundary conditions on the molecules can be analyzed by minimizing the free orientational energy of the system, and the reactions of the liquid crystal to disturbances such as an applied electric field can be analyzed in their dynamic return to the most stable energy orientation states. The analyses use the minimization techniques of the variational calculus with the orientational order S and directors $\boldsymbol{n}$ as parameters to calculate the orientation states of the liquid crystal, and thus the effect on incident light can be so parameterized.

All the presently used theoretical models of the operation of liquid crystal displays turn on the axes of the molecules' directors, and the effects of the electrical and fluid mechanical forces on the liquid crystal through that turning are manifested by the orientational order parameter through the mean electric dipole moments. So the theories must start off by modeling that critical dipole moment.

The Dipole Mean Field Is Born

Returning to the electric dipole moment described in Chapter 3, the molecules' charge separation ($\pm q$) produces an intrinsic dipole. Multiplying the charge by the distance between the charges (the lever vector $\boldsymbol{l}$) is just the intrinsic molecular electric dipole moment $\boldsymbol{p} = q\boldsymbol{l}$. Following the approach of models of ferromagnetism, the enormous number of individual dipoles form an average dipole moment orienting field that is felt by each molecule, but it is assumed that the molecules are not otherwise interacting.

Any vector field in space requires at least three vector components to describe, and the multivariable calculations thus becomes quite complicated, so a single variable scalar potential field is often used to represent the orienting field (confer the derivation of the Lorentz–Lorenz equation of Chapter 3). Conceptually, a given molecule will feel the potential much like a leaf floating in a potential sea, being subject to all the forces in the water causing it to bob up and down, but the leaf not being otherwise affected by neighboring leaves.

The first electric dipole scalar potential mean field model was proposed by the atomic physicist Max Born in 1916. His intrinsic electric dipole moment scalar potential mean field model was used to calculate the phase transitions in a liquid crystal. It successfully described the phase transition from isotropic liquid to anisotropic solid with temperature decrease, and although his model established the scalar potential mean dipole moment field model concept for liquid crystals, it was found later that the intrinsic dipole moment was not necessary for the description of the liquid crystal mesophase, and in fact a uniform nematic liquid crystal as a whole is nonpolar anyway (confer Chapter 5), so newer models supplanted the Born model, such as the Maier–Saupe theory still used today, to be described in Chapter 9 [5].*

References

[1] Pople, J.A., and Karasz, F.E. *J. Phys. Chem. Solids*, **18**, 28; **20**, 294 (1961); and Chandrasekhar, S. 1992. *Liquid Crystals*, 2nd edition. Cambridge, p. 17ff.
[2] Chandrasekhar, S. 1992. *Liquid Crystals*, 2nd edition. Cambridge, p. 37.
[3] Vuks, M.F. 1966. *Opt. Spektrosk.*, **20**, 644.
[4] Yang, D.K., and Wu, S.T. 2006. *Fundamentals of Liquid Crystal Devices*. Wiley, Chicester, p. 158.
[5] Born, M. 1916. *Sitz d. Phys. Math.*, **25**, 614; and Born, M. 1989. *Atomic Physics*. Dover, New York.

* An interesting fact is that the popular songstress of the late 1970's, Olivia Newton-John, is Max Born's granddaughter, exemplifying that both beauty and ultimate brains are possible within a family.

9

Maier–Saupe Theory

The mean field theory for liquid crystals was developed by the German physicist Wilhelm Maier and his student Alfred Saupe in the 1960s. While still employing the hard-rod molecule model, it included electrical effects by postulating induced dipole–dipole interactions between the molecules, which were treated as perturbations that engender attractive forces among the molecules that are responsible for the stability of the liquid crystal mesophase.

The Maier–Saupe theory successfully describes the observed mesophase to isotropic liquid phase transition at the clearing temperature as a discontinuous phase transition, and also provides an expression for the dielectric anisotropy as a function of the orientational order parameter, as will be seen below.

The Maier–Saupe theory first assumes that each hard-rod liquid crystal molecule is subject to a mean potential field arising from long-range induced dipole–dipole interaction forces; the Maier–Saupe potential field does not consider any short-range forces. From the cylindrical symmetry about the long axes of the liquid crystal molecules (often called the *preferred axis*) and the non-polarity of the liquid crystal as a whole, and based on the general physical truth that energy is the square of a field quantity, the orientational

Liquid Crystal Displays, First Edition. Robert H. Chen.

potential energy of the *i*th molecule is postulated to be proportional to the second order Legendre polynomial multiplied by the orientational order parameter variable *S*, as follows*

$$u_i \propto \frac{1}{2}(3\cos^2\theta_i)\cdot S.$$

where θ_i is the angle between the long axis of the ith molecule and the preferred axis of the liquid crystal. Maier and Saupe only considered the induced electrical dipole–dipole interaction between neighboring molecules producing anisotropic attractive forces; and no matter what the details of the interaction may be, the ith molecule is assumed as subject to an induced dipole potential given in simplified form as [1]

$$u_i = -\frac{A}{V^2}\cdot S\cdot\frac{1}{2}(3\cos^2\theta_i - 1),$$

where *A* is an empirical constant, and *V* is the molar volume. The denominator factor V^2 in the induced dipole moment potential equation above reveals that the attractive induced interaction between the dipole moments of neighboring molecules varies inversely as the 6th power of distance between them (l), that is, $l^3 = V$, so $V^2 = l^6$ and $u_i \sim l \times 10^{-6}$, seemingly small, but still considered a long-range molecular interaction relative to atomic and molecular dimensions.†

The Nematic to Isotropic Phase Transition Calculation

The Maier–Saupe theory can be used to calculate the mesophase to liquid phase transition-related thermodynamic parameters of the ordered state compared with the disordered state, that is, the excess entropy, Helmholtz free (orientational) energy, the latent heat of sublimation of the phase transition, the specific heat, and the isothermal compressibility.

The mean molar average orientational potential energy is calculated by taking the integral of the induced dipole potential (u_i) of the *i*th molecule

---

* Where energy is proportional to the square of a field quantity; for example, the kinetic energy is proportional to velocity squared and electrical energy to the square of the electric field vector.

† There is some controversy regarding the V^{-2} dependence; Cotter, M.A. 1977. *Mol. Cryst. Liquid Cryst.*, **39**, 173.

multiplied by a Boltzmann distribution of the angular orientation divided by the Boltzmann distribution,

$$U = \frac{N}{2}\frac{\int_0^1 u_i \exp(-u_i / k_B T) d(\cos\theta_i)}{\int_0^1 \exp(-u_i / k_B T) d(\cos\theta_i)} = -\frac{1}{2} N k_B T B S^2,$$

where k_B is the Boltzmann constant, T is temperature, N is the molar Avogadro number, $B = A / k_B T V^2$ is a therein defined thermodynamic coefficient, and the partition function (from the equipartition of energy) of a molecule is defined by the denominator as

$$f_i = \int_0^1 \exp(-u_i / k_B T) d(\cos\theta_i).$$

The statistical mechanical definition of entropy is just the Boltzmann constant k_B multiplied by the natural logarithm of all the possible orientational states of the system (cf. Chapter 6), so the excess entropy (excess of the unperturbed state relative to the perturbed) is just the above mean molar orientational potential minus the original orientational order entropy,*

$$\Omega = -N k_B \left[\frac{1}{2} B S (2S+1) - \ln \int_0^1 \exp\left(\frac{3}{2} B S \cos^2\theta_i \right) d(\cos\theta_i) \right]$$

The definition of Helmholtz free energy is (Chapter 6)

$$F = U - T\Omega,$$

where U is the molar potential given above, T is temperature, and $\mathbf{\Omega}$ is the entropy, so that from the above equations, Maier–Saupe's expression for the Helmholtz free energy (orientational energy) is

$$F = -N k_B T \left[\frac{1}{2} B S (S+1) - \ln \int_0^1 \exp\left(\frac{3}{2} B S \cos^2\theta_i \right) d(\cos\theta_i) \right]$$

---

* Of course entropy is usually denoted by S, but since that is used to denote the orientational order parameter in this book, "$\mathbf{\Omega}$" is used to denote entropy to avoid confusion.

Here the minimum orientational energy (the equilibrium state) can be obtained by using the elementary minimization technique of the differential calculus, setting the derivative to zero while V and T are held constant,

$$\left(\frac{\partial F}{\partial S}\right)_{V,T} = 0.$$

After taking the derivative, the necessary condition for the minimum orientational energy is

$$3S\frac{\partial \langle \cos^2 \theta_i \rangle}{\partial S} - 3\langle \cos^2 \theta_i \rangle + 1 = 0,$$

thus the consistency relation for orientational equilibrium is that the average orientational energy of the molecules is just the mean value of the orientational potential field, or in other words, the angle of the average orientation of the liquid crystal truly represents the mean angle of all the molecules; that is,

$$\langle \cos^2 \theta_i \rangle = \langle \cos^2 \theta \rangle.$$

The relation between the orientational energy F and the orientational order parameter S as calculated by Chandrasekhar and Madhusudana [2] using the Maier–Saupe equation for the Helmholtz free energy above is shown in Figure 9.1. The temperature T is the parameter, and T_c is the phase transition temperature at which the anisotropic nematic liquid crystal changes to the isotropic clear liquid (the clearing temperature). From the graph, one can see that for the curve $T \ll T_c$, at $S \approx 0.7$, there is a clear minimum that corresponds to the orientational equilibrium point for the nematic liquid crystal. For the curve $T < T_c$, the minimum point has moved to $S \approx 0.6$, where there still is some orientational order to speak of at the equilibrium point. For the $T > T_c$ curve, the minimum orientational energy is at $S = 0$, which of course represents the absence of orientational order of the isotropic liquid. However, for the $T = T_c$ curve, there are clearly two minima at $S = 0$ and $S \approx 0.43$, that is, two different orientational orders have the same minimum orientational energy equilibrium point. The Maier–Saupe theory thus can successfully calculate the first-order discontinuous meso-to-liquid phase transition of a liquid crystal, and the derived value of the orientational order parameter is not only within the range of the typical $S = 0.3 - 0.9$ experi-

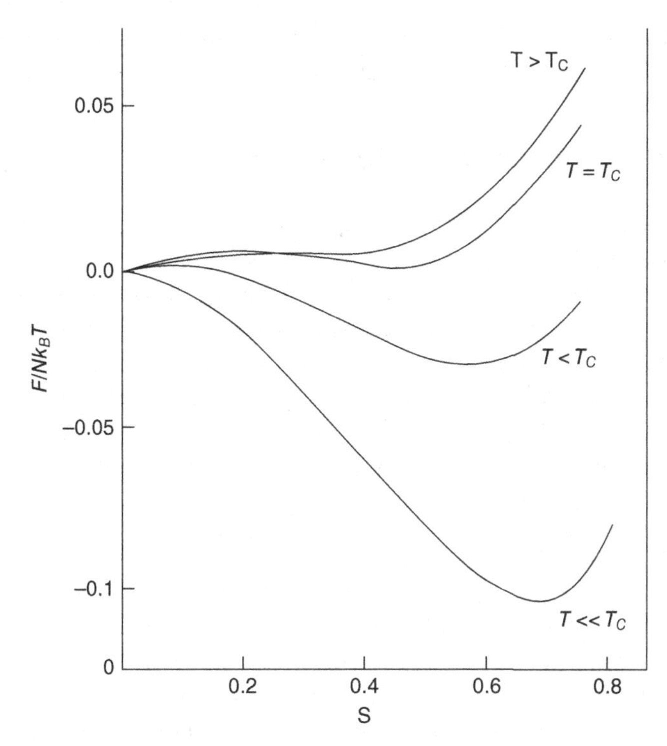

Figure 9.1 Maier–Saupe calculation of the relation between the Helmholtz free energy and the orientational order parameter *S*.

mental values for the orientational order parameter, but the $S = 0.43$ derived value is also close to the experimental value.

Values of the orientational order parameter S can be measured by applying a magnetic field and measuring the orientational order, or diamagnetic anisotropy of a liquid crystal sample. From the microwave radiation of the liquid crystal in a magnetic field, the uniaxial symmetry of the nematic phase is especially revealed by line splitting in the nuclear magnetic resonance (NMR) spectrum, but the lines are narrow, implying rapid molecular motion, which is characteristic of a fluid. The orientational order can be extracted from the NMR data as performed rather ingeniously by Saupe [3]. The Maier–Saupe theory has shown good agreement with the experimental results in some cases, so at least as far as the orientational order parameter is concerned, an induced dipole potential field perturbation model at the least has some merit. However, the Maier–Saupe calculations for the latent heat of sublimation, the specific heat, and the isothermal compressibility did not agree well with experimental results.

Dielectric Anisotropy Calculation

Any plausible model of a liquid crystal must be able to describe a liquid crystal's distinctive dielectric anisotropy as a function of an applied electric field. An external electric field (E) applied to the liquid crystal can be resolved into components ($E_\Updownarrow$ and $E_\perp$) parallel and perpendicular, respectively, to the principal long axis of the molecules, which axis is taken as representing the liquid crystal molecular orientation as a whole (also called the *axis of symmetry**). As described in Chapter 5, a low-frequency (particularly dc) signal causes the intrinsic molecular dipole to turn to align with the field, and a relatively minor displacement (polarization) of the atoms of the molecule, but there will also be an intermolecular electrical interaction inducing further molecular polarization, which is characterized by a *polarization field.*

Because of the difficulties of calculating the interactive polarization field effect on each individual molecule and then calculating their interactive cumulative influence, including the effects of the external field, it is expedient to assume for each molecule an *effective* induced dipole moment arising from the aggregate of induced dipole moments, and utilize the Vuks average index of refraction equation (Chapter 8) in a cylindrically symmetric hard-rod model in a mean dipole field. Then, in analogy with the theory of diamagnetic anisotropy, but adding on the intermolecular interaction polarization field in the form of *field factors*, the parallel and perpendicular components of the *effective induced dipole moments per molecule*, in analogy with diamagnetic anisotropy and adding terms for the polarization field, then can be written as[†]

---

* Previously, vector components were taken with respect to a molecule's orientation with respect to the liquid crystal director, but since this is a mean field theory, the external electric field and dipole moment components will be taken with respect to the principal axis (also often called the preferred axis and the *unique axis*) of the liquid crystal as a whole.

[†] The effective dipole moments are calculated using the simpler magnetic field susceptibility (because magnetic intermolecular interactions can be ignored as small); if z is the direction of the preferred axis, and η_1 and η_2 are the diamagnetic anisotropic susceptibilities in respect to the molecule's principal axis, then the average z-component of the susceptibility per unit volume is $\chi_z = n_V[\eta_1\langle\cos^2\theta\rangle + \eta_2\langle\sin^2\theta\rangle] = n_V[\bar{\eta} + \frac{2}{3}(\eta_1 - \eta_2)S]$, where n_V is the number of molecules per unit volume, and $\bar{\eta} = \frac{1}{3}(\eta_1 + 2\eta_2)$; similarly for the x- and y-components, $\chi_x = \chi_y = n_V[\bar{\eta} - \frac{1}{3}(\eta_1 - \eta_2)]S$, and $S = (\chi_z - \chi_x)/n_V(\eta_1 - \eta_2)$. Chandrasekhar, S. 1992. *Liquid Crystals*, 2nd edition. Cambridge, p. 39.

$$\bar{m}_{\Updownarrow} = \left(\bar{\alpha} + \frac{2}{3}\Delta\alpha \cdot S\right) RhE_{\Updownarrow}$$

$$\bar{m}_{\perp} = \left(\bar{\alpha} - \frac{1}{3}\Delta\alpha \cdot S\right) RhE_{\perp}.$$

where $h = 3\bar{\varepsilon}/(2\bar{\varepsilon}+1)$, is the *cavity field factor*, $R = 1/(1-\bar{\alpha}f)$ is the *reaction field factor*, $f = 4\pi N\rho(2\bar{\varepsilon}-2)/3M(2\bar{\varepsilon}+1)$, $\bar{\varepsilon}$ is the mean dielectric constant, $\bar{\alpha}$ is the mean polarizability, $\Delta\alpha$ is the polarizability anisotropy (there will be different charge polarizabilities in different directions), ρ is the density, M is the molecular weight, and S is the orientational order parameter.*

It can be seen that the Maier–Saupe effective induced dipole moment expressions include microscopic parameters, such as the dielectric constant and polarizability, and also macroscopic parameters, such as density, molecular weight, and the orientational order parameter.

The above is the effective *induced* dipole moment per molecule; long rod-shaped liquid crystal molecules' intrinsic charge polarizations structurally manifest *intrinsic* (permanent) dipole moments, and taking into consideration the intermolecular electrical interaction, an *intrinsic dipole moment vector* μ can be calculated as components parallel to the external electric field $\bar{\mu}_{\Updownarrow}$, and perpendicular to that external field $\bar{\mu}_{\perp}$ respectively, as follows.

In terms of a spatial coordinate system (x, y, z) and a molecular coordinate system (ξ, η, ζ), if their spatial relationship is given by the relative directions between the molecules and the liquid crystal as a whole (denoted by υ and υ'), and $\Psi(\theta)$ is the probability that a given molecule's polar orientation is within the angular range θ and $\theta + d\theta$; and if further for a low voltage applied electric field, the Boltzmann distribution of the orientational potential of the dipole moments can be approximated by a series expansion of the exponential function taking only the first two terms,†

$$\exp(-u_i/k_BT) \approx 1 + \mu hE/k_BT$$

then an effective mean value can be expressed by an integral over a volume in a spherical coordinate system wherein the μ_z and μ_x components of the dipole moment are integrated as follows to produce the effective parallel and perpendicular intrinsic dipole moments

* The h and f in the effective induced dipole moments per molecule expressions use mean dielectric constants, but strictly, they are affected by the anisotropy of the dielectric constant; this is ignored in this version of the Maier–Saupe theory.

† The series expansion is just $\exp(x) = 1 + x + x^2/2! + x^3/3! + \ldots$, taking the first two terms.

$$\bar{\mu}_{\Updownarrow} = \frac{\int_0^{\pi}\int_0^{2\pi}\int_0^{2\pi}[1+(\mu_z hE_{\Updownarrow}/k_BT)]\cdot\mu_z\cdot\psi(\theta)\sin\theta d\theta d\upsilon d\upsilon'}{\int_0^{\pi}\int_0^{2\pi}\int_0^{2\pi}[1+(\mu_z hE_{\Updownarrow}/k_BT)]\cdot\psi(\theta)\sin\theta d\theta d\upsilon d\upsilon'}$$

$$\bar{\mu}_{\perp} = \frac{\int_0^{\pi}\int_0^{2\pi}\int_0^{2\pi}[1+(\mu_x hE_{\perp}/k_BT)]\cdot\mu_x\cdot\psi(\theta)\sin\theta d\theta d\upsilon d\upsilon'}{\int_0^{\pi}\int_0^{2\pi}\int_0^{2\pi}[1+(\mu_x hE_{\perp}/k_BT)]\cdot\psi(\theta)\sin\theta d\theta d\upsilon d\upsilon'}.$$

If the angle between the intrinsic dipole and the principal axis of the liquid crystal molecules is given by β then

$$\mu_z = R\mu[\cos\beta\cos\theta + \sin\beta\sin\upsilon\sin\theta],$$

and

$$\mu_x = R\mu[\cos\beta\sin\upsilon'\sin\theta + \sin\beta(\cos\upsilon\cos\upsilon' - \sin\upsilon\sin\upsilon'\cos\theta)],$$

then the probability of molecules having angles between θ and $\theta + d\theta$ is given by the Boltzmann distribution as

$$\Psi(\theta) = e^{-u_i/k_BT}$$

where $u_i = \frac{1}{2}(3\cos^2\theta_i)\cdot S$, which is the ith molecule's orientational potential given previously. From the effective intrinsic electric dipole moment equations given above, then

$$\bar{\mu}_{\Updownarrow} = \frac{\mu^2}{3k_BT}[1-(1-3\cos^2\beta)S]hR^2E_{\Updownarrow}$$

$$\bar{\mu}_{\perp} = \frac{\mu^2}{3k_BT}\left[1+\frac{1}{2}(1-3\cos^2\beta)S\right]hR^2E_{\perp}$$

From the Vuks equations, and the definitions of f and F, adding the effective induced and intrinsic electric dipole moments $\bar{m}$ and $\bar{\mu}$ respectively, but because this is a mean field calculation, ignoring the different effects of $\varepsilon_{\Updownarrow}$ and $\varepsilon_{\perp}$ in $\bar{\varepsilon}$ on the average polarizability $\bar{\alpha}$, then

$$\frac{\varepsilon_{\Updownarrow}-1}{4\pi}E_{\Updownarrow} = \frac{N\rho}{M}(\bar{m}_{\Updownarrow} + \bar{\mu}_{\Updownarrow})$$

$$\frac{\varepsilon_\perp - 1}{4\pi} E_\perp = \frac{N\rho}{M}(\bar{m}_\perp + \bar{\mu}_\perp).$$

Now substituting the expressions for the effective induced $\bar{m}$ and intrinsic $\bar{\mu}$ dipole moments into the equations above, the parallel and perpendicular components of the dielectric constants are

$$\varepsilon_\Updownarrow = 1 + 4\pi \frac{N\rho R}{M}\left(\frac{3\bar{\varepsilon}}{2\bar{\varepsilon}+1}\right)\left\{\bar{\alpha} + (2/3)\cdot\Delta\alpha\cdot S + R\frac{\mu^2}{3k_BT}\left[1-(1-3\cos^2\beta)\cdot S\right]\right\},$$

and

$$\varepsilon_\perp = 1 + 4\pi \frac{N\rho R}{M} h\left\{\bar{\alpha} - \frac{1}{3}\Delta\alpha\cdot S + R\frac{\mu^2}{3k_BT}\left[1 + \frac{1}{2}(1-3\cos^2\beta)\cdot S\right]\right\}.$$

The dielectric anisotropy is defined as $\Delta\varepsilon = \varepsilon_\Updownarrow - \varepsilon_\perp$, so the Maier–Saupe mean field theory gives the expression for the dielectric anisotropy as

$$\Delta\varepsilon = \bar{\varepsilon}_\Updownarrow - \bar{\varepsilon}_\perp = 4\pi\frac{N\rho R}{M} h\left[\Delta\alpha - R\frac{\mu^2}{2k_BT}(1-3\cos^2\beta)\right]\cdot S.$$

The equation above gives the relationship of the dielectric anisotropy $\Delta\varepsilon$ to the intrinsic (μ), and the induced dipole moments (inherent in $\bar{\alpha}$ and $\Delta\alpha$), temperature, the angle between the intrinsic dipole moment and the principal molecular axis β, and in particular the orientational order parameter S.

The dielectric anisotropy relations calculated from the equation above generally agree with experimental results, as for example in the case of the synthesized liquid crystal compound 4,4′ p-azoxyanisole (PAA) to be discussed in later chapters.

The Maier–Saupe theory is a useful theoretical model of a complicated system, confirming the character of some intermolecular forces and providing guidance for further theoretical studies. It also can be used practically, for instance, in the guest-host compound liquid crystal to be discussed in later chapters, the changes in the orientational order as a result of the doping of a guest compound can be calculated using the Maier–Saupe theory, and thus the optimum guest-host mixtures can be derived for the best contrast values in liquid crystal display design.

The Maier–Saupe theory is still used today, however, although the theory can adequately describe the orientational order parameter in the discontinuous mesophase-to-liquid phase transition and in the liquid crystal dielectric anisotropy, it is not able to calculate other thermodynamic parameters that are in accord with experimental results, for example, the latent heat of sublimation of the phase transition, the specific heat, and the isothermal compressibility. As mentioned above, the problem lies in the inability of the hard-rod modeling of the molecules to handle the atomic level intermolecular interaction that produces short-range orientational order effects. The liquid crystal molecules are not really ideally cylindrical molecules, and there are flexible extremities that an oversimplified hard-rod model cannot adequately simulate. Thus, further studies of liquid crystal behavior require a deeper analysis into the short-range forces and a different model to include them.

Near Neighbor Correlation

When approaching and passing the temperature for the phase transition from anisotropic nematic liquid crystal to isotropic pure liquid, experiment has shown that the isotropic pure liquid evinces short-range attractive influences manifested in residual traces of anisotropic orientational order. In an attempt to model this short-range liquid crystal intermolecule interaction and its influence, Krieger and James applied Bethe's *molecular shell model* originally designed for binary metal alloys to the liquid crystal.*

Bethe's shell model postulates a paired orientational potential $E(\theta_{oj})$ to describe the short-range interaction between a jth molecule closest to a *central molecule* (denoted by the subscript "*o*"), and this neglects any interaction between any two of the other z neighbors surrounding the jth molecule, thereby simplifying the calculations immeasurably, but still considering a close intermolecular interaction. Every outer shell molecule j is also influenced by all the other z molecules through an angular interaction potential $V(\theta_j)$, which is just the long-range interaction. Thus, the angular distributions can still be described separately and respectively by Boltzmann distributions as follows,

$$f(\theta_{oj}) = e^{-E(\theta_{oj})/k_BT} \quad \text{and} \quad g(\theta_j) = e^{-V(\theta_j)/k_BT}.$$

The interaction potential of all the $z + 1$ molecules in an arbitrary state is

* Hans Bethe received the 1967 Nobel Prize for physics for his studies of the energy production of stars.

$$\prod_{j=1}^{z} f(\theta_{oj})g(\theta_j)$$

where the probability that the central molecule is in the orientational state (θ_o,φ_o) is given by the integral in spherical coordinates over all the z molecules,

$$\int\cdots\int\prod_{j=1}^{z} f(\theta_{oj})g(\theta_j)d(\cos\theta_j)d\varphi_j.$$

For application to liquid crystals, Krieger and James set the probability that the central molecule (o) and a neighboring molecule (1) are in the (θ_o,φ_o) and the (θ_1,φ_1) orientational states as

$$\psi(\theta_0,\varphi_0;\theta_1,\varphi_1) = f(\theta_{01})g(\theta_1)\prod_{j=1}^{z}\iint f(\theta_{0j})g(\theta_j)d(\cos\theta_j)d\varphi_j,$$

where $\psi(\theta_o,\varphi_o;\theta_1,\varphi_1)$ is the probability function. Because the central molecule and the other molecules should have the same probability of being in the given state (i.e., one could interchange them without affecting the calculation),

$$\psi(\theta_o,\varphi_o;\theta_1,\varphi_1) = \psi(\theta_1,\varphi_1;\theta_o,\varphi_o)$$

so that according to Chang's consistency relation [4],

$$\left[\iint f(\theta_{oj})g(\theta_j)d(\cos\theta_j)d\varphi_j\right]^z$$
$$= \iint f(\theta_{01})g(\theta_1)\left[\iint f(\theta_{oj})g(\theta_j)d(\cos\theta_j)d\varphi_j\right]^{z-1} d(\cos\theta_o)d\varphi_o,$$

the consistency condition for short-range liquid crystal interaction is

$$\frac{g(\theta_o)}{\left[\iint f(\theta_{0j})g(\theta_j)d(\cos\theta_j)d\varphi_j\right]^{z-1}} = const.$$

Then if the orientational order distribution for a liquid crystal follows the second-order Legendre polynomial as described in Chapter 5, higher even-order non-origin symmetric Legendre polynomials should be able to provide more detailed descriptions of the orientational order distribution. If the cylindrical symmetry (i.e, the energy has no relation to the azimuthal angle φ) is taken into account, and it can be assumed that the pair potential ($E(\theta_{oj})$) of the jth molecule is given by the second-order Legendre polynomial multiplied by a convenient new macroscopic thermodynamic coefficient (B^*)* similar to the Maier–Saupe $B = A/k_B T V^2$, and further if higher even-order Legendre polynomials (subscript $2n$) can describe the interaction potential $V(\theta_j)$, and the associated thermodynamic factor B^*_{2n}, the short-range pair potential and the long-range interaction potential then can be written respectively as

$$E(\theta_{oj}) = -B^* P_2(\cos\theta_{oj})$$

$$V(\theta_j) = -\sum_n B^*_{2n} P_{2n}(\cos\theta_j).$$

Calculations show that the higher the even order of the Legendre polynomial, the greater the self-consistency of the potential field of the z molecules system. For example, for $n = 6$ ($P_{12}[\cos\theta]$), practically total self-consistency results and thus the Krieger–James theory at the least can be said to be "extremely reasonable." Furthermore, and perhaps more importantly, the greater the value of z, for instance $z = 8$, the better the theory will match experimental measurements.

Following the Maier–Saupe molar mean orientational order potential calculation for the long-range interaction, and considering the short-range interaction potential as derived by Krieger and James, the mean orientational potential can be written as

$$U = -\frac{1}{2} N z B^* \langle P_2 \cos\theta_{oj} \rangle.$$

The result is that the calculated curve for U versus $1/T$ not only shows the mesophase-to-liquid phase transition's long-range interaction, the short-range interaction of the discontinuous first-order phase transition is also manifest. For instance, if T^* is assumed to be the temperature of the mesophase-to-liquid phase transition point for the short-range interaction

* B^* is an artificial coefficient contrived as a sole parameter for the Krieger–James near-neighbor correlation model; refer to Chandrasekhar, S. 1992. *Liquid Crystals*, 2nd edition. Cambridge, p.71ff.

represented by $P_2(\cos\theta_{oj})$, then if $z = 8$ (there are eight molecules in the ensemble), Krieger–James calculates a $(T_c - T^*)/T_c = 0.062$ that is closer to experimental values than Maier–Saupe. However, both theories are still far from the actual experimental value of 0.003 for the first-order mesophase-to-liquid phase transition. As for the orientational order parameter calculation, Krieger–James' inclusion of a short-range interaction results in a calculated value of $S = 0.40$, which is very close to the Maier–Saupe value of $S = 0.43$.

Thus, it turns out that although addressing the short-range attractive interaction in modeling the first-order mesophase-to-liquid phase transition, the Krieger–James temperature and order parameter calculations are in fact not much different from those of the Maier–Saupe theory that only considers long-range interactions. After ploughing through the complicated derivation, one might well ask "what's all the fuss?", but the Krieger–James theory aptly demonstrates that the short-range attractive interaction indeed is at work in Nature and can be addressed theoretically to produce the traces of orientational order in the mesophase anisotropic to isotropic transition inherent in liquid crystals.

References

[1] Chandrasekhar, S. 1992. *Liquid Crystals*, 2nd edition. Cambridge, p. 38ff.; and Collings, P.J., and Hird, M. 2004. *Introduction to Liquid Crystals*. Taylor & Francis, London, p. 250ff.

[2] Chandrasekhar, S., and Madhusudana, N.V. 1972. *Mol. Cryst. Liquid Cryst.*, **17**, 37.

[3] Saupe, A., and Englert, G. 1963. *Phys. Rev. Lett.*, **11**, 462; and de Gennes, P.G., and Prost, J. 1993. *The Physics of Liquid Crystals*, 2nd edition. Oxford, p. 43ff.

[4] Chang, T.S. 1937. *Proc. Camb. Phil. Soc.*, **33**, 524.

10

Phenomenological Theory

The molecular level mean statistical models introduced above, although simplifying the atomic interactions by mean field modeling of the liquid crystal, actually are still quite complicated, and the calculated thermodynamic parameters are at variance with experimental results. The problems lay with the oversimplification of the molecular structure and the neglect of intermolecular interactions, but further consideration of a more accurate molecular model and the intermolecule interaction would bring hopeless complications. Thus the French physicist Pierre-Gilles de Gennes finessed the complexities with a more direct and practical approach to the investigation of liquid crystals.

The phenomenological approach to the phase transitions is independent of the details of molecular structure and molecular interaction, and as such can be said to include the short-range as well as long-range interactions by a "synthesis" of the experimental observations with their concomitant thermodynamic parameters through the employment of coefficients that are adjustable within the confines of the logic of physics, such as symmetry and self-consistency. After accordingly formulating and fitting the coefficients arising from the synthesis, as will be seen, an elegant description of the liquid crystal nematic to isotropic phase transition and dielectric anisotropy is the result.

Liquid Crystal Displays, First Edition. Robert H. Chen.

The de Gennes phenomenological theory, based as it is on the symmetry inherent in long structures, and employing tensor analysis to mathematically describe the anisotropy, is applicable to any long-axis molecular structure, in particular liquid crystals, but also macromolecular polymers, and is a model of modern theoretical physics procedure.

For calculating the liquid crystal phase transition, the de Gennes phenomenological approach begins with careful analysis of the experimental data for the orientational order parameter S, including that from nuclear magnetic resonance spectroscopy, light scattering, and magnetic and electrically induced birefringence in liquids (the Kerr effect). de Gennes focuses on the domain below the clearing temperature T_c, where the fluctuations of the orientational order parameter S are relatively weak, with the greatest variation coming from changes of the orientation of the optical axis, and just above T_c, where the S fluctuations are larger, which can be treated by Landau's classical theory of the solid–liquid phase transition.*

de Gennes succeeds in extending the Landau theory to the more delicate phase transition in liquid crystals, deriving the first-order discontinuous mesophase-to-liquid phase transition. Then, in conjunction with the mean dipole moment construct as used by Maier–Saupe, an explicit expression for the orientational order parameter S is derived to calculate the liquid crystal birefringence [1].

The Nematic to Isotropic Phase Transition Calculation

The classical Landau phase transition theory, when faced with the vagaries of a nematic liquid crystal phase transition, was modified by de Gennes to take into consideration the special variations of the orientational order parameter S described above. The critical question addressed was that when the temperature reaches the clearing point, why should the isotropic liquid retain traces of orientational order? In the Landau theory, the isotropic liquid phase molecules should display no order, but a close inspection near the clearing temperature revealed that *locally*, the molecules are parallel to each other, and this order persists over a characteristic extent de Gennes called the *coherence length*, that is dependent on the temperature.

At the left in Figure 10.1 is a schematic depiction of the parallel ordering of a nematic liquid crystal. On the right in Figure 10.1 is the isotropic liquid crystal state at temperature near T_c that betrays traces of sectoral orienta-

* Lev Landau received the 1962 Physics Nobel Prize for studies of liquid helium.

Figure 10.1 Schematic depiction of the parallel ordering of a nematic liquid crystal (left), and the isotropic liquid crystal state at temperature near T_c betraying the nematic swarms (right).

tional order as suggested by discernible spherical droplets of molecules forming the sinister-sounding *nematic swarms*. The radius of the droplets is just de Gennes' coherence length, denoted by $\xi(T)$.

The conceptual orientationally ordered droplet of liquid crystal however carried within a mathematical liability; for the mathematics of a droplet model require a "surface" to indicate the boundary of the sphere, thus necessitating the existence of a discontinuity in the order parameter, and this would lead to exponential runaway at the droplet surface boundary. To circumvent the exponential abyss, de Gennes employed a continuous free energy power series expansion to "smooth away" the boundary discontinuities. That is, particularly given the typical smooth events in an orderly natural process, such as the flow of energy governed by the second law of thermodynamics, the free energy should be expressible by well-behaved continuous mathematical functions that can be approximated by power series expansion. Further, in the nematic liquid crystal phase transition to an isotropic liquid (the mesophase-to-liquid NI transition), the orientational order parameter S phenomenologically neither appreciably nor rapidly changes, so the Helmholtz free energy (in this case, the orientational energy) can be expanded in terms of the orientational order parameter power series as follows

$$F = \tfrac{1}{2}A(T)S^2 - \tfrac{1}{3}BS^3 + \tfrac{1}{4}CS^4 + \ldots.$$

To maintain the relative signs in the power series, it is assumed that both B and C are positive. The coefficients A, B, and C do not have any inherent physical significance themselves, and can be postulated through observation and conformance with known theories of molecular behavior; that is, they can be adjusted very practically to conform with what is experimentally observed. As for the factor $A(T)$ in the second-order term (the most significant term), because isotropic disorder ($S = 0$) will correspond with a minimum of F (lowest orientational energy) only if $A(T) > 0$, and the ordered phase ($S \neq 0$) will be in accord with a stable minimum of F only if $A(T) < 0$, thus the sign of $A(T)$ must change at a *critical temperature* T^*, and $A(T)$ can be written as

$$A(T) = a(T - T^*).$$

where a (>0) is an experimental constant. From the above, it can be seen that when $T > T^*$, then $A(T) > 0$, and when $T < T^*$, then $A(T) < 0$. Because the phase transition from the solid to the nematic mesophase involves an orientational order parameter $S \geq 0.4$, dropping the fourth order and above terms in the power series would amount to ignoring significant terms, so the truncated de Gennes' power series expansion can only be used for the weak (meaning very small S) first-order nematic to isotropic liquid phase transitions, that is, the area around T_c where there is not much orientational order.

The power series expansion of the Helmholtz free energy then is a function of the orientational order parameter S at an appropriate temperature (near the clearing point temperature T_c), and a function of the temperature T and the critical temperature T^*, and together with the adjustable coefficients a, B, and C, F, can be written as

$$F = \tfrac{1}{2}a \cdot (T - T^*)S^2 - \tfrac{1}{3}BS^3 + \tfrac{1}{4}CS^4 \dots.$$

To calculate the equilibrium orientation condition of the liquid crystal, the Helmholtz (orientational) free energy is minimized using the elementary minimization technique of the calculus, so the above expression is differentiated with respect to S and the expression set equal to zero,

$$\frac{\partial F}{\partial S} = a \cdot (T - T^*) \cdot S - BS^2 + CS^3 = 0.$$

The solution to this second degree equation is given by the quadratic equation as

$$S = 0 \text{ and } S = \frac{-B \pm \sqrt{B^2 - 4a \cdot (T - T^*)C}}{2C},$$

where $S = 0$ is the equilibrium state of complete lack of order of the isotropic liquid. The other two values of S from the plus and minus radical, however, indicate that there are indeed two orientational order values at the minimum (equilibrium) point. From this, de Gennes has very simply shown that the nematic liquid crystal to isotropic liquid phase transition involves a discontinuity resulting in two distinct values of the orientational order parameter, thereby manifesting the observed first-order phase transition.

A graph of the de Gennes calculation of the Helmholtz free energy as a function of the orientational order parameter is almost identical to that of the Maier–Saupe theory curve shown in Figure 9.1. The interesting relation between T^* and T_c furthermore can be just as well observed from the curves; however, the meaning of T^* is not simple, and can involve a liquid crystal's superheating and supercooling [2]. Thus, de Gennes' simple phenomenological theory can obtain the same nematic-to-liquid phase transition results as the far more complicated Maier–Saupe theory.

Old-line physicists may well note that the Maier–Saupe theory is based on statistical mechanics and the stolid molecular dipole moment, whereas the de Gennes phenomenological approach appears to be a mere coefficient-fitting exercise. However, it should be pointed out that the statistical mechanics is itself an extensive smoothing out of the actual atomic situation, and the dipole moment is also a simplification based on cylindrical symmetry. Perhaps more significantly, the continuous and symmetric nature of matter is perhaps a deeper and more substantial basis than a theory based on statistical mechanics, and an electric dipole moment, which itself embodies many simplifications of a very complex atomic and molecular interaction situation. Both approaches should be taken for what they can contribute to understanding of the liquid crystal, which may be considerable, and they are not incompatible in that the phenomenological theory employs the effective dipole moment to calculate birefringence.

Birefringence Calculation

The birefringence (double refraction, dielectric anisotropy) of liquid crystals arises from the intrinsic dipole moment and the induced electric dipole moment caused by an applied external electric field. The applied field will generate within the liquid crystal a mean orientational energy of the molecules $W(S)$. From the power series expansion of the Helmholtz (orientational) free energy above, adding on the $W(S)$ multiplied by Avogadro's number N, then gives the total molar free energy in the presence of the applied field as

$$F = \tfrac{1}{2}a\cdot(T - T^{*})S^2 - \tfrac{1}{3}BS^3 + \tfrac{1}{4}CS^4 \ldots + NW(S).$$

Now from the expressions for the effective induced dipole moments from Maier–Saupe in Chapter 9 and the electric field, and keeping in mind that energy is proportional to the square of a field quantity such as an electric field, the average orientational energy of the molecule due to the induced dipole moment can be written as

$$W_I(S) = -\tfrac{1}{3}\cdot\Delta\alpha\cdot R\left(\frac{3\overline{\varepsilon}}{2\overline{\varepsilon}+1}\right)^2 E^2\cdot S,$$

and the average orientational energy due to the intrinsic dipole moment can be written as*

$$W_P(S) = -\left(R^2\left(\frac{3\overline{\varepsilon}}{2\overline{\varepsilon}+1}\right)^2 \mu^2E^2/6k_BT\right)(3\cos^2\beta - 1)\cdot S,$$

where the terms are defined as in the Maier–Saupe theory for the dielectric anisotropy in Chapter 9. Because $W(S) = W_I(S) + W_P(S)$, substituting the above two expressions into the total orientational energy equation given above, the molar free energy is then

$$F = \tfrac{1}{2}a(T - T^{*})\cdot S^2 - \tfrac{1}{3}NR\left(\frac{3\overline{\varepsilon}}{2\overline{\varepsilon}+1}\right)^2 E^2\cdot\left[\Delta\alpha - (R\mu^2/2k_BT)\cdot(1 - 3\cos^2\beta)\right]\cdot S.$$

Using the elementary quadratic formula to solve the second-degree equation above, and neglecting the radical in that formula as insubstantial, the orientational order of the liquid crystal near the phase transition from nematic to isotropic liquid can be given explicitly as a function of temperature (T), and the angle between the molecular intrinsic dipole moment and the principal axis of the liquid crystal (β)†

$$S = \frac{NR\left(\frac{3\overline{\varepsilon}}{2\overline{\varepsilon}+1}\right)^2 E^2[\Delta\alpha - (R\mu^2/2k_BT)\cdot(1 - 3\cos^2\beta)]}{3a(T - T^{*})}.$$

* The calculation of the average orientational energy of the molecule due to the intrinsic dipole moment considers only even powers of the $\cos\theta$, where θ is the orientational angle of the molecules; this is in accord with the expression for the free orientational energy.

† The radical $\sqrt{(b^2 - 4ac)}$ in the quadratic formula can be calculated of course, but turns out to not have much influence.

Using a Lorentz–Lorenz/Vuks relation for the index of refraction as derived in Chapters 3 and 8, and substituting in de Genne's phenomenological theory equation above, the birefringence Δn can be expressed as a function of the charge polarization parameter $\Delta\alpha$ of the nematic liquid crystal, the mean dielectric constant $\bar{\varepsilon}$, the index of refraction n, molecular anisotropic polarizability $\Delta\eta$, and the molar volume V,

$$\Delta n = \frac{2\pi N^2 \Delta\eta (n^2+2)^2 R\left(\frac{3\bar{\varepsilon}}{2\bar{\varepsilon}+1}\right)^2 E^2[\Delta\alpha - (R\mu^2/2k_B T)\cdot(1-3\cos^2\beta)]}{27nVa(T-T^*)}.$$

The birefringence equation above can be confirmed as correct in principle by a simple calculation, which also demonstrates the theory's usefulness. Positive or negative birefringence is determined by the sign of the farthest right term in the equation above,

$$\pm[\Delta\alpha - (R\mu^2/2k_B T)\cdot(1-3\cos^2\beta)],$$

and the $\Delta\alpha$ for a cylinder is generally positive, so if

$$[(R\mu^2/2k_B T)\cdot(1-3\cos^2\beta)] > \Delta\alpha,$$

then $\Delta n < 0$, and the liquid crystal will demonstrate negative birefringence. The contribution of the dipole moment to birefringence depends on the value of the angle (β) between the intrinsic dipole moment μ and principal long axis of the liquid crystal molecules, if β is small, then $\Delta n > 0$, then the liquid crystal is positively birefringent; if β is sufficiently large, then $\Delta n < 0$, and the liquid crystal is negatively birefringent; that is, when the molecules turn such that the director angle with the intrinsic dipole moment is large, then the liquid crystal will show a negative birefringence. This was mentioned at the end of Chapter 5 with regard to the figure of the phenyl ring where the experimental value for birefringence sign reversal was $\beta = 55°$.

The left-hand side of the above equation $(R\mu^2/2k_B T)\cdot(1-3\cos^2\beta)$ is inversely proportional to the temperature T, and so according to the de Gennes theory, the birefringence of a nematic liquid crystal can change from a positive to a negative type with a change in temperature. Indeed, from values of a and T^* from the magnetic birefringence of the synthesized liquid crystal compound p-azoxyanisole (PAA), the change from positive to negative birefringence with temperature does occur.

Actually, a simpler analysis of the relation of birefringence to temperature can be gleaned by setting $\beta = 0$ (intrinsic dipole moment is absolutely aligned with the molecular principal long axis), then

$$\Delta n \propto (T - T^*)^{-1},$$

and so when the direction of the intrinsic dipole moment is substantially parallel with the molecular principal axis, ($\beta \sim 0$), there is a "strong" positive birefringence. For example, the birefringent nematic liquid crystal hexylcyanobiphenyl (HCB), no matter whether an applied field is electric or magnetic, demonstrates the inverse relation between birefringence and temperature as predicted by the de Gennes theory.

In summary, the Landau–de Gennes theory can calculate the isotropic to nematic reverse phase transition, as well as describe the molecular orientation of a weakly ordered isotropic liquid. Based on a Helmholtz free energy computation, de Gennes also produces a useful explicit equation for the orientational order parameter, and with temperature as a variable, also provides an expression for the liquid crystal birefringence.

The de Gennes theory can also be extended to study smectic liquid crystals and the macromolecular polymers important to the chemical technology product industries. Interestingly, the theory and methods used by de Gennes for the nematic to smectic-A phase transition are astonishingly similar to those used in the theory of low-temperature liquid helium superconductivity [3].

Of course, all of the theories of nematic liquid crystals must be confirmed by experiment. Today's high-precision measuring techniques, such as nuclear magnetic resonance and x-ray diffraction and scattering, have been used to measure the liquid crystal molecular structure and the various phase transitions, and it is no surprise, since the phenomenological theory is largely based on coefficient fitting to observations, later experiments have confirmed the basic attributes of the Landau–de Gennes theory.

With the market success of the notebook computer using a liquid crystal display screen in the 1990s, the technological brilliance behind liquid crystal technology was visibly evident to all, and so Pierre-Gilles de Gennes was recognized and received the 1991 Nobel Prize for Physics "for discovering that methods developed for studying ordering phenomena in simple systems can be generalized to more complex forms of matter, particularly liquid crystals and polymers" [4].

References

[1] de Gennes, P.G. 1971. *Mol. Cryst. Liquid Cryst.*, **12**, 193; Landau, L.D., and Lifshitz, E.M. 1980. *Statistical Physics*, Part I, 3rd edition. Pergamon, Oxford; and de Gennes, P.G., and Prost, J. 1993. *The Physics of Liquid Crystals*, 2nd edition, Oxford, p. 76ff.
[2] Chandrasekhar, S. 1992. *Liquid Crystals*, 2nd edition. Cambridge, p. 62; de Gennes, P.G., and Prost, J. 1993. *The Physics of Liquid Crystals*, 2nd edition. Oxford, pp. 78, 85; and Collings, P.J., and Hird, M. 2004. *Introduction to Liquid Crystal.* Taylor & Francis, London, p. 248.
[3] de Gennes, P.G., and Prost, J. 1993. *The Physics of Liquid Crystals*, 2nd edition. Oxford, p. 509.
[4] NobelPrize.org.

11

Static Continuum Theory

Up to now, the description of liquid crystals has been primarily in regard to its microscopic atomic and molecular structure and its interaction with light. However, when a liquid crystal is subject to a macroscopic external force, the strains exhibited in its splay, twist, and bend have dimensions of the order 10^{-6} m. In comparison with molecular sizes of about 10^{-10} m, it is clear that for many cases, specific molecular structure can be ignored when studying strain-induced changes in the director orientation of the liquid crystal, and the liquid crystal can be treated as a liquid continuum. The continuum theory of liquid crystals is derived from the well-developed classical theory of anisotropic fluids in continuum physics, and as in that theory, can treat the liquid crystal both statically and dynamically.

Basic Principles

In the static continuum theory, the liquid crystal's initial energy state equation is first determined, and the appropriate boundary conditions for the system under study are set; then the calculus of variations is employed to minimize the orientational energy of the liquid crystal (the Helmholtz free

Liquid Crystal Displays, First Edition. Robert H. Chen.

energy). The minimum orientational energy expressions then form a set of equilibrium equations, the solutions of which are the stable states of the system.

In applying an external force, because the liquid crystal is incompressible, its reaction will not be in a change of volume but rather an internal stress (before) and strain (after) within the liquid crystal. The stress and strain will alter the liquid crystal's orientational energy state and cause the system to transform to a different stable minimum energy state. Because the stresses and strains in the continuum of the liquid crystal act locally, the liquid crystal director (n), which was defined as the local average orientation of the long axes of the molecules, is the principal descriptive parameter used in continuum theory (rather than the orientational order parameter S used for the mean dipole moment models). Because of the elastic nature of a fluid, when the external force is released, the liquid crystal will tend to return to its original equilibrium state, which is its state of minimum orientational potential energy.

The continuum theory first assumes that the liquid crystal's restoring free energy density w can be expressed as a function of the unit vector director $\hat{n}$ and the distortion of that director $\nabla\hat{n}$ (which describes the distortions within the liquid crystal) [1],

$$w = w(\hat{n}, \nabla\hat{n}) w = w(\hat{n}, \nabla\hat{n}),$$

Since the natural state of the liquid crystal is its lowest energy state, the energy of any other state will be higher than the natural equilibrium state, so

$$w = w(\hat{n}, \nabla\hat{n}) \geq 0$$

Further, because of the cylindrical shape of the liquid crystal molecules, $\hat{n}$ and $-\hat{n}$ are geometrically the same, that is, because one-half of all the molecules can align in either direction, the intrinsically charge polarized molecules' energy density is symmetric

$$w(\hat{n}, \nabla\hat{n}) = w(-\hat{n}, -\nabla\hat{n}),$$

and w can be integrated to obtain the elastic free energy (which will be the orientational energy), so the Helmholtz free energy over the volume of the entire system is

$$F = \iiint w(\hat{n}, \nabla \hat{n}) dV,$$

But because liquid crystals are anisotropic, the different effects in different directions is best treated by tensors, a brief introduction to which will be given here.

If a force F is applied to the liquid crystal, it will affect the orientation of the director which represents the local average molecular direction within the liquid crystal. Just as in the case of the electric field vector of linearly polarized light incident on the solid crystal (Chapter 4), the force F will have different effects on the liquid crystal depending on the latter's different structure along the x-axis and the y-axis. That is, the liquid crystal's elastic structure begets responses differentially depending on its structure. For an arbitrary orientation of the liquid crystal, a force in the x-direction will produce a director response with x, y, and z components,

$$n_x = k_{xx} F_x, \quad n_y = k_{yx} F_x, \quad n_z = k_{zx} F_x$$

where k_{ij} is a proportionality constant with the i representing which component of n is involved and the j referring to the direction of the force. In the same way, for a force in the y-direction,

$$n_x = k_{xy} F_y, \quad n_y = k_{yy} F_y, \quad n_y = k_{zy} F_y$$

and for a force in the z-direction,

$$n_x = k_{xz} F_z, \quad n_y = k_{yz} F_z, \quad n_z = k_{zz} F_z$$

If the force is not along the x-axis, but in an arbitrary direction, the director will be given most generally by

$$n_x = k_{xx} F_x + k_{xy} F_y + k_{xz} F_z$$
$$n_y = k_{yx} F_x + k_{yy} F_y + k_{yz} F_z$$
$$n_z = k_{zx} F_x + k_{zy} F_y + k_{zz} F_z$$

The director behavior of the liquid crystal in response to an external force is completely described by the nine k_{ij} and this set of coefficients is called a *tensor*, which can immediately be seen to be best represented by a matrix.

If a different coordinate system is used, of course the components of the F and n vectors will be different, so all the k_{ij} will also be different. For any new coordinate system, the $n_{x'}$ will be a linear combination of n_x, n_y and n_z,

$$n_{x'} = aF_x + bF_y + cF_z$$

and similarly for the other components. Substituting for n_x, n_y, and n_z in terms of the F components will produce expressions for the n_i in terms of the components of F in the new coordinate system. Aligning the coordinate system with the symmetry of the liquid crystal can simplify the tensor for more convenient calculation, as will be done following. The curvature strain in a liquid crystal is more complicated than the response to the simple linear force in the example above, but the tensor principles are the same.

The elastic free energy density expression given above can be characterized as a strain relation described by curvature strain tensors (K_{ij}) and curvature strains (a_i) in a second-degree expansion,

$$w = \sum_{i=1}^{6} K_{ii}a_i + \frac{1}{2}\sum_{i=1}^{6}\sum_{j=1}^{6} K_{ij}a_i a_j.$$

The expression above shows that the orientation of the director is based on the three changes of the orientation in three spatial coordinate $(\Delta x, \Delta y, \Delta z)$, but owing to the anisotropic nature of a liquid crystal, it requires at least six elements of a curvature strain tensor (also called the curvature elasticity) to adequately describe the energy situation. Actually, the curvature strain tensor in the generalized Oseen–Zocher–Frank equation above originally had 81 components, but because of the cylindrical symmetry of the liquid crystal molecules, the principal long axis may be aligned in the z-direction, and by symmetry many of the elements vanish. Taking the partial derivative of the director with respect to each direction x ,y ,z, the anisotropic nature of the strain can be demonstrated; that is, based on the symmetry of the molecules' principal axis being aligned with the z direction, the director's z component does not change, so [2]

$$\frac{\partial n_z}{\partial x} = \frac{\partial n_z}{\partial y} = \frac{\partial n_z}{\partial z} = 0,$$

the six elements of the curvature strain and their expression as vector equations can be summarized as follows,

$$\frac{\partial n_x}{\partial x} = a_1, \frac{\partial n_y}{\partial y} = a_5(\text{splay}), \quad \nabla \bullet \hat{n} \neq 0$$

$$\frac{\partial n_y}{\partial x} = -a_4, \frac{\partial n_x}{\partial y} = a_2(\text{twist}), \quad \hat{n} \bullet \nabla \times \hat{n} \neq 0$$

$$\frac{\partial n_x}{\partial z} = a_3, \frac{\partial n_y}{\partial z} = a_6(\text{twist}), \quad \hat{n} \times \nabla \times \hat{n} \neq 0,$$

and the curvature strain tensor can be written in matrix form as

$$\left.\frac{\partial n}{\partial x}\right|_{ij} = \begin{bmatrix} \frac{\partial n_x}{\partial x} & \frac{\partial n_x}{\partial y} & \frac{\partial n_x}{\partial z} \\ \frac{\partial n_y}{\partial x} & \frac{\partial n_y}{\partial y} & \frac{\partial n_y}{\partial z} \\ 0 & 0 & 0 \end{bmatrix} = \begin{bmatrix} a_1 & a_2 & a_3 \\ -a_4 & a_5 & a_6 \\ 0 & 0 & 0 \end{bmatrix}.$$

The orientational energy density can be expressed as a strain tensor, and a uniaxial liquid crystal structure will not change around the principal cylindrical z-axis, so the curvature elastic coefficients can be expressed in a matrix as follows,*

$$K_{ij} = \begin{bmatrix} K_{11} & K_{12} & K_{13} \\ K_{21} & K_{22} & K_{23} \\ K_{31} & K_{32} & K_{33} \end{bmatrix}.$$

The diagonal elements K_{11}, K_{22}, K_{33} in the curvature elastic coefficients matrix are just those described previously in Chapter 5 as the splay, twist, and bend coefficients of elasticity, and recall that combinations of them can describe any deformation. Owing to the cylindrical symmetry of the liquid crystal molecules, the free energy orientational energy density equation is relatively simple, in the Cartesian coordinates vector notation, it is

* It should be noted that different subscripts notation schemes for the elastic coefficients are used in different texts and papers.

$$w_F = \sum_i \sum_j \sum_k \left[\frac{1}{2}(K_{11} - K_{22} - K_{24})\left(\frac{\partial n_i}{\partial x_i}\right)^2 + \frac{1}{2}K_{22}\left(\frac{\partial n_i}{\partial x_i}\right)\left(\frac{\partial n_i}{\partial x_j}\right) + \frac{1}{2}K_{24}\left(\frac{\partial n_i}{\partial x_j}\right)\left(\frac{\partial n_j}{\partial x_i}\right) + \frac{1}{2}(K_{33} - K_{22})n_j\left(\frac{\partial n_i}{\partial x_j}\right)n_k\left(\frac{\partial n_i}{\partial x_k}\right)\right],$$

and the orientational energy density (or free, Helmholtz, distortional, elastic, nematic, and so on various other names) in vector notation is

$$w_F = \frac{1}{2}K_{11}(\nabla \bullet \hat{n})^2 + \frac{1}{2}K_{22}(\hat{n} \bullet \nabla \times \hat{n})^2 + \frac{1}{2}K_{33}(\hat{n} \times \nabla \times \hat{n})^2.$$

The equation above is the *basic orientational energy density equation* that will be used in the analyses to follow; the total free energy is

$$F = \iiint w_F dV.$$

Static Continuum Theory Examples

Based on the principles outlined above, a few applications of the static continuum theory for practical liquid crystal display (LCD) problems will further understanding of how the theory is used, and provide insight into the physical processes at work.

A display still in common use is the twisted nematic liquid crystal mode using a rubbing compound that effectively anchors the rod-like molecules in a specific direction on each glass substrate. If the rubbing direction on the glass substrate is different, the molecules would then be forced to twist in the region between the glass substrates. The anchoring can be categorized as absent, strong, conical, and weak, altogether four distinct types. The examples below will use strong anchoring, wherein the molecules are either anchored to lie parallel to the plane of the glass substrate in the so-called *homogeneous mode*; or by anchoring their tails, making the molecules stand perpendicular to the glass substrate in the so-called *homeotropic mode*.

The Twisted Only

If the top and bottom glass substrates holding the liquid crystal are coated with rubbing compounds that strongly anchor the molecules in the vicinity of the substrates, and if the top and bottom rubbing directions are different

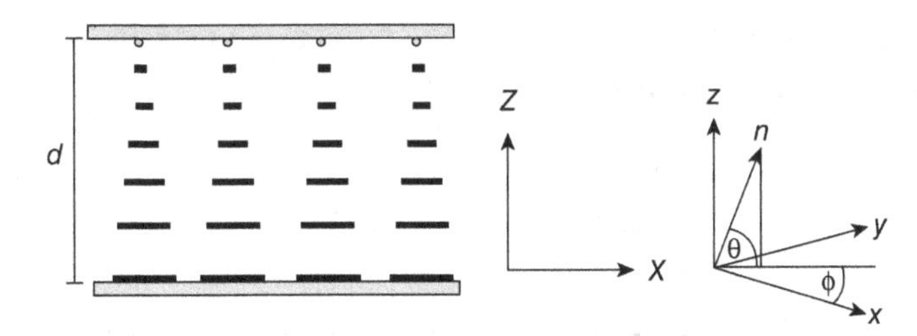

Figure 11.1 Twisted only liquid crystal from strong anchoring boundary conditions anchoring the molecules parallel to the substrate.

by 90°, the molecules in the liquid crystal will undergo a twist of 90° through the liquid crystal layer in the distance d if the Mauguin and Gooch–Tarry conditions are met (Chapter 16), as they are here by imposition of the boundary conditions and setting the cell gap d appropriately.

Figure 11.1 at left shows the molecules near the bottom substrate of the liquid crystal cell to be anchored parallel to the x axis (denoted by the —), and the molecules near the top substrate indicated by the dots (denoted by the ∘ ∘) are perpendicular to the x and z axes and parallel to the y axis, with the molecules in-between turning from the x direction towards the y direction (denoted by the decreasing perspective lengths —). Figure 11.1 at right depicts the director $\hat{n}$ in an x,y,z coordinate system and its orientation relative to the twist angle ϕ (azimuthal) and the tilt angle θ (polar) of the molecules in a polar coordinate system,

With the z axis as the axis of symmetry, the liquid crystal director $\hat{n}$ can be expressed in terms of the unit vectors $(\hat{x}, \hat{y}, \hat{z})$ as,

$$\hat{n} = \cos\phi(z)\hat{x} + \sin\phi(z)\hat{y} + 0\hat{z} = (\cos\phi(z), \sin\phi(z), 0),$$

where it should be noted that there is only twist in the xy-plane and no polar angle tilt; substituting the expression for the director $\hat{n}$ into the basic orientational free energy equation of the previous section, with $\phi(z)$ being the twist angle as a function of z, the *distortion energy as a function of twist angle* is (where K_{22} is the twist elastic coefficient)

$$W = \frac{1}{2} K_{22} \int_0^d \left(\frac{d\phi}{dz} \right)^2 dz$$

To find the lowest orientational energy state (the stable state), the Euler–Lagrange equation from the variational calculus introduced in Chapter 7 is used to minimize the orientational energy. For the stable state, ϕ must satisfy the Euler–Lagrange minimization condition

$$\frac{d}{dz}\left[\frac{\partial w}{\partial(d\phi/dz)}\right]-\frac{\partial w}{\partial\phi}=0.$$

Taking w from the integrand in the distortion energy as a function of the twist angle equation above, $w=(d\phi/dz)^2$, the Euler–Lagrange equation becomes just

$$\frac{d^2\phi}{dz^2}=0,$$

and the solution to this simple second-order differential equation is just

$$\phi(z)=az+b$$

With the directors near the top and bottom substrates set at a 90° relative angle, the boundary conditions for this case are

$$\text{at } z=0,\ \hat{n}_{\text{boundary}}=(1,0,0),\ \phi=0s$$

$$\text{at } z=d,\ \hat{n}_{\text{boundary}}=(0,1,0),\ \phi=\frac{\pi}{2}$$

thus $b = 0$, and

$$\phi(z)=\frac{\pi}{2d}\cdot z,$$

so the liquid crystal molecules twist in the xy-plane and execute a spiral going from the bottom to the top glass substrate and from the boundary conditions as set, when $z=d$, the twist angle is $\pi/2$ (90°). The energy per unit area of the substrate (W/cm^2) is easily calculated from the distortion energy equation as,

$$W=K_{22}\frac{\pi^2}{8d}.$$

In the same manner, the static continuum theory of liquid crystals can be used to calculate the tilt angle and the twist and tilt angle combination.

The Twist and Tilt

In this more general case, in addition to twist (ϕ), there is also a tilt (θ) out of the *xy*-plane toward the *z*-axis, in this case the director can be expressed as

$$\hat{n} = (\cos\phi(z), \sin\phi(z), \sin\theta(z)),$$

where now there is allowed a polar tilt in the *z*-direction. Based on the basic orientational energy density equation in the previous section, the orientational energy can be written in functional form as

$$w_F(\hat{n}, \nabla\hat{n}) = w(\theta, \phi, d\theta/dz, d\phi/dz),$$

where according to basic equation, the Helmholtz free energy is

$$\begin{aligned} w_F(\hat{n}, \nabla\hat{n}) = &\frac{1}{2}[K_{11}\cos^2\theta + K_{33}\sin^2\theta]\cdot\left(\frac{d\theta}{dz}\right)^2 \\ &+\frac{1}{2}[K_{22}\cos^2\theta + K_{33}\sin^2\theta]\cos^2\theta\cdot\left(\frac{d\phi}{dz}\right)^2, \end{aligned}$$

where K_{11}, K_{22}, and K_{33} in the equation above are respectively the splay, twist, and bend elastic coefficients. This more general orientational energy equation again can be minimized by the Euler–Lagrange equation of the variational calculus to find the stable states, but because the tilt angle variance has been added, the resulting stable state orientational energy condition is more complicated,

$$\begin{aligned} &2\cdot[K_{11}\cos^2\theta + K_{33}\sin^2\theta]\cdot\left(\frac{d^2\theta}{dz^2}\right) + \frac{d}{d\theta}[K_{11}\cos^2\theta + K_{33}\sin^2\theta]\left(\frac{d\theta}{dz}\right)^2 \\ &-\frac{d}{d\theta}[K_{22}\cos^2\theta + K_{33}\sin^2\theta]\cos^2\theta\cdot\left(\frac{d\phi}{dz}\right)^2 = 0 \\ &\frac{d}{dz}\left\{K_{22}\cos^2\theta + K_{33}\sin^2\theta]\cos^2\theta\cdot\left(\frac{d\phi}{dz}\right)\right\} = 0. \end{aligned}$$

The integration of the two equilibrium equations above provides two minimum orientational energy equations (with a and b as constants of integration),

$$[K_{11}\cos^2\theta + K_{33}\sin^2\theta]\cdot\left(\frac{d\theta}{dz}\right)^2 + [K_{22}\cos^2\theta + K_{33}\sin^2\theta]\cos^2\theta\cdot\left(\frac{d\phi}{dz}\right)^2 = a,$$

and

$$[K_{22}\cos^2\theta + K_{33}\sin^2\theta]\cos^2\theta\cdot\left(\frac{d\phi}{dz}\right) = b.$$

The equations above include both twist and tilt energies, but because the solution will depend on the relative values of the elastic coefficients, the solution will be different for different liquid crystals; that is, whether the twist or tilt is dominant will depend on the relative values of K_{11}, K_{22}, and K_{33}, so each solution must consider the different elasticities of different liquid crystals. However, an idea of the physical factors at play in the equations can be gained by limiting the analysis to only tilt orientations as a special example.

The Tilt Only

With no twist, there will be only the tilt of the molecules' directors involved; setting the boundary tilt angle with respect to the top and bottom glass substrates as θ_o, the boundary conditions are simply

$$\theta(0) = \theta(d) = \theta_o$$

$$\phi(0) = \phi(d) = 0.$$

Because K_{22}, K_{33}, $\cos^2\theta$, and $\sin^2\theta$ are all always positive, from the b orientational energy equation above, if $b = 0$, then $(d\phi/dz) = 0$, and $\phi(z) = \text{const.}$, but from the $\phi(z)$ boundary condition, that const. must equal zero; if $b \neq 0$, since all the factors multiplying the term $(d\phi/dz)$ in the b equation are positive definite, the sign of $(d\phi/dz)$ cannot change; then, applying the ϕ boundary condition, that is, for $z = 0$ to $z = d$, the sign of the slope of $\phi(z)$ cannot change from positive to negative, and, vice-versa, it follows that $\phi(z)$ must also be zero, and so the ϕ boundary condition mandates that there can be no twist. So if the twist angle is zero ($\phi(z) = 0$), and is substituted into the a orientational energy equation above, then

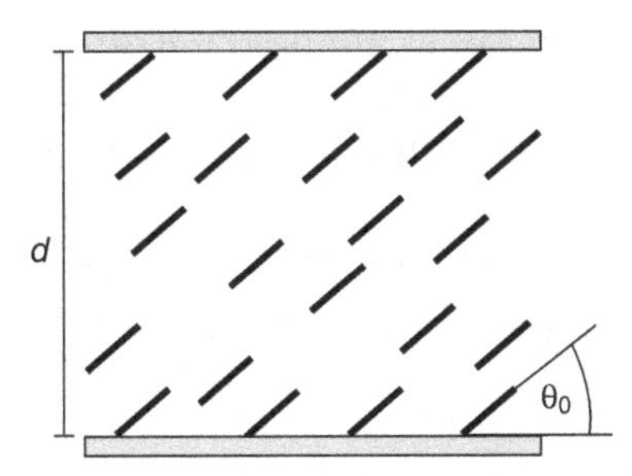

Figure 11.2 Tilt only liquid crystal from strong anchoring boundary conditions anchoring the molecules at initial tilt angle with the substrate.

$$[K_{11}\cos^2\theta + K_{33}\sin^2\theta]\cdot\left(\frac{d\theta}{dz}\right)^2 = a.$$

In the same fashion as for the b orientational energy equation, if $a > 0$, within the interval $0 \le z \le d$, $(d\theta/dz) \neq 0$, and the θ boundary conditions cannot pertain, so a must be equal to zero. Thus because the square-bracket term is positive definite, then the square derivative term must be equal to zero; that is, $(d\theta/dz) = 0$. Applying the θ boundary condition, $\theta(z) = \theta_o$, then within the interval $0 \le z \le d$, $\phi(z) = 0$ and $\theta(z) = \theta_o$, and the molecules' director tilt throughout the liquid crystal is a constant $\theta(z) = \theta_o$, as schematically depicted in Figure 11.2.

The situation devolves as described above principally because of the effects of the strong anchoring boundary conditions; this will change when the influence of an externally applied electric field is considered, as will be described in the following section.

The Freedericksz Cell

An external electric field interacting with the dipole moment of the liquid crystal molecules will cause those molecules to turn so that the dipole moment aligns with the applied field. The earliest experimental studies of the effects of electric and magnetic fields on liquid crystals were performed by the Russian physicist V.K. Freedericksz. He surrounded a liquid crystal with electrodes and electromagnets to generate electric and magnetic fields through a liquid crystal sample, and mounted various electrical and optical measuring instruments to monitor the liquid crystal's response to the fields; his apparatus in concept is still used today and is called a *Freedericksz cell.**

It is a fact that as opposed to the effects of the mutual electrical induction among the liquid crystal molecules, the relatively much smaller mutual magnetic induction in liquid crystal molecules can be ignored, so almost all the early experiments designed to probe liquid crystals utilized a magnetic field to make measurements and analysis simpler. However, since today's liquid crystal displays use electric fields to turn the molecules to control the transmission intensity, the examples described below will use external electric fields instead of the magnetic fields that are common to experimental work. It might be interesting to think about the question, however: Why don't LCDs use magnetic fields to re-orient liquid crystal molecules to polarize the light to control intensity?

The Splay Tilt

Because the polarizing effect is most easily observed when the tilt angle θ of the directors changes, Freedericksz first studied the effects of an external field on the splay of the liquid crystal molecules. As Figure 11.3 at left shows, the liquid crystal molecules initially are lying horizontally to the glass substrate, and applying an electric field ***E*** perpendicular to the glass substrate will engender a splay tilt as shown in Figure 11.3 at right.

A *z*-axis electric field vector can be expressed simply as,

$$\vec{E} = 0\hat{x} + 0\hat{y} + E\hat{z} = E(0,0,1).$$

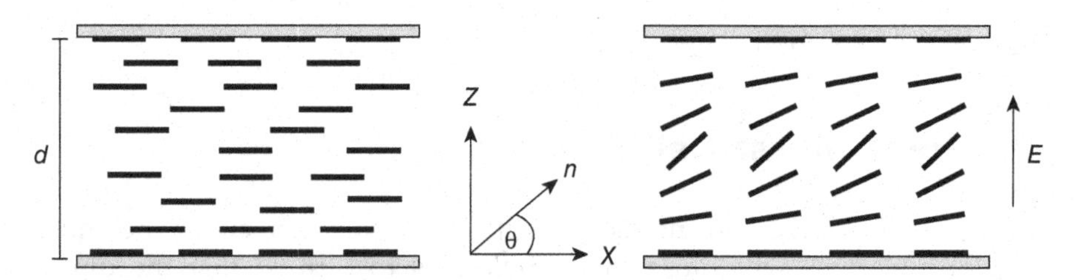

Figure 11.3 Splay tilt liquid crystal molecules lying horizontally with electric field perpendicular to the glass substrate.

* "Freedericksz" has also been spelled "Frederiks," Fréericksz, Freedericsz, apparently depending on the nationality of the writer invoking his name, the transliteration will be different.

As described in Chapter 5, if the liquid crystal is of positive dielectric anisotropy ($\Delta\varepsilon > 0$),its dipole moment is aligned along the principal long axes of the molecules, and the director $\hat{n}$ will tend to align with the electric field direction $\vec{E}$; $\hat{n}$ can be expressed as

$$\hat{n} = (\cos\phi(z), 0, \sin\theta(z)).$$

and the boundary conditions expressed as

$$\theta(0) = \theta(d) = 0.$$

From the principles section above, the basic orientational free energy density equation is

$$w_F = \frac{1}{2}K_{11}(\nabla \bullet \hat{n})^2 + \frac{1}{2}K_{22}(\hat{n} \bullet \nabla \times \hat{n})^2 + \frac{1}{2}K_{33}(\hat{n} \times \nabla \times \hat{n})^2.$$

Applying an external electric field adds an electrical energy expressed using the Maxwell displacement vector D for electric fields in material media as described in Chapter 3 (where ε_o is the electrical permittivity of free space),*

$$w_E = -\int_0^E D \bullet dE = -\frac{1}{2}\varepsilon_o\varepsilon_\perp E^2 - \frac{1}{2}\varepsilon_o\Delta\varepsilon(\hat{n} \bullet E)^2.$$

Because $\hat{n}$ will only turn in the xy-plane, the first RHS term in the equation above can be ignored (because that term includes $\varepsilon_\perp$, which is perpendicular to the xy-plane and is thus zero), then what remains is

$$w_E = -\frac{1}{2}\varepsilon_o\Delta\varepsilon(n \bullet E)^2 = -\frac{1}{2}\varepsilon_o\Delta\varepsilon E^2 \sin^2\theta.$$

The orientational free energy density adding on the external electric field energy density then becomes

$$W = \int_V w_F + w_E)dV = \frac{1}{2}\int_0^d [K_{11}\cos^2\theta + K_{33}\sin^2\theta] \cdot \left(\frac{d\theta}{dz}\right)^2 - \varepsilon_o\Delta\varepsilon E^2 \sin^2\theta]dz.$$

* As in Chapter 3, the Maxwell equations are written in Gaussian units (cgs), where $\varepsilon_o = 1$, but in the SI units system $\varepsilon_o = 8.854 \times 10^{-12}$ F/m.

Now again the Euler–Lagrange equation of the variational calculus can be used to find the stable minimum orientational energy state with the tilt angle θ as the independent variable, and from Euler–Lagrange, the necessary condition for the minimum energy state is

$$\frac{d}{dz}\left[\frac{\partial w}{\partial(d\theta/dz)}\right]-\frac{\partial w}{\partial\theta}=0,$$

where the w in the equation is now $w = w_F + w_E$, considering both the orientational free energy and the applied electric field energy. The equation for the stable energy state of the tilt angle of the director $\hat{n}$ is

$$[K_{11}\cos^2\theta+K_{33}\sin^2\theta]\cdot\left(\frac{d^2\theta}{dz^2}\right)+(K_{33}-K_{11})\cdot\left(\frac{d\theta}{dz}\right)^2\sin\theta\cos\theta$$
$$+\varepsilon_o\Delta\varepsilon E^2\sin\theta\cos\theta=0.$$

The equation above unfortunately is clearly nonlinear and has no closed-form solution in elementary mathematical functions. Of course using a computer to perform numerical analysis can provide solutions that can be compared with experiment. Also, certain special cases can reduce the problem so that solutions can be expressed in terms of some elementary functions.

To provide insight into the physical meaning of the equation, some simple approximation methods that are commonly employed in mathematical physics can be used here; for example, limiting the angle θ to small values can linearize the equation as follows: Since for small angles,

$$\sin\theta\approx\theta \text{ and } \cos\theta\approx 1,$$

the orientational free energy equation can be simplified to the differential equation

$$(K_{11}+K_{33}\theta^2)\left(\frac{d^2\theta}{dz^2}\right)+(K_{33}-K_{11})\cdot\left(\frac{d\theta}{dz}\right)^2\theta+\varepsilon_o\Delta\varepsilon E^2\theta=0,$$

and assuming θ is small, the θ^2 and $(d\theta/dz)^2\theta$ terms can be ignored, and the equation above becomes a tractable differential equation

$$K_{11}\left(\frac{d^2\theta}{dz^2}\right)+\varepsilon_o\Delta\varepsilon E^2\theta=0,$$

which, again using the differential equations solutions techniques that twice differentiating sinusoidal functions gives the original functions back, and that linear combinations of solutions are solutions as well, a solution of the above differential equation is just a well-known general sinusoidal solution of the form

$$\theta = A\sin\left(\sqrt{\varepsilon_o \Delta\varepsilon E^2 / K_{11}}\right)z + B\cos\left(\sqrt{\varepsilon_o \Delta\varepsilon E^2 / K_{11}}\right)z.$$

From the boundary conditions, $\theta(0) = \theta(d) = 0$, therefore $B = 0$, so

$$\sqrt{\varepsilon_o \Delta\varepsilon E^2 / K_{11}}\, d = m\pi \quad m = 1, 2, 3, \ldots$$

If the applied electric field is small, the molecules will only slightly tilt, and so the lowest frequency solution is applicable, and $m = 1$ is taken, and the electric field strength at which the molecules will begin to turn is the *threshold* electric field

$$E_{th} = \frac{\pi}{d}\sqrt{\frac{K_{11}}{\varepsilon_o \Delta\varepsilon}},$$

and the threshold voltage (from $V_{th} = E_{th}d$) is then the very useful expression

$$V_{th} = \pi\sqrt{\frac{K_{11}}{\varepsilon_o \Delta\varepsilon}}.$$

In the threshold voltage expression above, there is no dependence on the thickness of the liquid crystal layer (d); further, the value of the dielectric anisotropy ($\Delta\varepsilon$) for different liquid crystal materials can be found from the suppliers' data sheets, and so measuring the *Freedericksz transition* threshold voltage can be used to determine the splay elastic coefficient K_{11} for various liquid crystals using the above equation.

In-Plane Switching

Another molecule-turning regime is the so-called *in-plane switching* mode where the long cylindrically shaped molecules (denoted by ▬ in Fig. 11.4 at left) are lying flat in respect to the glass substrate, and the direction of the electric field ***E*** is perpendicular to the x- and y-axes and parallel to the y-axis going into the plane of the page (denoted by the $\otimes$). In Figure 11.4 at right,

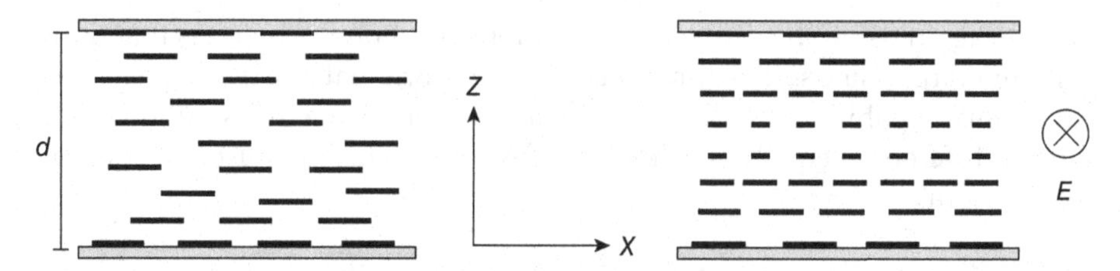

Figure 11.4 In-plane switching with liquid crystal molecules turning in the plane of the glass substrate.

the long axes of the cylindrical molecules are turning as denoted by the progression (▬▬) → (▬) → (-) and back again, so the boundary conditions are $\phi(0) = \phi(d) = 0$, and the director can be represented as a function of twist angle $\phi(z)$ only,

$$\hat{n} = (\cos\phi(z), \sin\phi(z), 0);$$

the electric field vector is given by

$$\vec{E} = 0\hat{x} + E\hat{y} + 0\hat{z} = E(0,1,0),$$

In the same manner as the previous example for the tilt angle orientational free energy density, the twist angle orientational free energy density differential equation is

$$K_{22}\left(\frac{d^2\phi}{dz^2}\right) + \varepsilon_o \Delta\varepsilon E^2 \sin\phi\cos\phi = 0,$$

and just the same as the tilt angle example, the threshold electric field and voltage values can be calculated and given by the simple expressions

$$E_{th} = \frac{\pi}{d}\sqrt{\frac{K_{22}}{\varepsilon_o \Delta\varepsilon}} \text{ and } V_{th} = \pi\sqrt{\frac{K_{22}}{\varepsilon_o \Delta\varepsilon}}$$

where now the twist (K_{22}) is the elastic coefficient of interest. This mode of molecular long-axis horizontal turning is called the *in-plane switching* mode (IPS), and is one of the major LCD modes, particularly for large-screen applications.

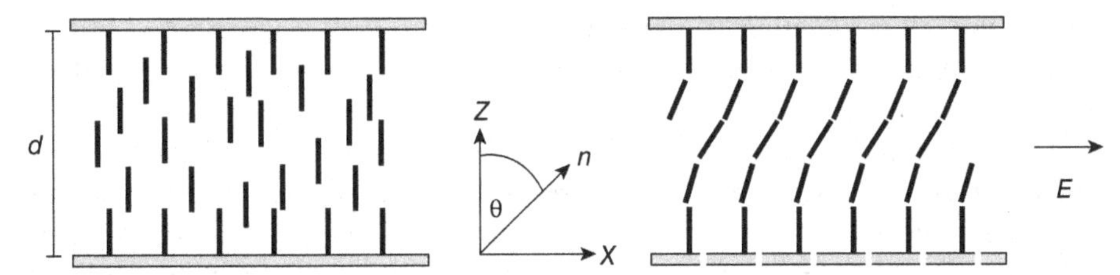

Figure 11.5 Bend perpendicular liquid crystal mode with electric field applied parallel to the *x*-axis.

The value for the dielectric anisotropy $\Delta\varepsilon$ can again be obtained from liquid crystal material suppliers, and the twist elastic coefficient K_{22} can thus be calculated from the above expressions from measurements of the threshold voltages needed to turn the liquid crystal molecules.

The Bend Perpendicular

The third kind of elastic distortion is the bend, represented by the bend elastic coefficient K_{33}. If the molecules in the Freedericksz cell are uniformly oriented perpendicular to the glass substrate, this is then a homeotropic mode. As shown in Figure 11.5, if the electric field is applied parallel to the x axis, and if the liquid crystal is positive anisotropic (dipole moment parallel to the molecular long axis), the mid-layer molecules will tilt in a direction making an angle θ with the z-axis.

The boundary conditions are $\theta(0) = \theta(d) = 0$, and the director vector is

$$\hat{n} = (\sin\phi(z), 0, \cos\theta(z)).$$

The electric field vector is

$$\vec{E} = E\hat{x} + 0\hat{y} + 0\hat{z} = E(1,0,0),$$

and similarly to the tilt angle example above, with K_{33} substituted for K_{11}, the threshold electric field and voltage expressions can be calculated as

$$E_{th} = \frac{\pi}{d}\sqrt{\frac{K_{33}}{\varepsilon_o \Delta\varepsilon}} \text{ and } V_{th} = \pi\sqrt{\frac{K_{33}}{\varepsilon_o \Delta\varepsilon}},$$

where the latter equation is just that given at the end of Chapter 5, and as in the other cases, the value of the bend elastic coefficient K_{33} can be calculated by measuring the threshold voltage.

Using rubbing compounds at the glass substrates to attract the ends of the molecules to make them stand perpendicular to the glass substrates and vertically throughout the liquid crystal, the molecules will be ordered in what is called the *vertical alignment* (VA) mode, which is a mainstream LCD mode for television panels.

For ease of manufacture, the electric field in VA panels is produced by electrodes that are plane parallel to the glass substrate, and so the electric field is oriented in the *z*-direction; then the long axis of the molecules of liquid crystals of negative dielectric anisotropy ($\Delta\varepsilon < 0$, dipole moments perpendicular to the molecular long axis) will tilt downwards away from the vertical just as if the electric field were oriented horizontally, as shown in Figure 11.5 at right.

In the Freedericksz transition, because the tilt angle θ measured from the vertical can have both positive and negative values, the liquid crystal can have two different minimum energy states. In order to promote one of the orientational equilibrium states, a so-called "pre-tilt" of the molecules in the preferred direction can be established in the liquid crystal by utilizing a strong anchoring compound disposed in the preferred direction to produce a tilt angle θ_o as in the example given in the above "Tilt Only" section.

The examples above all use strong anchoring; rubbing compounds also can generate weak or conical anchoring and the like, but the molecules near the glass substrates will then be more prone to influence by the electric field leading to rather complicated, but interesting calculations of tilt angle at the boundaries [3].

The Twisted Nematic

The commonly used twisted nematic (TN) LCD with tilt angle controlled by a vertical external electric field now can be analyzed as a Freedericksz transition. The situation is the same as described in the "Twist and Tilt" section above, but with more general boundary conditions and the addition of the effects of an electric field as shown in Figure 11.6. The molecules near the glass substrates are attracted to lie flat along the substrates, with a twist progressing between the substrates produced by aligning the rubbing compounds in different directions on the top and bottom glass substrates. The difference in alignment direction is given by $2\phi_o$ (the choice of the "2" will

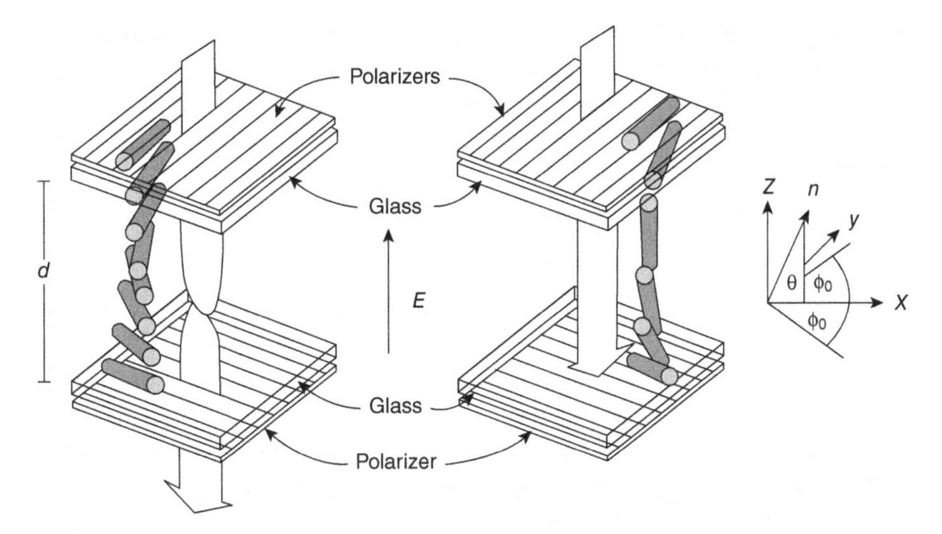

Figure 11.6 Twisted nematic liquid crystal display with applied electric field.

be apparent in the symmetry of the coordinate system), and the boundary conditions then are

$$\theta(0) = \theta(d) = 0$$

$$\phi(0) = -\phi_o \text{ and } \phi(d) = \phi_o,$$

and as Figure 11.6 on the left shows, the angular difference between the rubbing on the substrates is just $2\phi_o$.

A positive dielectric anisotropy ($\Delta\varepsilon > 0$) liquid crystal (with its dipole moment parallel to the molecular long axis) will tilt to align with the external electric field, so an applied electric field perpendicular to the glass substrates given by

$$\vec{E} = 0\hat{x} + 0\hat{y} + E\hat{z} = E(0,0,1),$$

will produce a tilt towards the z-axis. The most general expression for the director, considering both tilt (θ) and twist (ϕ) as functions of distance z through the liquid crystal, is

$$\hat{n} = (\cos\theta(z)\cos\phi(z), \cos\theta(z)\sin\phi(z), \sin\theta(z)).$$

The general expression for the twisted nematic orientational free energy from the basic energy density equation, in this case for both twist and tilt, and adding in the orientational energy contributed by the applied electric field given above, is

$$w = w_F + w_E = \left[K_{11}\cos^2\theta + K_{33}\sin^2\theta\right]\cdot\left(\frac{d\theta}{dz}\right)^2$$
$$+\frac{1}{2}[K_{22}\cos^2\theta + K_{33}\sin^2\theta]\cos^2\theta\cdot\left(\frac{d\phi}{dz}\right)^2 - \frac{1}{2}\varepsilon_o\varepsilon_\perp E^2 - \frac{1}{2}\varepsilon_o\Delta\varepsilon(\hat{n}\bullet E)^2.$$

In a positive dielectric anisotropy liquid crystal, adding the contribution of the applied electric field to the energy density, there are two angles of interest, tilt (θ) and twist (ϕ), that generally will be interdependent ("coupled"). The Euler–Lagrange minimization equations from the variational calculus for the tilt and twist angles respectively are,

$$\frac{d}{dz}\left[\frac{\partial w}{\partial(d\theta/dz)}\right] - \frac{\partial w}{\partial\theta} = 0,$$

and

$$\frac{d}{dz}\left[\frac{\partial w}{\partial(d\phi/dz)}\right] - \frac{\partial w}{\partial\phi} = 0,$$

which give the necessary conditions for the stable state orientational energy as

$$2\cdot[K_{11}\cos^2\theta + K_{33}\sin^2\theta]\cdot\left(\frac{d^2\theta}{dz^2}\right) + \frac{d}{d\theta}[K_{11}\cos^2\theta + K_{33}\sin^2\theta]\left(\frac{d\theta}{dz}\right)^2$$
$$-\frac{d}{d\theta}[K_{22}\cos^2\theta + K_{33}\sin^2\theta]\cos^2\theta\cdot\left(\frac{d\phi}{dz}\right)^2 + 2\varepsilon_o\Delta\varepsilon E^2\sin^2\theta\cos\theta = 0$$
$$\frac{d}{dz}\left\{K_{22}\cos^2\theta + K_{33}\sin^2\theta]\cos^2\theta\cdot\left(\frac{d\phi}{dz}\right)\right\} = 0.$$

Just as previously under the "Twist and Tilt" section for the static case, the integration of the above orientational energy differential equations gives the stable orientational states for the twisted nematic case. Here, however, the applied electric field also contributes to the orientational energy, but given a uniform electric field that does not vary with z, and the fact that the second differential equation above is with respect to changes in z, the effect of the applied electric field is just the additional term shown in the equation for a' below,

$$[K_{11}\cos^2\theta + K_{33}\sin^2\theta]\cdot\left(\frac{d\theta}{dz}\right)^2$$

$$+[K_{22}\cos^2\theta + K_{33}\sin^2\theta]\cos^2\theta\cdot\left(\frac{d\phi}{dz}\right)^2 + \varepsilon_o\Delta\varepsilon E^2\sin^2\theta = a'$$

$$[K_{22}\cos^2\theta + K_{33}\sin^2\theta]\cos^2\theta\cdot\left(\frac{d\phi}{dz}\right) = b.$$

Again, as discussed in the "Splay Tilt" section, because the equations above are nonlinear, and specific solutions depend on the relative values of the elastic coefficients (K_{11}, K_{22}, K_{33}) for each given liquid crystal material, there are no closed form general solutions for the above equations. However, if typical values for the elastic coefficients in a twisted nematic display are used,

$$2\phi_o = 90°,\ \Delta\varepsilon > 0,\ K_{33} - 2K_{22} \geq 0 \text{ or } K_{33} - 2K_{22} < 0,$$

then

$$\phi_o \leq \frac{\pi}{2}\sqrt{\frac{K_{11}}{2K_{22} - K_{33}}},$$

and a specific calculation of threshold voltage similar to those performed above can be made with the result [4]

$$V_{th} = \pi\sqrt{\frac{\left[K_{11} + \frac{1}{4}(K_{33} - 2K_{22})\right]}{\varepsilon_o\Delta\varepsilon}}.$$

The above expression for the threshold voltage of a twisted nematic display clearly shows that the bend, twist, and splay elasticities are all in play through their respective coefficients. Together with values for the dielectric anisotropy, the equation can be useed for designing twisted nematic liquid crystal displays.

In Memoriam

With the emergence of the twisted nematic, the liquid crystal display today has pervaded our visual space, having quickly become an indispensable and indeed beautiful adjunct to everyday life. The genesis and development of the liquid crystal display owes much to the eponymous Freedericksz cell and its progeny, the name commemorating the fundamental contributions to liquid crystal research made by Vsevolod Konstantinovich Freedericksz more than 80 years ago. His experiments are now clearly seen as brilliant and farsighted, performed well before their time [5]. But in 1936, just when Freedericksz was in the midst of conducting his fundamental studies, he was arrested and charged with inciting terrorism by the Soviet dictator Josef Stalin's secret police. Tragically, Freedericksz died in prison in 1944. Perhaps it was because of his Germanic family name that Stalin questioned his loyalty to the USSR, but whatever the reason for his fate, it was not until after the end of the War and three years after Stalin's death that the Freedericksz name was rehabilitated [6].

References

[1] Stewart, I.W. 2004. *The Static and Dynamic Continuum Theory of Liquid Crystals: A Mathematical Introduction*, Taylor & Francis, London; Chandrasekhar, S. 1992. *Liquid Crystals*, 2nd edition. Cambridge; and de Gennes, P.G., and Prost, J. 1993. *The Physics of Liquid Crystals*, 2nd edition. Oxford.

[2] Stewart, I.W. 2004. *The Static and Dynamic Continuum Theory of Liquid Crystals: A Mathematical Introduction*, Taylor & Francis, London, p. 14.

[3] Stewart, I.W. 2004. *The Static and Dynamic Continuum Theory of Liquid Crystals: A Mathematical Introduction*, Taylor & Francis, London, p. 95ff.

[4] Yang, D.K., and Wu, S.T. 2006. *Fundamentals of Liquid Crystal Devices*. Wiley, Chicester, p. 138ff.; Chandrasekhar, S. 1992. *Liquid Crystals*, 2nd edition. Cambridge, p. 106ff.; Stewart, I.W. 2004. *The Static and Dynamic Continuum Theory of Liquid Crystals: A Mathematical Introduction*. Taylor & Francis, London, p. 104ff.; and de Gennes, P.G., and Prost, J. 1993. *The Physics of Liquid Crystals*, 2nd edition. Oxford, p. 123

[5] Zocher, H., and Birstein, V. 1929. *Z. Physik. Chem.*, **142A**, 186; and Freedericksz, V.K., and Tsvetkov, V.N. 1935. *Compt. Rend. Acad. Sci. (USSR)*, **4**, 131.

[6] Castellano, J.A. 2005. *Liquid Gold*. World Scientific, Singapore, p. 4.

12

Dynamic Continuum Theory

The responses over time of a liquid crystal to external forces can be described by a dynamic continuum theory. The dynamic theory is founded on the classical fluid dynamics treatment of rotational and translational fluid flow. The liquid crystal, however, is special because its delicately ordered internal molecular orientation is significantly influenced by the internal translational motions within the liquid crystal, and the changing molecular orientations reciprocally affect the translational motions. In other words, any fluid-sliding motions will be coupled to the local dynamic orientational state of rotation, and the rotation in turn will affect the translation; the result is clearly observable effects on the polarization of light passing through, which is critical to the operation of liquid crystal displays.

In terms of hydromechanics, when an external force is applied, the liquid crystal molecules turn in a local but concerted manner, thereby producing a horizontal relative displacement of layers within the fluid, and the velocity gradient of the displacements produces a shear force that will in turn create a torque on the fluid, thereby generating further internal turning within the fluid.

Any mathematical theory describing the dynamic coupling of rotational and translational forces and motions of an anisotropic fluid at both the

Liquid Crystal Displays, First Edition. Robert H. Chen.

molecular and continuum fluid levels will of necessity be extremely complicated. The mathematical theory is further burdened by being split into different theoretical points of departure arising from basically different views of the fundamental dynamics. The predominant theories are the nominally macroscopic *Ericksen–Leslie dynamic continuum theory* and the nominally microscopic *Harvard correlation functions study.*

The Ericksen–Leslie theory is based on the velocity field of traditional continuum fluid mechanics, with the addition of a director to parameterize the local sector orientation of the liquid crystal to address its inherent anisotropy. The Harvard school, however, believed that the velocity field alone and its gradient are sufficient to functionally correlate the director orientation, that is, the director is not required as an independent variable, but accedes from the natural flow of the liquid crystal fluid.

This wielding of Occam's razor on the mathematical theory to cull it to the barest of necessary and sufficient postulate bones is at the very foundation of pure mathematical analysis and a difficult undertaking at best, but fortunately in liquid crystal research, the task is mundane, for it turns on the question of whether rotation of the microscopic directors can occur only by means of a macroscopic, nonuniform fluid flow. And this can be easily tested by experiment; for example, certain magnetic fields alone have altered the angle of the mid-layer molecules (as observed in a change of orientation of the optic axis). That is, in the absence of any macroscopic translation of the fluid, as evidenced by the stillness of a dust particle floating on the surface, the magnetic field caused the optic axis to observably rotate, thereby favoring the Ericksen–Leslie approach as requiring an independent director orientation. Notwithstanding the Ericksen–Leslie and Harvard disputes, de Gennes has demonstrated that the two *weltanschauung* were in fact consistent and compatible [1].

Both of the main dynamic nematic liquid crystal continuum theories ("nematodynamics")* are based on local slow spatial variations at low frequencies, the so-called *hydrodynamic limit*, as well as the typical incompressibility, isothermal, and strong anchoring basic assumptions. The following discussion of the dynamic continuum theory will adopt the Ericksen–Leslie approach, as it is the more generally-used theory in liquid crystal display research.

---

* de Gennes called the theories "nematodynamics," and although worms are cylindrically symmetric (though hardly stiff) just as liquid crystal molecules, images of squirming nematodes are conjured (maybe the erudite de Gennes was being sly).

Suffering an external force, the dynamic response of a liquid crystal continuum is first manifested as a strain that can be described by the traditional metrics of fluid dynamics, for example, the rate of strain tensor and the relative angular velocity. The equations describing the dynamic response of the liquid crystals in the Ericksen–Leslie theory can be formulated basically from the conservation of energy and momentum laws of physics. Then utilizing an asymmetric stress tensor to describe the anisotropic nature of the liquid crystal, the Ericksen–Leslie theory derives a set of constitutive relations from which the dynamic equations of the different physical states may be deduced, and the special characteristics of the liquid crystal analyzed. Among those special characteristics are the liquid crystal's viscosity, electric and magnetic anisotropy, different fluid flow modes, and its light scattering ability.

In the continuum theory, the response of the liquid crystals to electric and magnetic forces can be treated as a dynamic fluid motion. The general theory for electrical force effects is known as electrohydrodynamics, and if magnetic force effects are included it is firmly in the daunting field of the multiforce electromagnetohydrodynamics. The requirement of the tensor analysis to handle the liquid crystal's anisotropy in this dynamic environment makes the dynamic nematic continuum theory all the more mathematically forbidding. But feelings of mathematical inadequacy should not deter its study, for even the great fluid mechanics scholars Prandtl and Tietjens have noted [2]:

> Hydrodynamics. . . has little significance for the engineer because of the great mathematical knowledge required for it and the negligible possibility of applying its results. . . In classical hydrodynamics everything was sacrificed to logical construction. . . Theoretical hydrodynamics seemed to lose all contact with reality; simplifying assumptions were made which were not permissible even as approximations. . . [E]ach individual problem was solved by assuming a formula containing some undetermined coefficients and then determining those so as to fit the facts as well as possible. . . Hydraulics seemed to become more and more a *science of coefficients* (emphasis added).

Although famous scholars have so criticized the very fluid dynamics theories that they themselves developed, the basic principles at their heart are not indecipherable. Following is an introduction to those principles, a brief foray into the equations, and an example or two of the theory's application to liquid crystals, thereby offering a feel for this very mathematics-intensive discipline [3].

Conservation Principles

External electric or magnetic fields engender translational and rotational movements of the liquid crystal as a continuum. The motions are governed by the traditional conservation laws for mass, translational momentum, and angular momentum of the continuum liquid crystal, including the external, surface, and body forces and their moments:

Mass and the Equation of Continuity

$$\frac{d}{dt}\iiint \rho dV = 0$$

$$\frac{\partial \rho}{\partial t} + \nabla \bullet (\rho v) = 0$$

Translational Momentum

$$\frac{d}{dt}\iiint \rho \cdot \vec{v} dV = \iiint \rho \cdot \vec{f} dV + \iint \vec{t} dA$$

Angular Momentum

$$\frac{d}{dt}\iiint \rho(\vec{x} \times \vec{v}) dV = \iiint \rho(\vec{x} \times \vec{f} + \vec{k}) dV + \iint (\vec{x} \times \vec{t} + \vec{l}) dA.$$

In the four conservation equations above, V is the volume enclosed by a surface area A, ρ is the density, $\vec{v}$ is velocity, $\vec{f}$ is the external body force, $\vec{t}$ is the surface body force, $\vec{x}$ is position, $\vec{k}$ is the external body moment, and $\vec{l}$ is the surface body moment. Treating the liquid crystal as a continuum, the application of an external force will produce an external body force and a surface force, and the two forces will promote a linear translational stress and a resultant translation and potential energy. The external force will also produce a small velocity and its associated momentum.

The external body force $\vec{t}$ and the surface body moment $\vec{l}$ can be expressed as coupled stress tensors respectively as below, where $\tilde{n}$ is the unit vector perpendicular to the surface*

---

* Many texts and papers use v to represent the surface unit vector, but because it is so similar to the velocity v, its use can be confusing, so $\tilde{n}$ will be used instead, but note that $\tilde{n}$ is not the same as the director unit vector $\hat{n}$.

$$t_i = t_{ij}\tilde{n}_j \quad \text{and} \quad l_i = l_{ij}\tilde{n}_j.$$

In fluid dynamics, the Einstein summation convention generally is used, that is, within a given equation, any repeated subscripts imply a summation, so that for example (since j is repeated and i is not),*

$$t_i = t_{ij}\tilde{n}_j = \sum_j t_{ij}\tilde{n}_j.$$

The stress tensor denotes the influence of an external force acting on the surface i of the body's *jth* layer, and the respective components can represent the mechanical stress, for example, t_{11}, t_{22}, t_{33} are the normal stress components, and $t_{12}, t_{21}, t_{13}, t_{31}, t_{23}, t_{32}$ are the internal shear stress components. Using the familiar divergence theorem (Gauss theorem) introduced earlier in Chapter 2,

$$\iiint \frac{\partial t_{ij}}{\partial x_j} dV = \iint t_{ij}\tilde{n}_j dA,$$

the conservation of translation momentum above and the angular momentum (including the surface force angular momentum) can be expressed in tensor form. Then using the conservation of mass equation of continuity and the Reynolds transport theorem to "transport" the time derivative into the integrand, the translational momentum equation can be written as

$$\iiint \left(\rho \cdot \frac{dv_i}{dt} - \rho f_i - \frac{\partial t_{ij}}{\partial x_j} \right) dV = 0.$$

Since the volume V is arbitrarily chosen, the integrand in the above equation must be identically zero, so

$$\rho \cdot \frac{dv_i}{dt} - \rho f_i - \frac{\partial t_{ij}}{\partial x_j} = 0,$$

thus the translational momentum equation becomes

$$\rho \cdot \frac{dv_i}{dt} = \rho f_i - \frac{\partial t_{ij}}{\partial x_j}.$$

* In the following, the Einstein notation will be employed except when a summation symbol Σ may be used to impart particular emphasis or to avoid confusion.

In the same manner, the angular momentum conservation equation can be written in the Cartesian vector notation as

$$\left(\rho k_i + \varepsilon_{ijk} t_{kj} + \frac{\partial l_{ij}}{\partial x_j}\right) = 0,$$

where the alternator ε_{ijk} is the vector analysis point notation for components,

$$\varepsilon_{ijk} = \begin{cases} 1 & & i,j,k \quad \textit{unequal \& cyclical} \\ -1 & \text{if} & i,j,k \quad \textit{unequal \& noncyclic}. \\ 0 & & i,j,k \quad \textit{any} \cdot \textit{two} \cdot \textit{equal} \end{cases}$$

The essential difference between isotropic and anisotropic fluids can be highlighted and is manifest in that isotropic fluids do not require any consideration of external force moments ($\vec{k}$) or surface moments ($\vec{l}$), and the alternator in the above conservation of angular momentum equation for isotropic fluids is identically zero ($\varepsilon_{ijk} \equiv 0$); in other words, the stress tensor in the isotropic case can be said to have been completely divested by symmetry. In contrast and in great complication, the anisotropic liquid crystal has to be treated by means of the rarely encountered and daunting *asymmetrical stress tensor.*

The Leslie Work Hypothesis

How the external forces and their moments enter the liquid crystal to change the internal energy of the liquid crystal requires a hypothesis that the liquid crystal system as an entity absorbs the external forces and their moments, transforming the mechanical forces into orientational (nematic) energy or energy dissipation from viscous friction. After an exceedingly complex derivation, this *Leslie work hypothesis* in Cartesian vector form is [4]

$$t_{ij}\frac{\partial v_i}{\partial x_j} + l_{ij}\frac{\partial w_i}{\partial x_j} - \omega_i \varepsilon_{ijk} t_{kj} = \frac{dw_F}{dt} + D.$$

The ω in the above equation is the angular velocity of the director, w_F is the free energy density of the liquid crystal (as expressed in the orientational energy equations given previously in the static theory), and D is the liquid crystal's viscous dissipation resulting from the friction within the liquid crystal generated by the external forces. From the arbitrary choice of the volume (V) and according to the law of entropy (Ω) increase,

$$\frac{d}{dt}\iiint \Omega dV \geq 0,$$

D will always be greater than zero.

The turgid influence of the external force can be described by a strain tensor $\tilde{t}_{ij}$ which is in turn a function of the director orientation, changes of velocity, and rotational velocity, and where the rotational velocity of the director is a function of the relative angular velocity ω_i, the rate of strain tensor A_{ij}, and Leslie's so-called *co-rotational time flux* vector $\vec{N} = \vec{\omega} \times \hat{n}$ that represents the rotation of the director relative to the fluid. The strain tensor has been found experimentally to be amenable to representation as a linear function as follows,

$$\tilde{t}_{ij} = (A_{ij} + B_{ijk}N_k + C_{ijkp}A_{kp}),$$

where

$$\begin{aligned}
A_{ij} &= \mu_1\delta_{ij} + \mu_2 n_i n_j \\
B_{ijk} &= \mu_3\delta_{ij}n_k + \mu_4\delta_{jk}n_i + \mu_5\delta_{ki}n_j \\
C_{ijkp} &= \mu_6\delta_{kp} + \mu_7\delta_{ik}\delta_{jp} + \mu_8\delta_{ip}\delta_{jk} + \mu_9\delta_{ij}n_k n_p + \mu_{10}\delta_{jk}n_i n_p + \mu_{11}\delta_{ik}n_j n_p + \\
&\quad \mu_{12}\delta_{ip}n_j n_k + \mu_{13}\delta_{jp}n_i n_k + \mu_{14}\delta_{kp}n_i n_j + \mu_{15}n_i n_j n_k n_p,
\end{aligned}$$

and in the expressions, μ_i is an adjustable parameter involving the viscosity, n_i is the director, and δ_{ij} is the Kronecker delta, which is defined as

$$\delta_{ij} = \begin{cases} 1 & i = j \\ 0 & i \neq j. \end{cases}$$

Again using the tensor calculus, the commonly used expression for the viscous stress can be derived as

$$\tilde{t}_{ij} = \alpha_1 n_k A_{kp} n_p n_i n_j + \alpha_2 N_i n_j + \alpha_3 N_i n_j + \alpha_4 A_{ij} + \alpha_5 n_j A_{ik} n_k + \alpha_6 n_i A_{jk} n_k,$$

the α_i constants in the expression are the so-called *Leslie viscosity coefficients*, and after some tensor calculations,

$$\varepsilon_{ijk}\tilde{t}_{kj} = \varepsilon_{ijk} n_j \tilde{g}_k,$$

where

$$\tilde{g}_i = -\gamma_1 N_i - \gamma_2 A_{ip} n_p.$$

and

$$\begin{aligned} \gamma_1 &= \alpha_3 - \alpha_2 \\ \gamma_2 &= \alpha_6 - \alpha_5 \end{aligned}.$$

In the above expressions, γ_1 is the coefficient of rotational viscosity introduced in Chapter 5; γ_1 was ingeniously measured by the famous rotating magnetic field experiments of the Russian physicist Tsvetkov [5]. The other rotational viscosity coefficient γ_2 represents the gradient of the shear stress effect on the twist velocity gradient; it is called the *torsion coefficient* and represents the shear flow coupling between the director and the liquid crystal fluid continuum. Again to emphasize the extraordinary nature of liquid crystals, for common isotropic fluids, there is no need to consider the γ_1 and γ_2 viscosity coefficients; the complications they bring are only necessary in theories for anisotropic fluids such as liquid crystals.

The tensor analysis of anisotropic fluids, while indeed providing a formal analysis construct to the very difficult fluid dynamical study of liquid crystals, also intimates an extreme mathematical complication that itself may hinder comprehension. As the seventeenth-century French mathematician Blaise Pascal, one of the founders of fluid mechanics, observed [6],

> imagination tires before Nature....

It might be those innumerable subscripts in the daunting anisotropic electromagnetohydrodynamic theory that tired him.

Perhaps in response to all those subscripts, the constitutive equations of the Ericksen–Leslie theory have been reformulated in a simpler form to delineate the effects of viscosity. To begin, the $\tilde{g}_i$ in the equations above contain the viscosity coefficients and the viscosity and director tensors, and can express the spatial variation of the time variation of the viscous dissipation D with the director orientation as,

$$\tilde{g}_i = -\frac{1}{2}\frac{\partial D}{\partial(dn_i/dt)}.$$

The rate of change of the dissipation with respect to the spatial variation of the fluid velocity ($\vec{v}$) resulting from viscosity effects can be expressed by the viscous stress tensor $\tilde{t}_{ij}$ as

$$\tilde{t}_{ij} = -\frac{1}{2}\frac{\partial D}{\partial(dv_i / dx_j)}.$$

The derivation of the tensor equations above unfortunately also are so complicated that perhaps only fluid dynamics scholars can appreciate them, and the very number of coefficients makes the "science of coefficients" criticism of Prandtl and Tietjens all the more cogent, so they will not be derived here [7].

The aim of the description herein is to present the physical "flavor" of the dynamic tensor analysis of liquid crystals, and if detailed derivations of the reformulated constitutive equations are presented, the forest of subscripts and superscripts unfortunately may well overwhelm any single trees of understanding. What has been derived so far hopefully is sufficient to show the basic physical character of the dynamic constitutive equations, and the full derivation of all the individual equations probably will provide only limited further understanding. Therefore, only two of the more important constitutive equations will be presented below, the first is,

$$\frac{dv_i}{dt} = \frac{1}{\rho}\left(f_i + \frac{\partial t_{ji}}{\partial x_j} \right).$$

From the above equation, it can be seen that when the liquid crystal receives a body force (f) and a surface force's spatial change in surface position ($\partial t_{ji} / \partial x_j$), according to Newton's second law, there will be an acceleration (dv_i / dt), which will be inversely proportional to the density of the liquid crystal material ($1/\rho$).

The second constitutive equation to be discussed is

$$\frac{d^2 n_i}{dt^2} = \frac{1}{\rho_i}\left(G_i + g_i + \frac{\partial p_{ji}}{\partial x_j} \right).$$

where p_{ji} is the surface stress on the director, that is

$$p_{ji} = \beta_j n_i + \frac{\partial w_F}{\partial(\partial n_i / \partial x_j)},$$

Where β_j is an adjustable constant (yet another coefficient), and w_F is the orientational free energy density described previously. Again, according to Newton's second law this second constitutive equation is concerned with the acceleration of the director (d^2n_i/dt^2), as it reflects the external body force (G_i) and the internal body force (g_i), and the surface spatial variation of the surface stress of the director (which can be viewed as a form of pressure). The latter (p_{ji}) further reflects the interaction between the director and the free energy resulting from this surface stress. Finally, the acceleration of the director will be resisted by the density of the liquid crystal fluid.

Clearly, all these external, internal, surface, and body forces acting on an anisotropic fluid will result in an exceedingly complicated set of constitutive equations. Each flow, rotation, discontinuity, and electromagnetic influence can be described by the equations, but the price to be paid is a huge number of sub- and superscripts that together with the various notational conventions of summation and derivation present a daunting task of seeing through the equations to the physical reality.

Instead of ploughing through the equations, presented below are two simple but useful examples of the application of the Ericksen–Leslie theory to the dynamic response time analysis of liquid crystal displays, particularly important for television.

Turn-On Example

In order to study the response time of a liquid crystal display, an electric field in a Freedericksz cell is quickly turned on, and using the Ericksen–Leslie dynamic continuum theory considering molecular twist only, the influence of the viscosity and twist elasticity of the liquid crystal material on the response time of the display can be calculated. Then the electric field can be just as quickly turned off and the influence of the restoring elasticity and viscosity on the response of the liquid crystal can be calculated [8].

The dynamic response analysis at the outset follows the procedure of the static continuum theory examples presented above in Chapter 11. First assume that the director has no z component of translation and only turns in the xy-plane, then the only difference with the static continuum twist angle analysis is that there is a time variation (which is just the meaning of "dynamic"), so the director's twist angle now is a function of time, as shown in the expression below for the vector director

$$\hat{n} = (\cos\phi(z,t), \sin\phi(z,t), 0).$$

The direction of the applied electric field is parallel to the y-axis so that the E-field vector can be written as

$$E = (0, E, 0);$$

the strong anchoring boundary conditions are

$$\phi(0,t) = \phi(d,t) = 0.$$

The expression for the nematic orientational free energy density is the same as that for the static twist case,

$$w_F = \frac{1}{2} K_{22} \left(\frac{d\phi}{dz}\right)^2.$$

The Ericksen–Leslie theory is now applied through the effect of the coefficient of rotational viscosity on the energy dissipation in the dynamic system, in analogy with the orientational energy expressions above, as follows,

$$D = \gamma_1 \left(\frac{d\phi}{dt}\right)^2.$$

Again according to the translational motion and conservation of angular momentum for motion restricted to the xy-plane in the dynamic continuum theory, the rate of viscous dissipation can be represented by the coefficient of rotational viscosity γ_1 in the orientational energy differential equation with applied electric field E as

$$\gamma_1 \frac{\partial\phi}{\partial t} = K_{22} \frac{\partial^2\phi}{\partial z^2} + \varepsilon_o \Delta\varepsilon E^2 \sin\phi\cos\phi.$$

If an electric field greater than the threshold is suddenly turned on, the undisturbed value of the molecules' twist angle ($\phi = 0$) will be abruptly unstable, and the speed of the response of those molecules to the electric field will be reflected in the response time of the liquid crystal. Because this is a dynamic analysis, initial conditions must be applied, which are taken most simply where ϕ is very small initially ($\phi_o << 1$) and is the same at the top and bottom of the liquid crystal layer, as follows,

$$\phi(z,0) = \phi_o(z)$$

$$\phi_o(0) = \phi_o(d) = 0.$$

Using the small angle approximation that is adequate for the molecular twisting near the zero state, $\sin\phi \sim \phi$ and $\cos\phi \sim 1$, then the orientational energy with applied electric field equation above can be linearized as

$$\gamma_1 \frac{\partial\phi}{\partial t} = K_{22}\frac{\partial^2\phi}{\partial z^2} + \varepsilon_o \Delta\varepsilon E^2 \phi.$$

Setting $\tau = (K_{22}/\gamma_1)t$ and $c = \varepsilon_o\Delta\varepsilon(E^2/K_{22})$, the equation above can be written in a more convenient differential equation form as

$$\frac{\partial\phi}{\partial\tau} = \frac{\partial^2\phi}{\partial z^2} + c\phi.$$

Then using a canonical transformation, set

$$\phi(z,t) = \Phi(z,\tau)e^{c\tau},$$

and the orientational energy equation becomes a differential equation of a well-studied form

$$\frac{\partial\Phi}{\partial\tau} = \frac{\partial^2\Phi}{\partial z^2}.$$

The solution to the partial differential equation above is obtained from the standard method of separation of variables again, with the most general solution being a linear combination of sinusoidal functions in a Fourier series (which can represent any periodic function)

$$\Phi(z,\tau) = \sum_{n=1}^{\infty} A_n \sin(n\pi z/d)\exp[-(n\pi/d)^2\tau],$$

where A_n are the half-range Fourier coefficients,

$$A_n = \frac{2}{d}\int_0^d \phi_o(z)\sin(n\pi z/d)dz.$$

Recalling that as in the static case, the threshold voltage can be calculated using the small angle linearization technique as

$$E_{th} = \frac{\pi}{d}\sqrt{\frac{K_{22}}{\varepsilon_o\Delta\varepsilon}},$$

and returning to the original variables and substituting in the E_{th} expression above, the solution becomes

$$\phi(z,\tau) = \sum_{n=1}^{\infty} A_n \sin(n\pi z/d)\exp(-t/\tau_n)$$

where

$$\tau_n = \frac{\gamma_1}{\varepsilon_o\Delta\varepsilon(n^2E_{th}^2 - E^2)}.$$

If the liquid crystal is of positive anisotropy ($\Delta\varepsilon > 0$), then when $E < E_{th}$ and $\tau_n > 0$, from the above expressions, all the n Fourier modes will decay exponentially, so the initial twist angle remains zero ($\phi = 0$, the undisturbed stable state), and the liquid crystal will not undergo any distortion due to the externally applied electric field. However, when the applied electric field exceeds the threshold ($E > E_{th}$), then $\tau_n < 0$, and from the first equation above, the twist angle ϕ will increase exponentially with time (extremely rapidly), and the liquid crystal will be suddenly "turned on," with the molecules quickly turning to align with the applied field. In this case, τ_n is the *switch-on time*, denoted by τ_{on}, which is just the switch-on response time of the liquid crystal. The switch-on time can be expressed in the following different forms, used in textbooks and the literature; it can be seen that the expressions provide the relation among the parameters and material coefficients to allow analysis and adjustment to optimize the switch-on response time.

$$\tau_{on} = \frac{\gamma_1}{\varepsilon_o\Delta\varepsilon(E_{th}^2 - E^2)} = \frac{\gamma_1 d^2}{\varepsilon_o\Delta\varepsilon(V_{th}^2 - V^2)} = \frac{\gamma_1 d^2}{K_{22}\pi^2\left[(V^2/V_{th}^2) - 1\right]}.$$

If now the applied voltage is suddenly switched off, the *switch-off time* can be calculated in a similar manner. but now with the initial condition being the disturbed twist angle $\phi(z,0)=\phi_o$; so when $E > E_{th}$, the switch-off time (τ_{off}) is

$$\tau_{off} = \frac{\gamma_1}{K_{22}}\left(\frac{d}{\pi}\right)^2.$$

From the switch-on and switch-off time expressions above, the dynamic character of liquid crystals can be understood in terms of the parameters and material coefficients of the display. For example, the on/off time is directly related to the viscosity (γ_1) and thickness (d) of the liquid crystal layer, but inversely proportional to its twist elasticity (K_{22}); that is, high viscosity liquid crystal materials will inhibit the molecules from turning quickly and cause the response time consequently to be slow, but a liquid crystal material with high elasticity will promote a fast response time because of a fast restoration due to high elasticity snap-back. Further, the thinner the liquid crystal layer, the faster the response, the thickness ("cell gap") being a pronounced effect of the second degree (d^2).*

The solutions to the nonlinear dynamic constitutive equation above were obtained by using a small angle approximation linearization technique, and were derived for illustrative purposes; of course the analysis is not entirely proper for the larger twist angles. There are other more rigorous, but also more complicated approximation methods, as well as the brute-force computer numerical analysis that can produce more exact solutions, but the results of interest are not that different from the simple small-angle approximation used above, and that approximation can provide physical insight because of the relatively simple, but very useful, closed form function solutions that can be obtained [9].

The example above, for purposes of relational simplicity and intuitive understanding, only considered a twist; in the more general situation, the splay and bend of the liquid crystal molecules will of course also influence the response time. That is, the switching on and off of the electric field also generates splay and bend, as well as translational flow, all of which will influence the dynamic response of the liquid crystal fluid.

Quickly turning on the field will not cause the velocity field of the liquid crystal at the edges of the cell or in its middle layer to change much, but in other sectors such as the middle regions, because of the rapid turning of the

---

* Switch-on time is also called *rise time* or *reaction time*. Switch-off time is also called *decay time* or *relaxation time*.

molecules, the liquid crystal will first flow in one direction and then abruptly reverse direction, creating what is aptly called a *backflow*. The result is that the flow generated in the Freedericksz cell will reduce the viscosity effect so that the switch-on time actually will be faster than the predictions of the simple twist nematic mathematical model.

And conversely, the sudden switching off of the electric field will cause the splay (tilt) angle θ to go from the initial 90° tilt to increase past the original 90° by about 20° to 110°, and then gradually relax to the 0° tilt. This phenomenon is called *kickback* or *bounce,* and causes the switch-off time to be slower than originally calculated, resulting in significant changes in polarization during operation of twisted nematic liquid crystal displays, with consequent adverse effects on display quality [10].

Based upon the equations derived from the theories, the various parameters of interest may be calculated, and the influences of each of the factors can be analyzed with the relevant effects optimized to design the best response times for the liquid crystal display.

Furthermore, the Ericksen–Leslie theory and the Freedericksz transitions have been extensively utilized to study the dynamic flow characteristics of liquid crystals and the associated parameters under the special conditions of laminar, Poiseuille, shear, and Couette flow [11].

From the rapid small-angle rotations of the liquid crystal molecules, a light scattering regime is present in liquid crystals that is much more pronounced than in other materials; and thus scattering observations can be utilized to study the characteristics of liquid crystals. The dynamic continuum model together with the classical theories of light scattering can produce a set of light-scattering intensity equations, wherein the solutions of the orderly eigenmodes and eigenfrequencies and light spectral line width analysis can be used for instance to derive values for the elasticity and viscosity coefficients [12].

Hydrodynamic Instability

In a negatively anisotropic liquid crystal ($\Delta\varepsilon < 0$), the electric dipole moment is perpendicular to the liquid crystal director, and when an electric field is applied, the electric dipole moment that turns to align with the external field will cause the long axis of the molecules to turn towards the perpendicular to the direction of the electric field. But it was observed that for 4,4 p-azoxyanisole (PAA) that is negatively anisotropic, the molecules instead rotated to be parallel with the applied field direction.

Searching for the reasons behind the anomaly, it came to be believed that positive ions in the liquid crystal were also influenced by the electric field in a manner opposite to that of the electrons, and this would produce a horizontal polarization space charge generated by the separation ions and electrons. The space charge would then produce a horizontal electric field, and the applied vertical electric field together with this horizontal space charge-induced electric field would then cause the ions to flow, and the shear force generated by the flow would create a torque in the liquid crystal. From the Ericksen–Leslie theory, the liquid crystal fluid then will exert a frictional torque moment, and altogether the sundry multidirectional forces will produce a disorderly hydrodynamic instability in the liquid crystal.

Although hydrodynamically unstable, the intensity of the light scattering resulting from this instability in fact can be controlled by an applied electric field. It was just this PAA liquid crystal compound in a Freedericksz cell that was to become the world's first liquid crystal display.

Conclusion

With the advent of fast computers and numerical analysis programs, based on the physical theories introduced above, from the four variables of position, orientation, velocity, and angular velocity, adding on an external force's effect on the molecules, utilizing Newtonian mechanics, it is theoretically possible to calculate the mechanical variables and their next instantaneous value in time. Therefore, organizing all the factors and performing brute force calculations in principle can mathematically reproduce the physical characteristics of the nematic liquid crystal.

But the sheer number of variables and equations in any realistic system still present difficulties, and in the early days, no such computations were possible anyway, so the molecular dipole moment mean field and the static and dynamic fluid continuum models were developed to model the liquid crystal systems. The theories produced elegant differential equations and used sophisticated approximation techniques to arrive at closed-form solutions that provided both physical insight and useful engineering formulas. As is common in applied physics, the solutions further provided confirmation of physical effects and relations, and served as guidelines for further investigation. Later, as in so many fields, computer-based numerical analysis based on the theories are used by scientists and engineers in simulation studies for the design and analysis of liquid crystal displays.

The inherent attractiveness of the beautiful and delicate liquid crystal seemingly should promote great interest and rapid advancement of the science. However, after the early scientific discoveries in Europe, history showed that rather than from the staid progress of scholarly investigation, the flowering of liquid crystal research came from an accidental discovery of an anomalous scattering produced by an unstable flow. This startling discovery in the New World coming some 80 years after Europe's first look at liquid crystals, led to a rapid development of the technology of the liquid crystal display and an increased scrutiny of the liquid crystals themselves.

References

[1] de Gennes, P.G., and Prost, J. 1993. *The Physics of Liquid Crystals*, 2nd edition. Oxford, p. 198ff.

[2] Prandtl, L., and Tietjens, O.G. 1934. *Fundamentals of Hyro- and Aeromechanics*. McGraw-Hill, New York; and Choudhuri, A.R. 1998. *The Physics of Fluids and Plasmas*. Cambridge, p. 17.

[3] Stewart, I.W. 2004. *The Static and Dynamic Continuum Theory of Liquid Crystals: A Mathematical Introduction*. Taylor & Francis, London, p. 133ff.; Chandrasekhar, S. 1992. *Liquid Crystals*, 2nd edition. Cambridge, p. 85ff.; (for the entropy and energy dissipation approach theory) de Gennes, P.G., and Prost, J. 1993. *The Physics of Liquid Crystals*, 2nd edition. Oxford, p. 198ff.; (for the basic classical fluid mechanics) Batchelor, G.K. 1974. *Fluid Dynamics*. Cambridge; and Choudhuri, A.R. 1998. *The Physics of Fluids and Plasmas*. Cambridge.

[4] Stewart, I.W. 2004. *The Static and Dynamic Continuum Theory of Liquid Crystals: A Mathematical Introduction*. Taylor & Francis, London, p. 140ff.; and de Gennes, P.G., and Prost, J. 1993. *The Physics of Liquid Crystals*, 2nd edition. Oxford, p. 201ff..

[5] Freedericksz, V.K., and Tsvetkov, V.N. 1935. *Compt. Rend. Acad. Sci. (USSR)*, **4**, 131; Stewart, I.W. 2004. *The Static and Dynamic Continuum Theory of Liquid Crystals: A Mathematical Introduction*. Taylor & Francis, London, p. 170ff.; and Chandrasekhar, S. 1992. *Liquid Crystals*, 2nd edition. Cambridge, p. 144ff.

[6] Mandelbrot, B.B. 1991. How long is the coast of Britain? In *World Treasury of Physics, Astronomy, and Mathematics* (Ferris, T., ed.). Little Brown, New York, p. 449.

[7] Chandrasekhar, S. 1992. *Liquid Crystals*, 2nd edition. Cambridge, p. 97ff.

[8] Stewart, I.W. 2004. *The Static and Dynamic Continuum Theory of Liquid Crystals: A Mathematical Introduction*. Taylor & Francis, London, p. 218ff.

[9] Stewart, I.W. 2004. *The Static and Dynamic Continuum Theory of Liquid Crystals: A Mathematical Introduction*. Taylor & Francis, London, p. 218ff.; and Chandrasekhar, S. 1992. *Liquid Crystals*, 2nd edition. Cambridge, p. 161ff.

[10] Stewart, I.W. 2004. *The Static and Dynamic Continuum Theory of Liquid Crystals: A Mathematical Introduction.* Taylor & Francis, London, p. 223ff.
[11] Stewart, I.W. 2004. *The Static and Dynamic Continuum Theory of Liquid Crystals: A Mathematical Introduction.* Taylor & Francis, London, p. 176ff.
[12] Stewart, I.W. 2004. *The Static and Dynamic Continuum Theory of Liquid Crystals: A Mathematical Introduction.* Taylor & Francis, London, p. 236ff.); Chandrasekhar, S. 1992. *Liquid Crystals*, 2nd edition. Cambridge; and de Gennes, P.G., and Prost, J. 1993. *The Physics of Liquid Crystals*, 2nd edition. Oxford.

13

The First Liquid Crystal Display

Nearly 80 years after Reinitzer's discovery of naturally occurring liquid crystals in carrots, in 1962 Richard Williams found himself at the Radio Corporation of America Laboratories ("RCA Labs"), studying the synthetic negative anisotropy liquid crystal p-azoxyanisole (PAA), and wondering why the molecules tended to align with the applied electric field rather than having their perpendicular dipole moment turn them away, as theory predicted. While playing with the Lab's Freedericksz cell, he noticed that turning the field on would produce an orderly black and white pattern as shown in Figure 13.1 at left, and turning the field off would result in a uniformly lighted pattern as shown at the right in the picture (electrode plates were used on the left half of the liquid crystal cell only, so the field-on and field-off domains could be compared side-by-side). Williams was perplexed, first with the anomaly of the reverse turn of the molecules, and then with the deeper question of just how does the electric field alter the light transmission through the liquid crystal when there is no net charge? [1]

As for the first question, Williams surmised that within the PAA sample, there were positively and negatively charged impurity ions that were being separated by the applied electric field, and owing to the long rod shape of the liquid crystal molecules that created a preferred axis, the ions would

Liquid Crystal Displays, First Edition. Robert H. Chen.

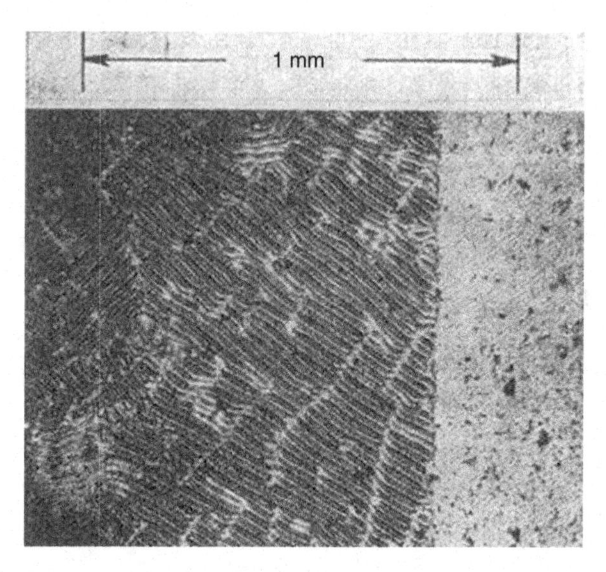

Figure 13.1 Williams domain field-on striated pattern (left) and field-off uniform lighted pattern (right). From Kawamoto, H. 2002. *Proceedings of the IEEE*, **90**, 4, 465, April.

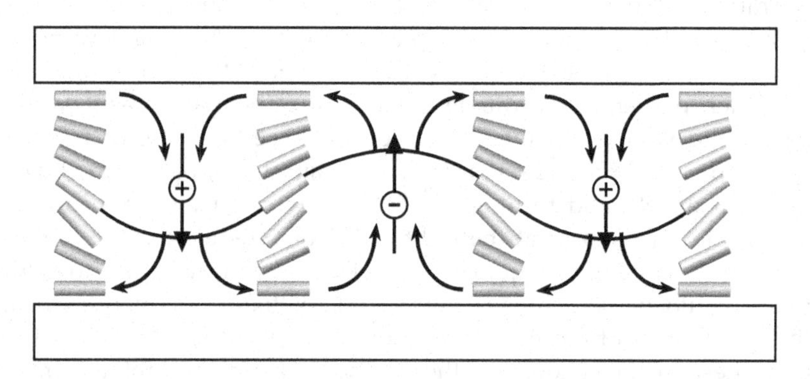

Figure 13.2 Schematic of convection cell in the negatively anisotropic liquid crystal.

preferably move in directions parallel to the long molecular axis rather than perpendicularly to it, thereby creating a "positive conductivity" to oppose the negative anisotropy of the molecule. This resulted in a flow of ions in a direction along the long axis of the liquid crystal molecules, as shown schematically by the curved arrows in Figure 13.2. It is this flow of ions that gives the negatively anisotropic PAA liquid crystal molecules a positive anisotropy bent, and causes them to anomalously align with the electric field as observed [2]. So there was an explanation for the first riddle.

The answer to the second question was to prove more significant (and rather more difficult to explain). Williams surmised that the application of an external electric field causes the positive and negative impurity ions to separate, producing a space charge that induces the liquid crystal molecules to flow in the manner of a convection cell along the direction of the curved arrows in Figure 13.2. The rotational flow then engenders a periodic change of orientation of the liquid crystal directors (denoted by the wavy curve in Fig. 13.2) that alters the polarization state of linearly polarized incident light in a spatially periodic fashion, and the resultant "focusing" of the light into stripes, apparently by the orderly changing of the optic axis because of the polarization, produces the pattern observed, later to become known as the *Williams domain*, the first observation of electrically controlled light intensity in a liquid crystal [3].

Colorful events were in the offing. As a dichroic dye that absorbs light polarized in the direction of the long axis of its molecules and otherwise transmits light (the linear polarizer), so a *pleochroic* (multicolor) dye can exhibit different colors when viewed from different directions because of the different absorption of light depending on the orientation of the incident light's electric vector relative to the dye's long axis. Just 2 years after observations of the Williams domain, RCA's George Heilmeier enlisted the turning power in an applied electric field of the strong intrinsic dipole moment of the "host" butoxy benzoic acid liquid crystal molecules to control the orientation of a "guest" dopant pleochroic dye, and demonstrated that relatively small electric fields could turn the characteristic dye colors on and off in a display.* An excited Heilmeier saw an RCA-produced flat panel liquid crystal color television just around the corner [4].

Dynamic Scattering

The young Heilmeier's enthusiasm was understandable, but there was still a long way to go before RCA could produce a full-fledged liquid crystal color

* The phenomenon of different colors produced by varying the strength of the electric field in a guest-host liquid crystal compound has been used in a wristwatch television produced in 1982 with an active matrix addressing system, but now the mainstream LCDs all use color filters or LEDs for color.

Figure 13.3 Heilmeier demonstrating the DSM display projecting a standard television test pattern. From Kawamoto, H. 2002. *Proc. IEEE*, **90**, 4, 467, April.

television set. Most importantly the host liquid crystal lacked chemical stability. To address the problem, Heilmeier tried a newly synthesized liquid crystal, anisylidene para-aminophenylacetate (APAPA).

This time, when a stronger electric field (5000 V/cm, about 6 V for a 12-μ cell gap) was applied to APAPA, the liquid crystal suffered a strong electrohydrodynamic instability, or field-induced turbulence. This turbulence from the more stable liquid crystal produced intense forward light scattering, creating a white surface, and increasing the applied field increased the brightness [5]. Figure 13.3 shows Heilmeier demonstrating his *dynamic scattering mode* (DSM) display using a backlight to project a standard television test pattern. The RCA liquid crystal group, believing the DSM to be a major breakthrough in television technology, hosted a press conference to demonstrate the world's first commercial liquid crystal display, a digital clock.*

The Liquid Crystal Display Calculator

Just as the research on liquid crystals was moving along smoothly and productively, RCA's research management abruptly transferred Heilmeier's

* In Europe, the DSM is also called the "Carr–Helfrich effect" named after rival researchers.

liquid crystal display research team to RCA's Semiconductor Group. This was unfortunate because the liquid crystal team immediately came under the sway of the "semiconductor culture" and the aegis of the Semiconductor Group R&D managers. The Semiconductor Group saw organic chemistry research, in comparison to solid state physics, as essentially "dirty, disorderly, and difficult," and deemed liquid crystal study as unlikely to lead to anything of use. The managers summarily terminated all research activity for liquid crystal displays.

To say that a research was not to be pursued because it was ostensibly dirty, disorderly, and difficult of course cannot be a reason for its cessation; any ground-breaking physiochemical research is by necessity all of the above. The real reasons likely were founded in business and politics. The dominant RCA product at the time was the transistor-based cathode ray tube television with the vaunted RCA color shadow mask as the flagship proprietary technology. The Semiconductor Group probably was concerned that development of a rival liquid crystal display technology would undermine the hugely successful and highly profitable CRT television business and the handsome royalties received from the shadow mask licenses.

As has happened often in the history of liquid crystal display development, in 1968, just as RCA was in the process of winding down their liquid crystal display work, the Japanese broadcast network NHK came to RCA to film on-site part of the documentary called "Companies of the World, Modern Alchemy." In the film there was a scene of Heilmeier operating his precious DSM display, but during the filming, albeit having abandoned the endeavor, RCA still circumspectly made sure that all the bottles of liquid crystal compound samples in the background were turned around with their labels facing the wall to avoid disclosure of trade secrets to competitors.

NHK's documentary showed RCA that contrary to the Semiconductor Group's view, apparently there were many who were interested in liquid crystal displays, not least among them the Japanese. Further, soon after the filming, America's Defense Department and National Aeronautics and Space Administration (NASA) also weighed in with questions about liquid crystal displays. Their interest of course was not commercial television sets, but rather robust, lightweight displays for use in jet fighters and spacecraft. Liquid crystal displays were not only light and thin, they were less affected by magnetic fields and the strong vibrations inherent in aircraft. After all, in air combat or space exploration, the accuracy and durability of the display devices held life-and-death significance.

About at this time, eager to enter the high technology electronics business, Japan's Sharp Corporation was aggressively developing a pocket-sized calculator. Wada Tomio was in charge of making displays for the Sharp calculator, which at the time employed the integrated circuit central processing units from America's defense giant Rockwell Corporation. The cost of the Sharp calculator was a hundred thousand Japanese Yen, but even that princely sum was not the main obstacle to success. The power requirement, size, and weight were the major impediments to a viable consumer product, and the prime culprit was the staid old energy-hungry, cumbersome, and heavy vacuum tube display. When Wada found out about the RCA press conference and the DSM, and later viewed the NHK documentary, he excitedly met with his supervisor, the legendary technologist Sasaki Tadashi, suggesting that the Sharp calculator should use the DSM display. Eager to know more, Sasaki immediately set off with Wada to New Jersey to observe the DSM apparatus in person. After a demo, an impressed Sasaki quickly proposed a cooperative effort with RCA to develop a DSM display for calculators [6].

Albeit remorseful after almost killing its liquid crystal research, RCA unexpectedly turned Sasaki down flat, condescendingly pointing out that DSM was not suitable for calculators because of its slow response time; RCA preferred to develop DSM solely for clocks and watches. Chagrined, Sasaki and Wada returned to Japan, resigned to the fact they would have to develop a calculator liquid crystal display by themselves, armed only with a patent license from RCA that cost Sharp a cool three million American dollars.

Because RCA's DSM liquid crystal compound was secret (here one might wonder what the three million dollar license was for), it was incumbent upon Wada to begin the tedious process of gathering all the known liquid crystal compounds available and testing each and every one for use in a calculator display; of particular concern was the temperature stability of the nematic phase. Sharp's attempts to copy the DSM involved the investigation of more than three thousand liquid crystal compounds, but happily, fate smiles on the diligent, and Sharp finally found a suitable compound based on the liquid crystal compound research by the British chemist George Gray (see Chapter 14). It was nematic in the broad temperature range of –20° to 63°C; it was an aromatic Schiff bases named*

N-(4-methoxybenzyliden)-4-n-butylaniline (MBBA).

* The following three synthesized mixtures were used at one time or another: N-(4-methoxybenzyliden)-4-n-butylaniline (MBBA), n-(4-ethoxybenzyliden)-4-n-butylaniline (EBBA), and n-(4-butoxybenzyliden)-4-n-butylaniline (BBBA).

Although the nematic temperature problem apparently was resolved by this synthesis, the Schiff base compounds are known to be hydrolytically unstable, that is, on contact with water, they spontaneously decompose, thus necessitating a tedious water-tight sealing step in the manufacturing process that would be a severe hindrance to efficient mass production of the calculators. It was a first problem in need of immediate resolution for the commercial success of Sharp's liquid crystal display calculators.

There was a second problem as well: Providing a uniform electric field to control the liquid crystal molecular orientation required that the electrodes distributing the field through the liquid crystal must be both electrically conducting and transparent to allow light from the backlight to pass through the liquid crystal. Fortuitously, the research arm of the Japanese government's Ministry of International Trade and Industry (MITI) just at this time was developing an indium tin oxide (ITO) electrode that just happened to be transparent. Upon hearing of this, Sharp immediately dispatched engineers to the MITI Osaka laboratory to learn about the new ITO electrodes.

Technology suitable for creating and controlling the electric field to orient the liquid crystal molecules also was advancing rapidly. Since the energy requirements for the liquid crystal display were much lower than that for vacuum tube displays, the pocket calculator could use the newly developed complementary metal oxide semiconductor (CMOS) invented by RCA in 1962 and first mass produced by Toshiba. The *clocked pulse* CMOS, in addition to providing the central processing unit for the calculator, would later also serve as the switch for controlling the electric field applied to the liquid crystal in the soon-to-be-developed active matrix addressing system (Chapter 17). The CMOS could also conserve energy, that is, the addressing system of the display could use CMOS transistors and their clocked pulse, switching on only when there is a clock signal, and storing its charge state otherwise, thereby not requiring voltage to maintain a state, and so the CMOS used much less electric power.

Furthermore, the CMOS transistors could be laid out in integrated circuits that could be fabricated directly on glass surfaces. This was ideal for the liquid crystal display, which used glass plates to hold the liquid crystal. The CMOS transistor, the thin film transistor, and semiconductor fabrication will be discussed in greater detail in Chapters 20 and 21.

The timely development of new technology, its incorporation into new products, and the skill and ingenuity of its engineers allowed Sharp to quickly and efficiently gather together the basic technology, and in 1973, Sharp announced the world's first commercial pocket calculator with a liquid crystal display. The new *Elsi Mate EL-805*, pictured in Figure 13.4, was

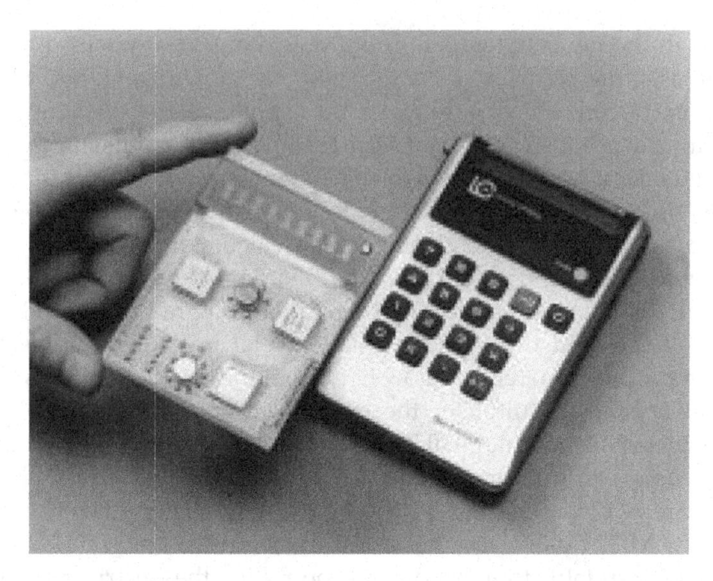

Figure 13.4 Sharp's breakthrough *Elsi Mate EL-805* liquid crystal display calculator. From Kawamoto, H. 2002. *Proc. IEEE*, **90**, 4, 470, April.

only 2.1 cm thick, which was 12 times thinner than anything on the market, weighed only 200 g, which was lighter by 125 times, and used only 2 mW of power, which was an astounding 1/9000 factor lower energy consumption. It was a remarkable engineering achievement, and the critical new component allowing the technological breakthroughs was the calculator's dynamic scattering mode (DSM) liquid crystal display.

The rapid development of a marketable liquid crystal display product perfectly demonstrated Japan's industrial prowess and the skill and dedication of its engineers. Many years later, the Director of the famed RCA Sarnoff Research Laboratories, James Tietjen, while taking in the Sharp Museum's pioneering product displays, resignedly noted [7],

> Two technologies—LCDs and CMOS circuits—had started RCA Laboratories, but ended up at Sharp Corporation.

Notwithstanding the technical breakthroughs, because of the DSM Schiff base MBBA's hydrolytic instability and a small operating temperature range of 0–40°C, Sharp's pocket calculator would not operate reliably under extreme environmental conditions, and at lower temperatures, the display figures were slow to appear and change. The next generation of calculators would have to obey General Manager Asada Atsushi's edict [8]:

> a water-resistant, indoor/outdoor wide temperature range calculator that is clearer and brighter, with faster response and capable of displaying Chinese characters.

References

[1] Kawamoto, H. 2002. The history of liquid-crystal displays. *Proc. IEEE*, **90**, 4, pp. 464–471; and Castellano, J.A. 2005. *Liquid Gold*. World Scientific, Singapore, p. 16ff.

[2] Collings, P.J. 2002. *Liquid Crystals*. Princeton, p. 45.

[3] Chandrasekhar, S. 1992. *Liquid Crystals*, 2nd edition. Cambridge, p. 178ff.; Collings, P.J. 2002. *Liquid Crystals*. Princeton, p. 44ff.; and Collings, P.J., and Hird, M. 2004. *Introduction to Liquid Crystals, Chemistry and Physics*. Taylor & Francis, London, p. 218ff.

[4] Heilmeier, G.H., and Zanoni, L.A. 1968. *Appl. Phys. Lett.*, **13**, 3, 91–92; and Castellano, J.A. 2005. *Liquid Gold*. World Scientific, Singpore, p. 20ff.

[5] Heilmeier, G.H., Zanoni, L.A., and Barton, L.A. 1968. *Proc. IEEE*, **56**, 1162–1171.

[6] Kawamoto, H. 2002. The history of liquid-crystal displays. *Proc. IEEE*, **90**, 4, pp. 464–471; and Johnstone, B. 1999. *We Were Burning*. Basic Books, New York.

[7] Kawamoto, H. 2002. The history of liquid-crystal displays. *Proc. IEEE*, **90**, 4, p. 470.

[8] Kawamoto, H. 2002. The history of liquid-crystal displays. *Proc. IEEE*, **90**, 4, p. 471.

14

Liquid Crystal Display Chemistry

The Williams domain PAA liquid crystal could be electrically controlled to produce dark and light, but had a nematic range of 117° to 134°C; Heilmeier's DSM, although demonstrating that the APAPA liquid crystal could indeed serve as a display medium, also has a nematic liquid crystal phase melting temperature (T_m) of 116°C, requiring heated platforms, which made both liquid crystals clearly impractical for use in consumer products. Furthermore, they were not chemically stable when exposed to water and ultraviolet light, and they had relatively high viscosities.

These problems, together with Asada's order to the Sharp engineers to develop a practicable calculator together formed an acknowledgment that the liquid crystal in a general use liquid crystal display must satisfy at least the following four criteria:

1. below zero to body temperature operating range;
2. hydrolytic and chemical stability;
3. low viscosity; and
4. colorless.

The four requirements can be abbreviated collectively as the "TSVC" (temperature-stability-viscosity-color) conditions. The objective was clear,

---

Liquid Crystal Displays, First Edition. Robert H. Chen.

but the search to find and develop a liquid crystal that could satisfy those criteria was to be long and arduous.

The Aromatic Compounds

The molecular structure of an organic compound is described by the *line/angle formula,* wherein the endpoints or intersections denote the presence of a carbon atom, and the associated hydrogen atom is denoted by an "H," or its presence is implied by the apex of the angle between two lines. The most stable configuration is that of the ubiquitous natural aromatic benzene ring, wherein the six carbon and six hydrogen atoms (C_6H_6) are arranged in a regular hexagon having a carbon-to-carbon bond length of 1.39 Å (10^{-10} m). Depictions of the benzene ring shown in Figure 14.1 use lines or a circle inside the ring to indicate the characteristic conjugated double bond. In many organic reactions in solution, the radical C_6H_5 phenyl ring occurs as an intermediate structure, and it turns out that the linearly linked ring units in display-use liquid crystals have a phenyl ring core.

Linearly linking benzene rings in the central portion of the molecule combined with flexible hydrocarbon chains at the end of the central benzene rings will produce the long-rod/flexible extremity liquid crystal molecules described in Chapter 5. The type of central linking group between the benzene rings determines many of the features of the liquid crystal, and so is commonly used as the name of the liquid crystal compound. Different length hydrocarbon chains added to the central portion constitute a series of liquid crystal compounds.

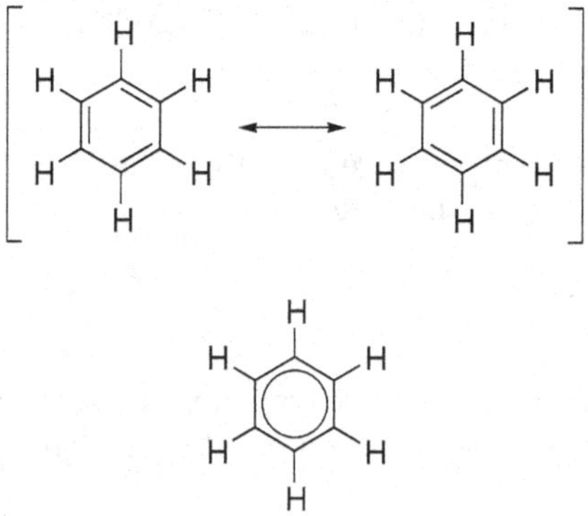

Figure 14.1 Benzene ring structure using lines or a circle inside the ring to indicate the conjugated double bond.

For example, the Williams domain p-azoxyanisole (PAA) has an oxygen atom connecting the hydrocarbon chain with the benzene ring, so the central linkage between benzene rings is called the *azoxy* group, with the different numbers of carbon and hydrogen atoms in the hydrocarbon chains at the extremities forming the series. The anisylidene para-aminophenylacetate (APAPA) used in Heilmeier's DSM has an *imine* linking group *Schiff bases* (–CH=N–) connection forming the central portion of the series. The liquid crystal phase type and temperature range vary for different members of each of the different series.

For many liquid crystal series, the stable nematic temperature decreases as the length of the hydrocarbon chains increases, and the longer hydrocarbon chains naturally promote the layering that is characteristic of the smectic phase (confer Chapter 5). Thus if the lamellar packing of the molecules producing the layered structure were inhibited, this would be conducive to formation of the nematic phase, so it was reasonable to guess that compounds made up of choices from nonlayered molecular structures could produce nematic liquid crystals.

It is the central portion of the molecule that determines the type of liquid crystal phase (for displays, hopefully nematic) and the manner in which the hydrocarbon chains are attached to the center can be varied to change the nematic temperatures (hopefully lowering the temperature and widening the range).

The German chemist Hans Kelker tried a Schiff bases central linkage group, which has an oxygen atom linking one hydrocarbon chain to the benzene ring but no oxygen atom linking the other hydrocarbon chain; a series then can be formed by adding more carbon and hydrogen atoms to the oxygen-linked chain. The resultant so-called MBBA demonstrated significantly lower phase transition and nematic temperatures down to room temperature. The process is outlined in Figure 14.2; the

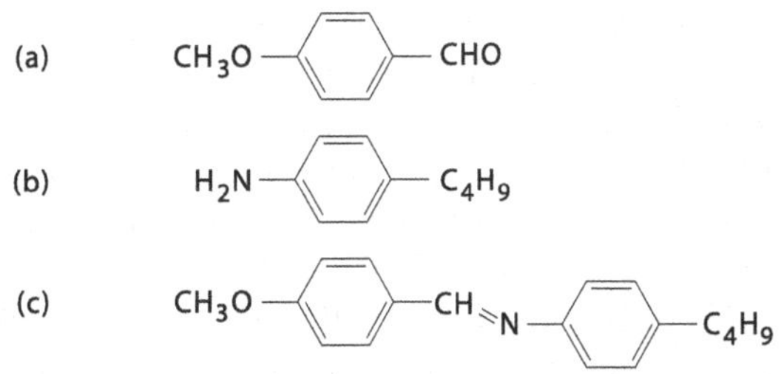

(a) CH_3O — benzene — CHO

(b) H_2N — benzene — C_4H_9

(c) CH_3O — benzene — CH=N — benzene — C_4H_9

Figure 14.2 Molecular structures of (a) p-n-methoxybenzaldehyde and (b) p-n-butylaniline to produce (c) MBBA (plus water).

reaction is between (a) p-n-methoxybenzaldehyde and (b) p-n-butylaniline, that produces (c) p-methoxybenzylidene-p′-n-butylaniline (plus water). To promote the reaction between (a) and (b) requires a solvent, and the resulting chemical compound must also remove water as well as the solvent, after which a filtering process can produce the unrefined MBBA, (c) which must further undergo recrystallization for synthesis of the final MBBA liquid crystal material [1].

Although the procedure as just described seems simple and straightforward, the actual implementation is fraught with difficulties, for example, even just the procurement of the basic chemicals was problematic, and eventually most of them had to be made from scratch by the researchers themselves. The synthesis procedure itself was also very complicated and prone to contamination, so the final product often lacked sufficient purity for further experimental use, which then required repeating the synthesis and further analysis and processing.

The room-temperature MBBA was indeed a significant breakthrough, and it was the liquid crystal that Sharp used in its calculator described in Chapter 13. However, as criticized by Asada, the central linking group of the Schiff bases would break apart upon exposure to water, and this hydrolytic instability was an impediment both to efficient manufacture and use in consumer products requiring robustness.

The Search for a Robust Display Liquid Crystal

The purely scientific investigations of liquid crystals and the later development of display prototypes, for the hundred years from inception, occurred only on the European mainland and in America. But in the late 1960s, a new player to the scene emerged in Great Britain. It all began in 1967, when the British Minister of State for Technology John Stonehouse, on a visit to the Royal Radar Establishment, was astounded by a report by the Aeronautical Display Division; to wit, the British television industry's royalty payments to RCA for the color television shadow mask were greater than Britain's entire expenditure for the (very expensive) Concorde supersonic passenger jet aircraft. Disquieted, the Minister the very next day ordered the Royal Radar Establishment to develop at once a flat-panel color television that could obviate the RCA shadow mask. After accepting the order and upon being asked whether there was in fact any hope of ever doing such a thing, the Head of the Aeronautical Display Division Cyril Hilsum candidly replied, "none whatever."

Nevertheless, a new interdisciplinary committee to study new forms of color television was created at the Royal Radar Establishment, and so in 1968 Hilsum found himself as the newly named committee head, and was engaged in making preparations for a conference inviting experts to report on the feasibility of liquid crystal displays replacing the CRT and shadow mask.

The conference started off inauspiciously and proceeded slowly downhill. At the conclusion of one of the final presentations, Hilsum politely requested questions from the audience. In the ensuing awkward silence, Hilsum performing his chair duty, tried to break the ice by off-handedly asking the speaker why the light from the slide projector when reflected by the speaker's liquid crystal sample would cast some curiously shaped patterned curves on the screen. Starting to answer, then realizing that what he was saying didn't make much sense, the speaker began to fumble through his notes and books for the answer, and in his anxiety caused them to fall to the floor in disarray. As he kneeled down to gather the papers, the audience began murmuring in empathetic concern and the meeting seemed destined for an embarrassing end, when from the back of the room a voice quietly and calmly asked, "I wonder if I can help".

Immediately recognizing possible release from a distressing situation, Hilsum gratefully replied "I'd be most obliged if you would". Whereupon his rescuer, a professor from Hull University named George Gray, proceeded to explain that perhaps the curves and patterns were caused by light scattering from transient turbulence causing changes in orientation of the liquid crystal directors in the sample.*

Several days later, Hilsum when asked about the meeting and whether there were any significant conclusions reached, after thinking a bit, Hilsum replied "Yes, we must put the man from Hull on a contract" [2].

On RRE contract, Gray's hitherto purely scholarly liquid crystal research now found itself face-to-face with the practical realities of synthesizing a liquid crystal material that satisfied all of the TSVC (temperature-stability-viscosity-color) requirements for commercial displays and particularly for television.

Gray surmised that if the hydrocarbon chains were connected directly to the benzene ring, there then would be less layering (avoiding the smectic) and different liquid crystal features would be produced in a new series

* Gray's explanation has been lost with the passage of time and attempts to re-create the scene from personal accounts proved fruitless; the author offers the most plausible scenario from circumstantial evidence.

of nematic compounds. So Gray eliminated the central linkage group by connecting the benzene rings together, forming biphenyl and terphenyl compounds that were nematic at room temperature, and, most importantly, chemically and hydrolytically stable. This was the breakthrough to high-volume commercial liquid crystal display production.

After reading *"Gray's Anatomy"* (of liquid crystals) [3], Joseph Castellano at RCA widened the nematic temperature range by implementing the idea that although different nematic compounds have different nematic temperature ranges, some mixtures can exhibit greater nematic temperature ranges than those of their individual constituents, and this depended on the number of carbon atoms in the terminal side chains.

Armed with this idea, and through the use of three-dimensional phase diagrams such as that shown schematically in Figure 14.3, the different nematic state phase temperatures in the different compounds (A, B, C, in Fig. 14.3) could be delineated, and thus different mixtures of liquid crystal

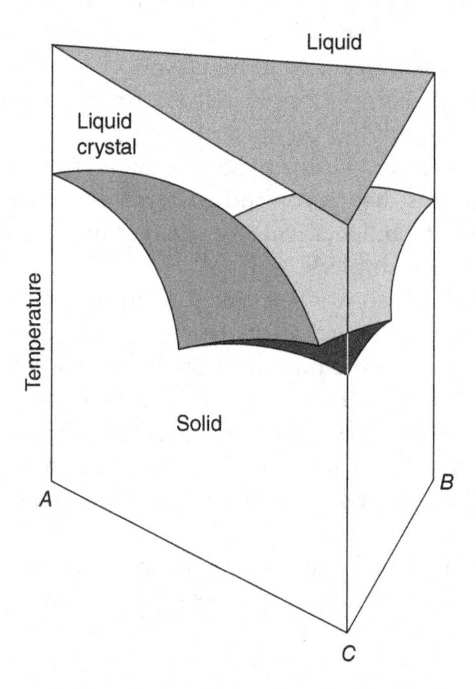

Figure 14.3 Representative three-dimensional phase diagram showing the different nematic state phase temperatures in different compounds (*A, B, C*).

compounds might be found, which were stable in nematic temperature ranges that were lower than the nematic temperature ranges of the individual constituents.

Because of the great number of different aromatic compounds possible for mixture and the difficulties of synthesizing them, three-dimensional phase state diagrams were extremely useful in narrowing and systematizing the search for stable room-temperature liquid crystal compounds, and avoiding tedious wild goose chases.

After trying first binary then ternary mixtures of different nematic aromatic Schiff bases derivatives, Castellano successfully synthesized a compound having a wider nematic temperature range. Later syntheses would widen the nematic temperature range even more, for example,

35% p-n-pentyl-p`-cyanobiphenyl (5CB) nematic range 24°–35°C
plus
65% p-n-octyloxy-p`-cyanobiphenyl (80CB) nematic range 67°–89°C
becomes
cyanobiphenyl (CB) a mixture with nematic range 5°–50°C.

Figure 14.4 shows the molecular structures of the cyanobiphenyls and terphenyls (E-7) used extensively in the 1980s for liquid crystal displays; by weight, the mixture is A: 16%, B: 51%, C: 25%, D: 8%.

After years of sustained effort, a cyanobiphenyl was finally synthesized, having the temperature range, chemical and hydrolytic stability, low viscosity, and no color; that is, it satisfied the TSVC requirements.

Another significant development came from the Royal Radar Establishment. Investigating a sample of cyanobiphenyl from Gray, Peter Raynes found that the molecules in the sample sometimes turned oppositely to the preferred direction under application of an electric field. This adverse reverse turn produced an inapposite phase retardation resulting in light leakage in the display (refer to Chapter 23), so Raynes herded the molecules in the right direction by doping the cyanobiphenyl with a chiral nematic having a twist in the desired turning direction. Because of the mutual intermolecular attractive forces, the chiral dopant succeeded in promoting the desired turn in the liquid crystal. This was portent of the twisted nematic display to come, for later chiral dopants would be widely used in the twisted nematic that ushered in the age of the liquid crystal display.

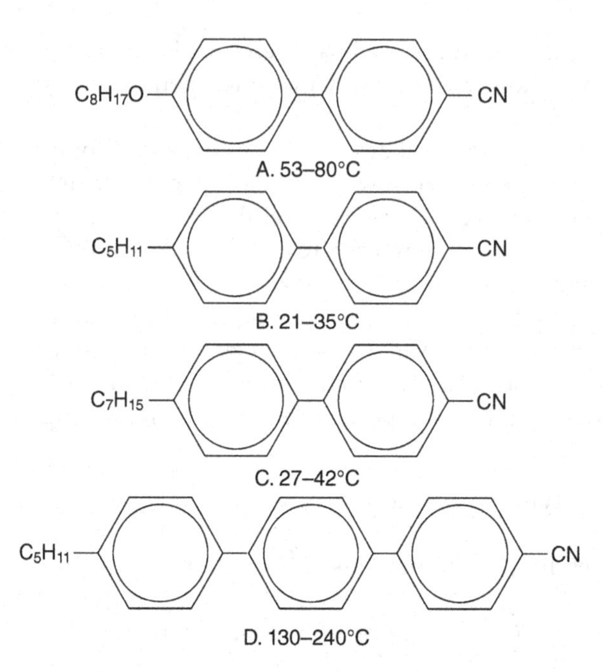

Figure 14.4 Molecular structures and nematic temperature ranges of the cyanobiphenyls (E-7) used in early watch and calculator displays.

The TSVC were essential for robust display operation, hydrolytic stability was for ease of manufacture and durability. That is, in manufacture, the glass substrates holding the liquid crystal now did not require the tedious glass frit sealing procedure to keep out moisture; instead, liquid crystal cell manufacture could simply utilize an epoxy sealant to join the glass substrates. This seemingly innocuous advance made mass production of liquid crystal displays viable because it not only reduced the processing time and cost, it improved the quality and thus the reliability of the finished product [4].

Refinements of Gray's cyanobiphenyl came in rapid succession. The E-7 mixtures were manufactured by the British Drug House, later to become BDH Chemicals, part of the giant Glaxo group, but later sold to the major liquid crystal supplier Merck in Germany.

Merck's liquid crystal research however was directed not for calculators or digital watches but rather for computer screens. Thus Merck's research objective centered on increasing the response speed of the liquid crystals. In 1976, Rudolf Eidenschink took the positive-cyano ester in the relatively high

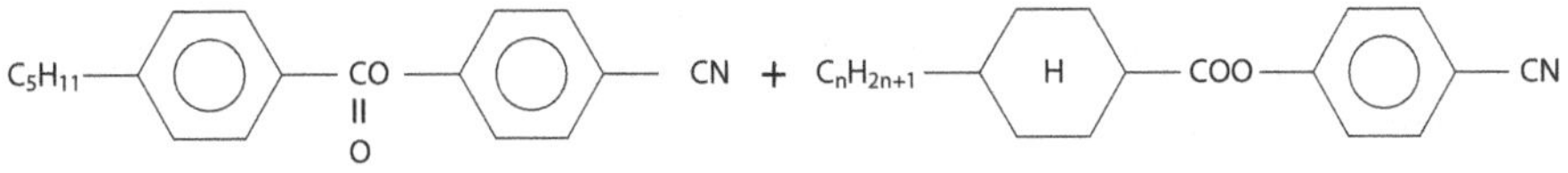

Figure 14.5 High viscosity positive-cyano ester (left) added to the cyclohexane (right).

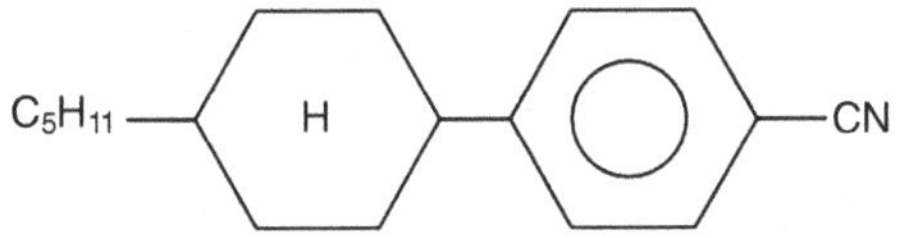

Figure 14.6 Structure of the low viscosity phenyl cyclohexane (PCH).

viscosity digital watch display liquid crystal material structure shown in Figure 14.5 at left, and added it to the cyclohexane at right with the aim of further reducing the viscosity.

The result of the mixture was that the -COO- link in the middle of the cyclohexane ester of Figure 14.5 was replaced in the cyanobiphenyl by the cyclohexane without the -COO- link, resulting in a very low viscosity phenyl cyclohexane (PCH) as shown in the structure schematic of Figure 14.6.

Most of the liquid crystal materials developed thereafter were based on this very low viscosity (hence fast response time) molecular structure. The development process culminating in the principal liquid crystals used in displays can be summarized as follows:

phenyl → biphenyl → cyanobiphenyl (CB) → chiral dopant
→ phenyl-cyclohexane (PCH) → cyanophenyl-cyclohexane (CPCH)

Figure 14.7 shows the organic chemical structures of the cyanophenyl cyclohexane (CPCH) series compounds and their individual nematic temperature ranges. Different proportions of each compound can be mixed to achieve different desired design capabilities for different liquid crystal display uses [5].

The cyanophenyl cyclohexane families and their derivatives gradually became the mainstream liquid crystal material. Today the principal commercial suppliers are Germany's Merck, Switzerland's Hoffmann-LaRoche, and Japan's Chisso Chemical and Dai Nippon Ink.

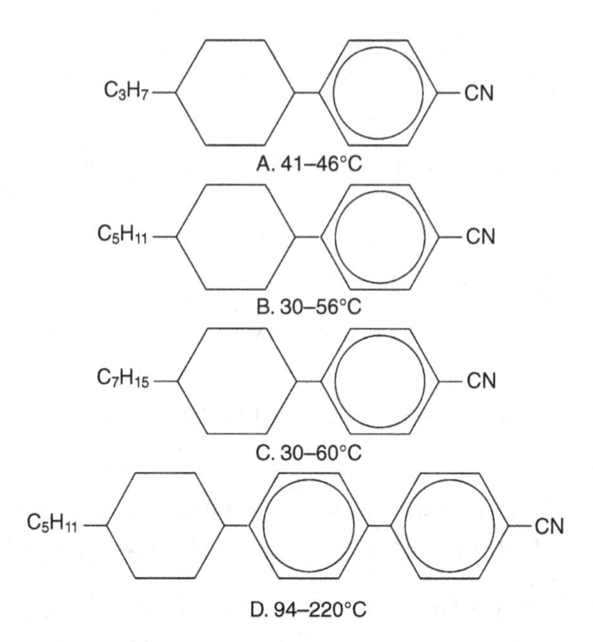

Figure 14.7 Cyanophenyl cyclohexane (CPCH) series compounds used in displays and their individual nematic temperature ranges.

References

[1] Collings, P.J. 2002. *Liquid Crystals*, 2nd edition. Princeton, p. 115.

[2] Hilsum, C. 1984. The anatomy of a discovery—biphenyl liquid crystals. In *Technology of Chemical and Materials for Electronics* (Howells, E.H., ed.). Ellis Horwood, Chichester, p. 43ff.; and Kawamoto, H. 2002. The history of liquid-crystal displays, *Proc. IEEE*, **90**, 4, p. 477ff.

[3] Gray, G.W., Harrison, K.J., and Nash, J.A. 1973. *Electron Lett.*, **9**, 6, 130-131.

[4] Castellano, J.A. 2005. *Liquid Gold*. World Scientific, Singapore, p. 13ff.

[5] Castellano, J.A., and Harrison, K. 1980. Liquid crystals for display devices. In *The Physics and Chemistry of Liquid Crystal Devices* (Sprokel, G.J., ed.). Plenum Press, New York, p. 263; Kawamoto, H. 2002. The history of liquid-crystal displays, *Proc. IEEE*, **90**, 4, p. 482ff.; Collings, P.J., and Hird, M. 2004. *Introduction to Liquid Crystals*, Taylor & Francis, London; and Collings, P.J. 2002. *Liquid Crystals*, 2nd edition. Princeton, p. 51ff.

15

The Twisted Nematic

The liquid crystal display prototypes developed at RCA interested Wolfgang Helfrich, a German researcher in Heilmeier's Group, who was engaged in applying the Ericksen–Leslie dynamic continuum theory to the operation of the display prototypes. While investigating whether a shear stress torque was necessary to produce the Williams domain pattern, he asked himself a cogent question: Is it the twisted stream of molecules in the convection cell that changes the polarization of the incident light that subsequently modulates the intensity? And would it be easier to modulate the light intensity if there were a natural or induced twist in the liquid crystal?

Continuing this train of thought, he decided to apply a thin film compound on the glass substrate and used a cotton swab to trace parallel troughs in the film; because of the long cylindrical shape and soft, flexible ends of the nematic liquid crystal molecules, the rubbed compound troughs would cause the long axes of the molecules to adhere to the glass surface, thereby anchoring them at a predetermined angle. Now if the angle of the rubbing on the top and bottom glass substrates were different, the stream of molecules, anchored as they were at those angles at the top and bottom substrates, would undergo a twist in the middle layers. At left in Figure 15.1 is a micrograph of the anchoring compound as applied, the middle micrograph shows

Liquid Crystal Displays, First Edition. Robert H. Chen.

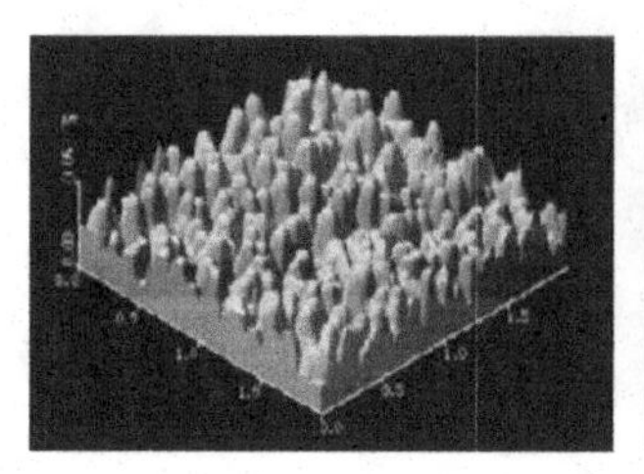

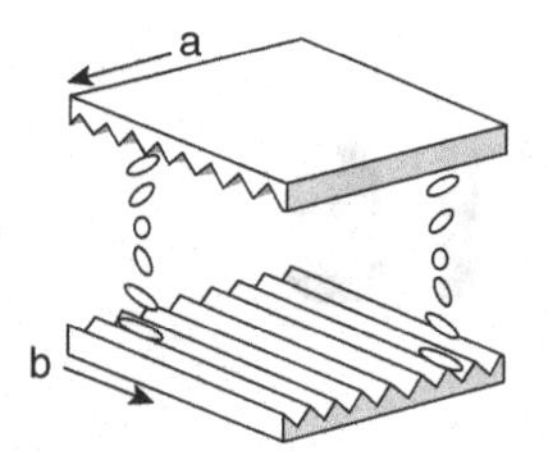

Figure 15.1 Micrograph of the anchoring compound (left), rubbed parallel troughs (middle), and schematic drawing of the 90° twist of liquid crystal molecules (right). From Kawamoto, H. 2002. *Proc. IEEE,* **90**, 4, 476, 472, April.

the parallel troughs wrought by "rubbing" the compound, and the right schematic drawing shows the 90° twist that the stream of molecules undergo in the middle layers.

If *crossed polarizers* (deposed with transmission axes at a relative 90° angle) are fitted outside the glass substrates, the light from the backlight will first be linearly polarized by the first polarizer, and then just as for the natural chiral optical activity in cholesteric liquid crystals described in Chapter 5, the plane of polarization of the linearly polarized light will rotate with the induced twist angle of the liquid crystal molecules, so that upon reaching the second polarizer, the plane of polarization will be parallel to the polarizer's optical axis and the light will pass through unimpeded. As the light undergoes no phase retardation, nor are any components of the light blocked, at the second linear polarizer, it is the brightest possible upon exit from the second linear polarizer, and since the definition of contrast is just the brightest divided by the darkest, this results in a very good contrast ratio, that is, Helfrich's *twisted nematic display* was an advance towards Asada's desire for a "clearer and brighter image." This so-called *white state* is the "normal state" for the twisted nematic, and since this requires no voltage to maintain, the *normally white* twisted nematic display also conserves power.

In operation, an applied electric field could precisely control the orientation of the twisted nematic liquid crystal molecules' dipole moment, which orientation produces a phase retardation in the light passing through from the backlight that changes the polarization state, and in conjunction with a second polarizer, modulates the intensity of the light. This precision control of the phase retardation producing different light intensities provides a fine *gray scale* for image display. Helfrich's twisted nematic display prototype certainly appeared to portend an extremely attractive new product.

However, when Helfrich presented the results of his research to his supervisor Heilmeier, the latter was not only unmoved, he criticized the prototype as "requiring tedious additional steps of applying and rubbing the aligning compound," and pointed out that the addition of two polarizers would not only increase the cost, but also reduce the intensity of the image (the highest transmissivity of any polarizer is 50%). Heilmeier clearly was not interested in Helfrich's twisted nematic display, but his coldness perhaps was founded in a desire to protect his own DSM display, or perhaps more insidiously, it was because RCA management had just decided to once again disband the liquid crystal research group and terminate all related research.

A saddened and frustrated Helfrich thereupon left RCA to join the pharmaceutical conglomerate Hoffmann-LaRoche in Switzerland. Barely one year later in 1970, Helfrich and Hoffmann-LaRoche's liquid crystal research colleague Martin Schadt published a paper in the journal *Applied Physics Letters* [1]. This paper purportedly was the first publication of the twisted nematic display technology, at once a report of scientific significance, but concomitantly also an event of legal import; for they prosaically also applied for what they claimed to be world's first twisted nematic display patent based on that report [2]. In order to prove the feasibility of their invention, and in preparation for a demonstration to persuade Hoffmann-LaRoche's management that commercialization of the TN display should begin immediately, Schadt hastily cobbled together the 3.5-inch twisted nematic prototype display shown in Figure 15.2.

Notwithstanding the dramatic technological breakthrough that the precision-controlled twisted nematic display proved to be, and the in spite of a workable prototype and even patent protection, due to personnel

Figure 15.2 Twisted nematic prototype display. Courtesy of M. Schadt.

changes in management, just as at RCA, Hoffmann-LaRoche's liquid crystal research group was disbanded, and in 1972 Helfrich once again found himself out of a job. Helfrich's twisted odyssey symbolized the early years of liquid crystal display development, an extremely promising technology, seemingly forever unrequited.

But the flickering light of the twisted nematic somehow seeped through again to Japan. Barely one year after Hoffmann-LaRoche's decision to let Helfrich go, out of the blue, Sony contacted Hoffmann-LaRoche with an offer to purchase the twisted nematic display patent. Chastened, Hoffmann-LaRoche's research managers now came to realize that the polyimide-rubbed alignment layered, polarizer-heavy twisted nematic device might have some value after all. They hastily ordered Schadt to summon the exiled Helfrich back from unemployed limbo, and together they reestablished the liquid crystal research group.

Hoffmann-LaRoche's business was chemicals, so the company had no interest in manufacturing the electronic displays themselves. But the displays needed liquid crystals which were indeed chemicals that Hoffman-LaRoche could produce in volume, and the more liquid crystal display manufacturers there were, the more sales they could make. So Hoffmann-LaRoche embarked on a vigorous licensing campaign, and soon in particular, many Japanese electronics manufacturers and some interested American companies signed up. Thus, through Hoffman-LaRoche's self-interested broad licensing, the commercial development of the new twisted nematic liquid crystal display apparently was on the verge of a development unfettered by legal obstacles that proprietary patents often imposed on new products.

A Twist of Fate

But while Helfrich was contemplating the twisted nematic at Hoffmann-LaRoche, James Fergason, formerly at Kent State University's Liquid Crystal Institute, and then principal of his own company Ilixco, had developed a twisted nematic liquid crystal from the departure point of his researches into the helical twist and temperature sensitivity of cholesteric liquid crystals. Fergason illustrated his ideas by means of a clever model comprising a bundle of sticks connected by rubber bands pictured in Figure 15.3. In the field-off state, the rubbing alignment on the glass substrates promotes a gradual helical twist, but when a field of about 5 V is applied, the mid-layer molecules will align with the field, and as the twist coefficient of elasticity is weaker than the splay and bend elasticities, so those molecules nearby

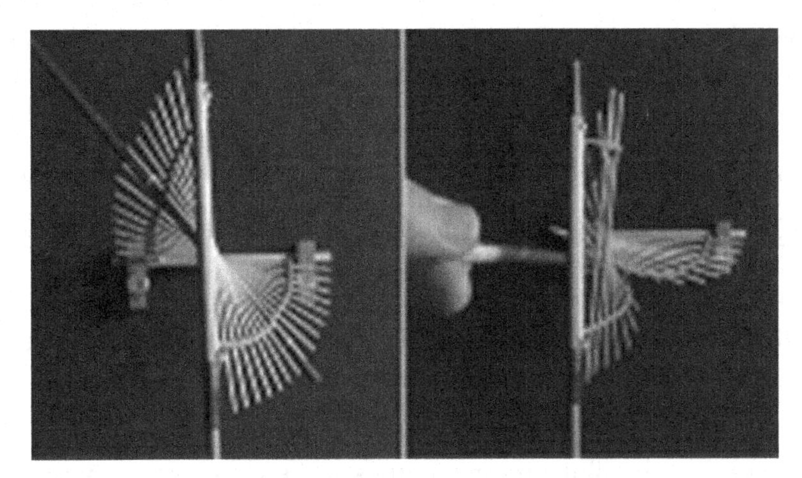

Figure 15.3 Twisted nematic stick and band model showing mid-layer collapse made by Tom Harsch of the Kent State Liquid Crystal Institute. Courtesy of Fergason Intellectual Properties.

also will turn abruptly towards the direction of the field (as shown in Fig. 15.3 at right), and the twist suddenly "collapses" to a perpendicular state, just as predicted in the theoretical section in Chapter 12 on threshold voltage response time. Now in the field-on state, because there is no optical activity (the plane of polarization of the linearly polarized light no longer follows a helical twist of the molecules), the linearly polarized backlight light will be completely blocked by a crossed polarizer at exit from the liquid crystal, producing a dark state. This experiment at the least demonstrated that the twisted nematic could perform the function of a fast optical switch [3].

In the spirit of the professional inventor that he was to become, Fergason quickly applied for two twisted nematic patents [4].

The Gathering Patent Storm

In the modern industrial age, a successful new commercial technology swiftly developing from disparate quarters invariably engenders disputes regarding primacy of conception, and the twisted nematic was no exception. A complicated intellectual property case was on the horizon.

First off was the question of ownership of Helfrich's twisted nematic invention. In almost all technology companies, unless otherwise agreed upon, the ownership of patent rights belongs to the company where an individual is employed to do technical work and not to the individual himself. Helfrich was at RCA when he came up with the twisted nematic idea, but it was Hoffmann-LaRoche who applied for the patent. Helfrich contended that he was employed by RCA not to invent but merely to do theoretical research (e.g., studying the Ericksen-Leslie dynamic continuum theory). Helfrich's rather strained argument was contradicted by his erstwhile RCA colleague Castellano, who averred that Helfrich invented the twisted nematic at RCA and took the idea to Hoffmann-LaRoche.

While hotly contesting this issue, the two sides discovered that only five months after the date of Schadt and Helfrich's patent application, Fergason had filed his two cholesteric theory-based twisted nematic patent applications. Although his application priority dates were later than Hoffmann-LaRoche's, Fergason argued that he had invention primacy because of prior conception. Unlike all the other countries in the world, America espouses the first-to-invent principle of patents whereby evidence of conception of an invention trumps a mere administrative filing for a patent. Fergason produced evidence that as early as 1968 at Kent State, he had conceived the idea for the twisted nematic and he had published an article on that very subject before Schadt and Helfrich's patent application date and the *Applied Physics Letters* article publication date. This dispute then was ripe for what is called in the Patent Office a *patent interference proceeding*.*

---

* Interestingly, the inventorship dispute in the United States delayed Hoffmann-LaRoche's patent issuance in Japan until May 1986, thereby giving Japanese companies a 14-year grace period to pursue liquid crystal product research. Based on the 20-year lifetime of a patent in Japan, commencing from the filing date, the basic twisted nematic patents expired in 1993, coincidentally just when the mass production of notebook computers using twisted nematic liquid crystal displays took off. So the fundamental twisted nematic patents missed their great licensing realization. As an aside, perhaps illuminating the role of the Japanese government in industrial research, the Japanese patent application examination system in those days often took all of six years to complete, and an apparent anomaly in the context of the relatively efficient operations of Japanese government agencies. Thus it is difficult to avoid the suspicion that such a slow examination system is part of Japan's industrial policy, that is, the early publication system used in Japan requires all patent applications to be made public 18 months after filing; an examination period of six years leaves four and a half years of free use. It could be expected that Japanese companies would make the reading of the early publications in Japan's Patent Gazette part of research and development's standard operating procedure, and this probably explains why Sharp contacted RCA about DSN and Sony offered to purchase Hoffmann-LaRoche's TN patent, both events occurring "out of the blue." In the 1980s, just when America was under severe pressure from Japanese consumer products, the situation became so serious as to warrant a bill in Congress accusing the Japanese Patent Office of unfair competition. It is not possible to prove direct cause-and-effect, but not long thereafter, Japan did indeed shorten its examination period, but every patent attorney knows that it still is always the Japanese patent that is the last to issue.

While the twisted nematic invention ownership dispute between RCA and Hoffmann-LaRoche was raging, Hoffmann-LaRoche brought in Fergason to join the patent interference fray ostensibly to weaken RCA's conception argument, whereupon Kent State now jumped in to sue Fergason on the basis of the latter's own evidence of prior conception of the twisted nematic at Kent State, that is, the university, and not Fergason, should own the patent. It did not help matters that there was overt, continuing personal animosity among the parties, in particular between his former colleagues at Kent State and Fergason.

The complex court case and its ramifications plus the enormous legal fees debilitated Fergason's start-up company and Ilixco soon faced bankruptcy. As a solo inventor up against the giant Hoffmann-LaRoche, Fergason ultimately grudgingly sold all his intellectual property rights to Hoffmann-LaRoche. In the settlement negotiated among the parties to the dispute, the royalties for the twisted nematic display were to be divided in equal 30% shares among Hoffmann-LaRoche, Fergason, and Hoffmann-LaRoche's subsidiary Brown Boveri, with Kent State receiving the remaining 10% [5]. It is not clear why RCA was left out in the cold.

Watches and Calculators

With the patent controversy settled, the twisted nematic display could take off, and in 1972 Taiwan's Sun Lu at Texas Instruments produced purportedly the world's first field-effect twisted nematic liquid crystal digital watch. The market appeared so hot that many entrepreneurial-minded TI researchers who worked on the TN watch subsequently left to form start-up companies supplying liquid crystal displays for big brand watchmakers such as Japan's Seiko, and America's giant Timex acquired the liquid crystal divisions of both RCA and Intel in preparation for a full-scale liquid crystal onslaught on the digital watch market.

Besides TI and Timex, other big companies such as National Semiconductor, Motorola, Beckman Instruments, and defense giants Rockwell, Hughes Aerospace, Honeywell Aerospace, as well as Hewlett-Packard, Bell Labs, and Commodore Computers all jumped headlong into the liquid crystal sea.

But they were all too late, for by the time they took the plunge, high-volume manufacture of digital watch and calculator displays had already moved to the high-efficiency producers in Japan. Large quantities of the latest liquid crystal materials from Merck and Hoffmann-LaRoche had been swept up by Sharp and Seiko, and Hitachi, Citizen, Epson, Sanyo, and NEC also began liquid crystal display manufacturing. The Japanese companies were pouncing on the advances in liquid crystal materials to storm the market for products using liquid crystal displays; and indeed by 1974, the world was awash with liquid crystal display watches and calculators, substantiating the appeal of the twisted nematic.

The new twisted nematic liquid crystal watches and calculators had demonstrated supreme market appeal, but almost any electronic device requires an information display capability. The prospects for other products, such as displays for notebook computers, naturally were on the minds of all the companies, but for the new displays to be created, they must be first designed and engineered. At this juncture, it is necessary to delve into the technology of the twisted nematic display in order to understand developments, and this requires some basic tools for analysis, to be introduced in the following Chapter 16.

References

[1] Schadt, M., and Helfrich, W. 1971. Voltage-dependent optical activity of a twisted nematic liquid crystal. *Appl. Phys. Lett.*, **18**, 4, 127–128.

[2] Swiss Patent No. 532,261, filing date Dec. 4, 1970, to Martin Schadt and Wolfgang Helfrich.

[3] Fergason, J.L. "Birth of a Technology," as told to A.L. Berman, unpublished manuscript provided by Fergason Intellectual Properties.

[4] USP 3,731,986, filing date April 22, 1971, and USP 3,918,796, filing date Jan. 24, 1973.

[5] Kawamoto, H. 2002. The history of liquid-crystal displays. *Proc. IEEE*, **90**, 4, 465, 474; and Castellano, J.A. 2005. *Liquid Gold*, World Scientific, Singapore, p. 77.

16

Engineering the Liquid Crystal

To review the fundamental operation of a liquid crystal display pixel cell: Light from a backlight source is linearly polarized by a first linear polarizer. Upon entering the liquid crystal medium, an applied electric field changes the orientation of the liquid crystal molecules, which engenders a phase retardation that changes the polarization state of the light. Then passing through a color filter and various compensation films to improve viewing angle, contrast, and color quality, in conjunction with a second linear polarizer, the light's intensity has been modulated pixel by pixel to simulate the light from the scene being imaged.

From the perspective of physical optics, if the polarization state of the light traversing each optical element in the display system can be characterized mathematically, then appropriately designed optical elements may be combined to produce and improve the desired image. This characterization of the polarization state can be achieved by the methods of mathematical physics, such as the following.

Liquid Crystal Displays, First Edition. Robert H. Chen.

Poincaré Sphere

The Poincaré sphere* is a purely geometrical construct to describe polarization, wherein a point on the sphere represents a specific polarization state, and a curve on the surface of the sphere represents the traversal of the light through the optical system from the different polarization states. The Poincaré sphere approach leads to a formulation of components of the polarizaton vectors representing light intensity, the so-called *Stokes parameters*, that have been used primarily to handle partial polarization in optical systems [1].

Refractive Index Ellipsoid

Another way to mathematically describe polarization is to constitute the material body's index of refraction in a refractive index ellipsoid. Recalling that elliptical polarization is the most general polarization state (whereby the electric field components resolve to form an ellipse), then if a body's indices of refraction are represented by the semi-major and semi-minor axes of a solid ellipse (an ellipsoid), this refractive index ellipsoid then can represent the polarization state of the light passing through the body. Changes in the state of polarization, through ellipses of different ellipticity and the circular and linear special cases, are exhibited in changes of the values of the semi-major and semi-minor axes of the refractive index ellipsoid. Conversely, the refractive index ellipsoid's semi-major and semi-minor axes can represent the material body's influence on the polarization state of the light passing through, so that the interaction between the light and the material body can be demonstrated by changes in the shape of the refractive index ellipsoid [2].

Jones Vector

In the Jones calculus method, a two-component vector called the *Jones vector* represents the polarization state of the light, and an optical element (such as a polarizer, wave plate, or the liquid crystal itself) is represented by a 2×2 matrix, called the *Jones matrix*. The Jones vector method has been advantageously used for the analysis of polarization states in liquid crystal displays. In particular, light passing through a nonhomogeneous medium is encoun-

* Poincaré is the French mathematician and physicist who famously proposed the Poincaré conjecture in 1904, which was finally solved in 2003 nearly one hundred years later.

tering molecular directors that are changing orientation, making calculation of the polarizing effects difficult. If, however, the medium is divided into thin slabs, each of which can be considered isotropically homogeneous, then the phase retardation can be calculated for each slab using matrix analysis. The Jones matrix of the medium as a whole can then be obtained by multiplying all the matrices in sequence [3].

Actually, Jones matrix methods are based in the fundamental quantum matrix mechanics descriptions of matter in the form of eigenstate vectors, including, significantly for the optics of liquid crystal displays, the polarization eigenstates. The Jones vector, more than just a mathematical construct to describe an optical system's polarization, is also a fundamental physics formulation of matter [4]. More mundanely, the Jones vector can also provide useful optical parameters and conditions for design of the twisted nematic display using numerical simulation.

The basic Jones vector technique will be illustrated in the following derivations of the very useful *phase retardation parameter*, the *Mauguin parameter*, and the *Mauguin* and *Gooch-Tarry conditions*, as well as in the analysis of the interesting twisted nematic *waveguiding* phenomenon (also called *adiabatic following*).

The Phase Retardation Parameter

To begin, the polarization state of light traversing a given slab of liquid crystal can be described by a Jones vector in the form of an ordered pair of the light's electric field vector components

$$E = \begin{pmatrix} E_x \\ E_y \end{pmatrix}.$$

As described previously, anisotropic materials have a fast and a slow axis which produce the double refraction (birefringence) of the light passing through. Figure 16.1 shows the relative positions of the fast and slow axes of the light in relation to the xy-coordinate system on the body as expressed by the azimuthal angle Ψ.

From simple trigonometry, the coordinate transformation to the fast and slow axes system is given by a matrix equation as

$$\begin{pmatrix} E_{slow} \\ E_{fast} \end{pmatrix} = \begin{pmatrix} \cos\Psi & \sin\Psi \\ -\sin\Psi & \cos\Psi \end{pmatrix} \begin{pmatrix} E_x \\ E_y \end{pmatrix}.$$

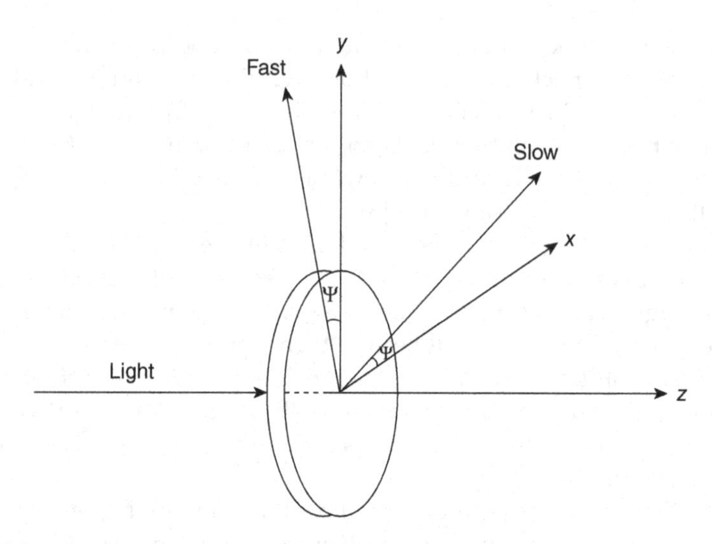

Figure 16.1 The relative positions of the fast and slow axes of incident light in relation to the *xy*-coordinate system on the body as expressed by the azimuthal angle Ψ.

Because one of the light electric vector components is slower than the other, a phase retardation appears, changing the polarization state of the incident light. The refractive index can also be characterized as fast n_{fast} or slow n_{slow}; which of the indices of refraction, ordinary n_o or extraordinary n_e will be "fast" or "slow" depends on whether the liquid crystal anisotropy is positive or negative.

From Chapter 2, the sinusoidal solutions to Maxwell's equations are,

$$E_x = a_x \cos(\omega \cdot t + \delta_x)$$
$$E_y = a_y \cos(\omega \cdot t + \delta_y),$$

where ω is the angular frequency, and $\delta_y - \delta_x$ is the phase difference between the two components of the electric vector. Passing through an optical element of thickness d, and because $\omega = 2\pi/\lambda$ (where one cycle is 2π radians and λ is the wavelength of the light), from Euler's formula,

$$\exp(-i\Psi) = \cos\Psi - i\sin\Psi,$$

and the polarization state of light passing through an optical element, for example, a uniaxial liquid crystal, can be expressed in the Jones matrix form "fast-slow" coordinate system as,

$$\begin{pmatrix} E'_{slow} \\ E'_{fast} \end{pmatrix} = \begin{pmatrix} \exp(-in_{slow} 2\pi d/\lambda) & 0 \\ 0 & \exp(-in_{fast} 2\pi d/\lambda) \end{pmatrix} \begin{pmatrix} E_{slow} \\ E_{fast} \end{pmatrix},$$

where the exponential arguments in the diagonal terms of the Jones matrix are combined to define an oft-used parameter for display design, called the *phase retardation parameter,*

$$\Gamma = \frac{2\pi}{\lambda}(n_{slow} - n_{fast}) \cdot d.$$

A liquid crystal layer with a birefringence of $\Delta n = n_{\text{slow}} - n_{\text{fast}}$ will then phase retard light of wavelength λ as it passes through a distance z within the liquid crystal layer by

$$\Gamma = \frac{2\pi}{\lambda} \Delta n \cdot z,$$

but this is just the phase difference discussed in Chapters 2 and 4, and for a liquid crystal layer of thickness d, a very useful expression for the phase difference is just

$$\delta = \frac{2\pi}{\lambda} \Delta n \cdot d.$$

The Mauguin Condition

As early as 1911, Charles Mauguin performed experiments on molecular helical twisting in liquid crystals and published his results in French and German journals (thereby probably anticipating the inventions of Helfrich and Fergason, making all their disputes moot), but his experimental results were not acknowledged as fundamental contributions until more than 60 years later. Today, his research is commemorated in a basic twisted nematic design index, called the *Mauguin parameter* for optical analysis, and for engineering design, it is called the *Mauguin condition.*

The Mauguin parameter is usually denoted by u, and can be easily derived from the phase retardation parameter Γ that is derived using the Jones vector as above. If the desired angle of twist is 90° ($\phi = \pi/2$ radians), in a positive anisotropic liquid crystal (where $\Delta n = (n_e - n_o) > 0$), with the extraordinary (slow) refractive index given by $n_e = n_{\text{slow}}$, and the ordinary (fast) refractive index given by $n_o = n_{fast}$, the birefringence is $n_e - n_o$; and with the wavelength

λ, and the thickness of the liquid crystal layer d, the Mauguin parameter then for a 90° twisted nematic liquid crystal design is

$$u = \left(\frac{\Gamma}{2\phi}\right) = \frac{2}{\lambda}(n_e - n_o)\cdot d.$$

In order to design a twisted nematic liquid crystal cell with a normally white state, for a twist angle of $\phi = \pi/2$ radians = 90°, the thickness of the liquid crystal layer d, the birefringence $\Delta n = n_e - n_o$, and the wavelength of the light, all must be in accord with the Mauguin condition which is,

$$\phi << \frac{2\pi}{\lambda}(n_e - n_o)\cdot d.$$

The Gooch–Tarry Condition

The transmittance of normally incident light of a twisted nematic cell is $T_\perp = |M|^2$, where M is given by Jones matrices as

$$M = \begin{bmatrix}\cos\beta & \sin\beta\end{bmatrix}\begin{bmatrix}\cos\phi & -\sin\phi \\ \sin\phi & \cos\phi\end{bmatrix}\begin{bmatrix}\cos X - i\dfrac{\Gamma}{2}\dfrac{\sin X}{X} & \phi\dfrac{\sin X}{X} \\ -\phi\dfrac{\sin X}{X} & \cos X + i\dfrac{\Gamma}{2}\dfrac{\sin X}{X}\end{bmatrix}\begin{bmatrix}-\sin\beta \\ \cos\beta\end{bmatrix},$$

where ϕ is the twist angle, β is the angle between the axis of polarization and the directors at the surface of the liquid crystal, $\Gamma = 2\pi d\Delta n/\lambda$ is just the phase retardation parameter with cell gap d, and $X = \sqrt{\phi^2 + (\Gamma/2)^2}$.

After some algebra, the transmittance (for normal incidence in the field-off state) can be written as:

$$T_\perp = |M|^2 = \left(\frac{\phi}{X}\cos\phi\sin X - \sin\phi\cos X\right)^2 + \left(\frac{\Gamma}{2}\frac{\sin X}{X}\right)^2 \sin^2(\phi - 2\beta).$$

For a conventional 90° twist, $\phi = \pi/2$ and the transmittance expression is simplified as

$$T_\perp = \cos^2 X + \left(\frac{\Gamma}{2X}\cos 2\beta\right)^2 \sin^2 X.$$

When X is an integer multiple of π ($X = m\pi$); that is, $T_{\perp} = 1$, so the Gooch–Tarry condition for 100% transmittance in a 90° twist TN cell with no applied electric field is

$$\frac{d\Delta n}{\lambda} = \sqrt{m^2 - \frac{1}{4}}.$$

For the first minimum condition $m = 1$, $d\Delta n / \lambda = \sqrt{3} / 2$, and for the second minimum condition $m = 2$, $d\Delta n / \lambda = \sqrt{15} / 2$. For a given wavelength, the larger the cell gap d can be, the easier to manufacture, so the Gooch–Tarry second minimum can be used for low-end LCD displays, such as digital watches and calculators. Generally, the first minimum must be met for higher-end products, such as notebook screens, where a smaller cell gap will provide faster response times.

Twisted Nematic Waveguiding

The particularly interesting feature of the twisted nematic cell is its "waveguiding" (or optical activity) capability, and the mathematical demonstration of the twisted nematic waveguiding also can serve as an example of Jones vector analysis. If the plane of polarization of the linearly polarized incident light is parallel to the surface director of the liquid crystal, the so-called *E-mode*, then the Jones vector expresses the polarization state as follows,

$$E = \begin{pmatrix} E_x \\ E_y \end{pmatrix} = \begin{pmatrix} E_e \\ E_o \end{pmatrix} = \begin{pmatrix} 1 \\ 0 \end{pmatrix},$$

where again E_e and E_o are respectively the extraordinary and ordinary components of the electric field vector of the incident light. Upon exit from the liquid crystal layer, the polarization state is represented by the E' equation given previously in the phase retardation parameter section above, and utilizing the Euler formula,

$$\begin{pmatrix} E'_e \\ E'_o \end{pmatrix} = \begin{pmatrix} \cos X - i\Gamma \sin X / 2X \\ -\phi \sin X \end{pmatrix}.$$

where again Γ is the phase retardation parameter, and $X = \sqrt{\phi^2 + (\Gamma / 2)^2}$.

In an approximation, taking only gradual changes of the twist angle in a nematic liquid crystal, the twist angle can be assumed to be much smaller than the phase retardation parameter ($\phi << \Gamma$), and $X \approx \Gamma/2$ and $E_o' \approx 0$, the expression above can be simplified as

$$\begin{pmatrix} E_e' \\ E_o' \end{pmatrix} = \begin{pmatrix} \cos(\Gamma/2) - i\sin(\Gamma/2) \\ 0 \end{pmatrix}.$$

Since the ordinary wave electric field component at exit from the liquid crystal layer E_o' is zero in the equation above, just as it is zero upon entrance to the liquid crystal layer, the expression shows that the electric vector of light while passing through the liquid crystal layer will maintain an orientation parallel with the director of the liquid crystal, that is, without any external energy influence, the electric vector follows the twist of the liquid crystal molecules, and that is why this phenomenon is also called *adiabatic following*.*

If the plane of polarization of the incident light is perpendicular to the liquid crystal directors at the surface, the so-called *O-mode* (use "orthogonal" as mnemonic), from the same Jones vector analysis, the plane of polarization of the light exiting the liquid crystal will remain perpendicular to the liquid crystal director, demonstrating the general 90° rotation of the plane of polarization through the twisted nematic liquid crystal [5].

The Twisted Nematic Cell

A schematic drawing of a twisted nematic liquid crystal cell used in liquid crystal displays is shown in Figure 16.2. The plane-parallel upper color filter and the lower thin film transistor glass substrates sandwich a nematic liquid crystal layer. Applied to the inside surface of each glass substrate are rubbed alignment layers that strongly anchor the liquid crystal molecules so that if the angles of rubbing alignment are different on the different substrates, the mid-layer molecules will twist to form a helical pattern as shown, the degree of twist being a function of the different angles of rubbing direction. In addition to the relative rubbing alignment angles, the birefringence (Δn) and thickness (d) of the liquid crystal layer must adhere to the Mauguin and Gooch–Tarry conditions, and the wavelength of the incident light must be

* Strictly speaking from the point of view of an outside observer, the light is undergoing elliptical polarization as its plane of polarization is rotated in traversing through the liquid crystal.

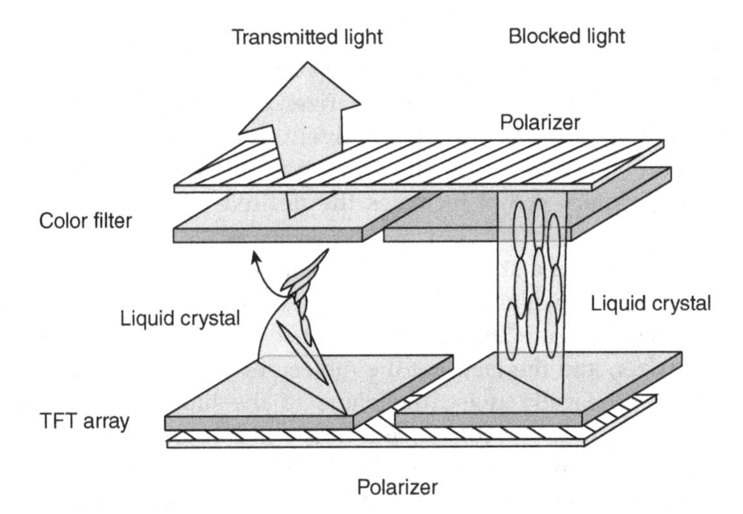

Figure 16.2 A schematic drawing of a twisted nematic liquid crystal cell used in liquid crystal displays.

within the conditions as well (visible light wavelengths work). If the conditions are satisfied, then the incident light will be waveguided 90°, and the normally white state can be achieved.

When light from a backlight unit is passed through a first linear polarizer, the light is linearly polarized, then passing through the liquid crystal, in the absence of an applied electric field, the plane of polarization of the linearly polarized light will follow the twist of the molecules (optical activity) so that upon reaching the second linear polarizer, the plane of polarization is exactly parallel to the transmission axis of that polarizer, and the light passes through with the plane of polarization rotated 90°. This is the normally white state (NW), and when there is no applied electric field, the light undergoes neither phase retardation nor blocking at the second linear polarizer, and is thus the brightest possible upon exit from the second linear polarizer, providing optimum contrast. This contrast was the essential feature of the twisted nematic that furthered the commercial applications of the LCD.

Applying an electric field perpendicular to the glass substrates will cause the dipole moments of the liquid crystal molecules to tend to align with that field in the mid-layers of the liquid crystal, and positive anisotropy molecules will tilt towards the applied field direction, providing different phase retardations and subsequent changes in the polarization state of the light from the backlight. The new polarization states, upon incidence on the second polarizer, will not pass through unaltered, and the light intensity will change (just like rotating polarized sunglass lenses), thereby providing gray scale. But before reaching the second polarizer, the polarization-altered light passes through a color filter that respectively transmits the primary colors red, green, and blue, and blocks all other colors; the mixing of the different intensities of primary colors recreates the desired color of the scene (see Chapter 22).

Upon application of a stronger electric field, the molecules will substantially align with the field, and a plane vertical ordering of the molecules will result, as shown in Figure 16.2 at right. Because of the linear polarization by the first polarizer, and the fact that the molecules of the liquid crystal midlayer no longer helically twist, the plane of the linearly polarized light remains the same (does not rotate), and thus will be blocked by the perpendicular transmission axis of the second linear polarizer, creating a *dark state*.*

At this juncture, the understanding of the physics, organic chemistry, and optics of the twisted nematic display appeared to be more than sufficient to produce a good commercial liquid crystal display, and in fact the technology to this point has been shown to be certainly good enough for displaying the numerals for digital watches and calculators. But for the display of more complicated words and images, and particularly moving pictures, the physics and chemistry described above will require assistance from the practical arts of electrical engineering.

References

[1] Born, M., and Wolf, E. 1999. *Principles of Optics*, 7th edition. Cambridge, p. 32ff.
[2] Huard, S. 1997. *Polarization of Light*. Wiley, Chichester, p. 51ff.
[3] Huard, S. 1997. *Polarization of Light*. Wiley, Chichester, p. 86ff; Yeh, P., and Gu C. 1999. *Optics of Liquid Crystal Displays*. Wiley, New York, p. 103ff.
[4] Dirac, P.A.M. 1972. *The Principles of Quantum Mechanics*, 4th edition. Oxford, p. 4ff.; Baym, G. 1990. *Lectures on Quantum Mechanics*. Westview/Perseus, New York, p. 1ff.
[5] Yeh, P., and Gu, C. 1999. *Optics of Liquid Crystal Displays*, Wiley, New York, p. 121ff.

* The anchoring alignment layers commonly use a polyimide compound. The twist angle is sometimes called the *spiral azimuthal angle*; the transmission axis is also called the *axis of polarization* or the *optical axis*.

17

The Active Matrix

The liquid crystal display digital watch and pocket calculator were tremendous commercial successes, perhaps too much so, for they attracted so many producers that competition to gain market share required rampant price cutting. The lower prices of course meant lowering manufacturer costs, and just as in many other electronics industries at the time, this engendered an inexorable shift to Asia, where the unskilled and skilled labor costs could be cut to one-fourth or less. In Japan, manufacturing quality and efficiency ruled at first, but when prices began rapidly falling and the market expanded to lower-priced products, the electronics manufacturers in Hong Kong and Taiwan sniffed out the new opportunities, and digital watch and calculator plants soon began sprouting like bamboo shoots after a spring rain. This proliferation of new, lower-cost manufacturers, using the ever more standardized manufacturing processes, while succeeding in cutting costs, exacerbated the price wars, and the prices fell even more precipitously. Once pricey digital watches and calculators quickly were available at discount retailers in the suburbs and on the streets and in the alleys of big cities at one-tenth the former prices. The losses for the American and Japanese manufacturers soon became debilitating.

Liquid Crystal Displays, First Edition. Robert H. Chen.

The older electronic companies knew that competing with the Hong Kong and Taiwan manufacturers was futile, and so realized that the only way to gain profitability was to add value and expand the market into new areas. Now coming to this conclusion requires no great insight, but the logic was sound: identical numerals-only products could not compete, differentiated new products were needed to move the market into new product areas like word processors and computer monitors. That meant that the liquid crystal screens would have to be able to display multiple lines of Arabic numbers, Roman words, Japanese script, Chinese characters, graphics, and even moving images; only then would the market grow larger and the prices of the products employing the nematic liquid crystal displays increase. The objective was clear: the liquid crystal display manufacturers had to develop new capabilities in their nematic displays for use in new products or die.

Matrix Addressing

In the watches and calculators, only numerals were necessary, and numerals can be displayed simply using the familiar "Figure Eight" with seven sectors, as shown in Figure 17.1, with activation by the wire connections (including a period) as shown. But to display words (particularly for example, complex Chinese characters), the seven sector scheme was clearly inadequate, and more and denser aggregates of sectors were necessary. For example, as shown in Figure 17.2, a matrix of 35 picture elements (pixels) has much more latitude to express a character than the standard "8" configuration. However, the multitude and density of the pixels required complicated, entangling wire connections, as can be imagined in connecting up the matrix in Figure 17.2, to address and activate each element to form the word.

The early pixel matrices addressing method was to first turn on (send voltage to) all the pixels in the first column containing the desired pixel, then activate all the pixels in the row that contains the desired activated pixel; the intersecting pixel would then be chosen (addressed) and lit up. In this way, one column–row for a pixel and the next column–row for the next pixel would be sequentially lit up, and so the desired character would be "written" and appear as the conglomeration of the chosen pixels. This manner of simultaneously addressing and writing is known as *time multiplexing*, and is a form of *passive matrix addressing*; and the process cycle of activating columns and rows to light up a single pixel is called the *duty cycle*.

An image is composed of *frames* to be displayed; if there are N rows and M columns, then there are $M \times N$ pixels in the frame, and so each pixel can

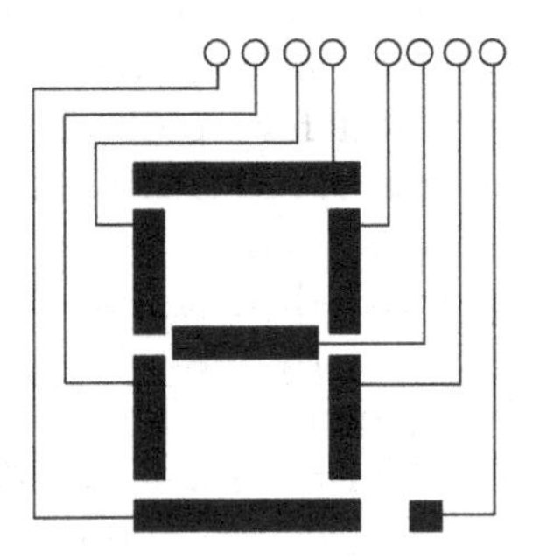

Figure 17.1 Seven-sector numeral.

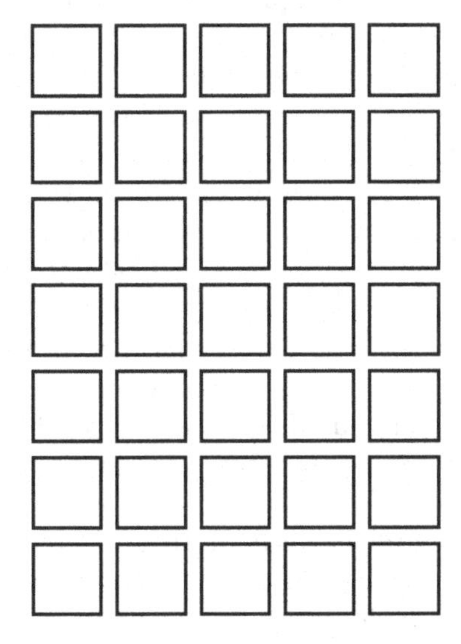

Figure 17.2 Thirty-five-sector picture element matrix.

only use $1/N$ of the time needed to display the frame. That is, the duty cycle is equal to $1/N$, and if the frame is driven simultaneously in two directions, up and down, the duty cycle is $2/N$.

Complicated textual or image displays require many columns and rows, the total number and density of which determines the resolution of the display. The large numbers of columns and rows for the more complicated characters require more and more operations in a duty cycle, with the result that soon the data signals sent will not have sufficient time to be completely displayed within the time for a frame display, and the frame will be incomplete.

A solution would be to increase the time to complete the frames, but then the display of the changing images will be slowed. For watches and calculators, this is not a major problem, but for rapidly changing words and numbers and particularly moving images, speed is critical.

Furthermore, since the address/write voltage must pass through neighboring pixels, there will be so-called *crosstalk* between the pixels, wherein a pixel's voltage will be disturbed by the voltage on a neighboring pixel. Because of the multitude and frequency of signals in a complicated image, crosstalk between pixels will be even more serious, resulting in very poor contrast (intensity distinction between neighboring pixels), and other undesirable effects such as inaccurate intensities.

The response time of nematic liquid crystal displays at the time was indeed inadequate for uses in displays that required characters other than numerals. For example, in 1982, a 480×128, $1/64$ duty cycle matrix-addressed twisted nematic display could only display very low-resolution words, but if the rows and columns were increased to improve resolution, the response time, viewing angle, and contrast were miserable, to the extent that even simple words were sometimes difficult to distinguish. Increasing the voltage in principle would cause more molecules to turn faster and improve the response time, but higher voltages, among other things, would produce more crosstalk, which would decrease contrast and accuracy between neighboring pixels. There were many attempts at adjusting parameters, introducing new and complicated driving schemes, changing liquid crystal materials, and so on, but all the efforts proved futile; the display was stuck at low resolution and could not move out of the blurry morass of indistinguishable words.

At this desperate juncture, IBM's Paul Alt and Peter Pleshko, using graphical analysis, came up with a solution. A transmission/voltage graph (*t/v* curve) from Schadt and Helfrich's twisted nematic device shown in Figure 17.3 displays the transmissivity of the liquid crystal (left ordinate) as a func-

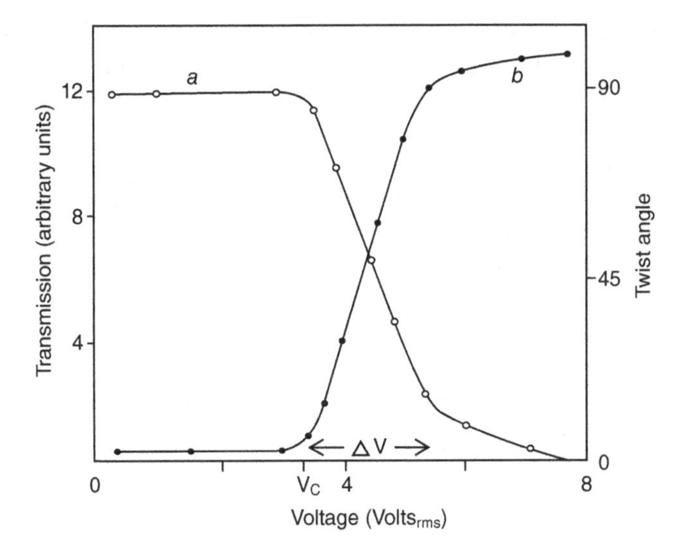

Figure 17.3 Transmission/voltage graph (*t*/*v* curve).

tion of applied root-mean-squared (rms) voltage V_c (abscissa) and the twist angle (right ordinate) [1]. The normally white curve (a) clearly can be seen to have a high transmissivity, and when a threshold voltage of about 3.5 V is applied, the transmissivity decreases sharply. Conversely, the normally black curve (b), when there is no applied field, has a nearly zero transmissivity (the second polarizer's transmission axis is perpendicular to the light's linear polarization direction and completely blocks the light). When there is a voltage applied exceeding the threshold, the transmissivity increases sharply. The steep decrease and increase of transmissivity with applied voltage is denoted by ΔV, and IBM's researchers using signal bursts in a simple calculus extrema calculation found that the maximum number of column and rows (N_{max}) was completely dependent on the ratio between ΔV and V_{th}, which they called P, that is, the maximum resolution could be simply expressed as [2]

$$N_{\max} \sim \left(\frac{1}{P}\right)^2 \quad \text{and} \quad P = \frac{\Delta V}{V_{th}}.$$

Clearly then, to increase the number of rows and columns, P has to be small, and P can be small if either the voltage range ΔV is compressed or the threshold voltage V_{th} is increased. Because a low-operation voltage can save energy, decrease crosstalk, and reduce heat, raising the threshold voltage is generally not a good option. So the solution was clear, to focus hard on compressing ΔV; that is, find a liquid crystal material that possessed a very steep *t/v* curve.

There were several different approaches taken to finding steep *t/v* curve materials, among which were to investigate the ratio of bend to splay elastic coefficients (K_{33} / K_{11}) in various liquid crystal materials (since this was a twist mode display, adjusting the K_{22} twist coefficient might adversely influence display operation). The reasoning was that a faster elastic snap-back response should increase the slope of the *t/v* curve. There were some salutary effects from different choices of liquid crystal, but after numerous experiments, the adjustment of elastic coefficient trials could at best only allow an N_{max} of 89, which was still far from the requirements of a high resolution display.

The Super Twisted Nematic

A different and novel approach to increase the slope of the *t/v* curve came from the Royal Radar Establishment researchers Colin Waters and Peter Raynes, who used their dye guest-host mixture to promote an additional twist to increase the speed of the liquid crystal response. The dye could push the twist to a 270° "super twist," and apparently because of an increased tendency to spring back, the response speed did indeed increase, or perhaps it was the constructive interference of the 270° super twist that increased the transmissivity *t*. Whatever the reason, the t/v slope was slightly steeper; however, because the dye guest viscosity was rather high, the response speed did not increase much, and the dye caused the liquid crystal to be colored, and there was significant light leakage. Nonetheless, the Royal Radar Establishment's concept of a *super twist* to increase the slope of the *t/v* curve was born, and the idea at least provided a sound basis on which to proceed.

Following the RRE concept, Brown Boveri's Terry Scheffer and Jurgen Nehring instead used a naturally twisting chiral nematic as dopant to instigate the 270° super twist and its concomitant steeper t/v slope. With the larger N value possible, they succeeded in producing a high-resolution, high-contrast, wide-viewing angle super twisted nematic (*STN*) display. The sharper image contrast was a result of the 270° constructive interference, since the brights could be brighter and the darks darker (the ratio of which is the definition of "contrast"). Further, after tuning the twist angle by changing the concentration of chiral dopant, introducing a pre-tilt, and adjusting the angles of the polarizers, the viewing angle could also be improved (rf. Chapter 23).

The dream of a high-resolution twisted nematic liquid crystal display was rapidly becoming a reality. In the 1985 Society for Information Display (SID) annual meeting, Scheffer and Nehring proudly demonstrated their *540 × 270* matrix, *1/135* duty cycle *12 cm × 24 cm* STN display. Attending the meeting and particularly impressed by the prototype demonstration were several engineers from Sharp. Soon after the meeting, Sharp invited Nehring to Osaka to specially demonstrate his display. Within a year, in addition to Sharp, Hitachi and Seiko also demonstrated their own STN displays at that year's All-Japan Electronics Show.

The polarizing effect of liquid crystals will always be distorted by random thermal atomic motion, but since the STN does not waveguide like the TN (which requires only a 90° twist), there is more pronounced wavelength dispersion resulting in light leakage of different colors (so the early STN displays were always in shades of light green, blue, or yellow, wavelengths chosen for a more pleasing shade). One method used to reduce this light leakage is to use an oppositely opposed optical element that cleverly retracts the leakage. That is, add a reversely oriented optical element to the STN display, such that the distortion defects will be exactly canceled (confer Chapter 23 on wide-view compensation).

Sharp's "back-to-back" apparatus added a positive 240° and an oppositely 240° liquid crystal elements to form the so-called *double super twisted nematic* (DSTN) display, which came on the market in 1988. However, although the DSTN could resolve the light leakage problem, the price to pay was that the display imposed high-precision alignment requirements in the manufacturing process, and this delicate procedure rendered production difficult and time consuming. Further, an entire canceling second optical unit not only complicated the production process, it also increased the cost and weight, and decreased the transmission intensity; all-in-all relegating the DSTN to the fate of a merely interesting technological workpiece.

To get around the alignment and weight problems, Seiko designed a birefringent film to replace the second panel in the Sharp back-to-back apparatus. The reverse film was developed by Seiko with the two specialty manufacturers Nippon Electric and Sumitomo Chemical. A lighter display called the *film super twisted nematic* (FSTN) was produced, but because there were difficulties in maintaining film uniformity (which affects color) over large screen surface areas, FSTN has been limited to use in small displays.

Although Sharp and Seiko each had their own products, the two companies cooperated to develop the *triple super twisted nematic* (TSTN) display using a third film to compensate for coloring defects over large screens. In fact, many of the displays used in products commonly called "STN," are actually either FSTN or TSTN, with the former used in smaller displays, and the latter primarily for use in larger ones. Generally, a twist of 240° or 260° and a pre-tilt angle 4°~5° or 6°~8° has become standard. But because the transmissivity of STN depends on the constructive wave interference, the displays were still sensitive to color so much so that a green display will still leak light in the other primary colors, blue and red.

The inherent color was still a problem, but the twisted nematic passive matrix addressing technology succeeded in expanding the uses of liquid crystals for displays from the simple digital watch and calculator numeric displays to slightly more complex displays, and the STN expanded applications further to the more complicated and dense displays for use in multipurpose calculators, word processors, and small computers.

Now, following in the path of the cathode ray tube, attention turned to the development of liquid crystal displays for color television. Having more or less satisfied the conditions of temperature, viscosity, and stability, the principal problem remaining for a color display was the inherent color of the liquid crystal itself. The liquid crystal must be completely colorless (and the other optical components as well), that is, the liquid crystal and optical components should not affect the color of the light passing through, the color was to be provided solely by a color filter mounted after the liquid crystal layer (see Chapter 22).

With the resolution improved by the advent of STN, for television, the liquid crystal now had to be capable of presenting fast, dynamic images. If the effort was successful, John Stonehouse's seemingly reckless order to the Royal Radar Establishment to develop a flat-panel color television to make the RCA shadow mask obsolete finally would have a chance at realization.

Active Matrix Addressing

The super twisted nematic displays were suitable for calculator and simple word processing displays, but the passive matrix addressing system described above was not capable of providing the even higher resolution, fast response, and timely overall control of the individual pixels required for more advanced computers and color television. In a continuing theme, the first instance of a resolution of the problem once again came from RCA. The DSM invented at RCA in 1971 actually utilized an electric circuit connection to the individual pixels to form a prototypical active matrix addressing system. And in principle, the active matrix addressing can avoid the duty cycle and crosstalk problems outlined above by separating the addressing and writing operations, as described following.

Separating addressing and writing requires connecting the signal line to the gate of a transistor switch that opens the desired pixels in one instance (addressing), then the image information signals are sent by way of multiple-line multiplexing to the transistor's source electrode to rapidly form the image frame in the next instance.

Bernard Lechner and Frank Marlowe of RCA did this by mounting transistors at each pixel of a matrix addressing system, and added a capacitor to hold the charge so that the signals could be timed to occur simultaneously. The addressing signal would open up the designated pixels, and then the data signal would be written into the capacitors of those opened pixels. Because the transistors act as passive, binary, nonlinear devices, a low voltage is sufficient for operation, thus saving power, reducing heat, and also reducing crosstalk. Furthermore, because transistor switching is fast, more lines and columns can be addressed and written in a single duty cycle, thereby increasing resolution. All the problems of resolution and response time were in principle solvable by this active matrix design using transistors [3].

The realization of the active matrix addressing scheme for liquid crystal displays, however, required a suitable transistor that could be mounted on the glass substrates used in liquid crystal displays.

It turns out that as early as 1962, Paul Weimer, while designing the RCA television camera, had already researched a glass-mountable transistor, which he called the *thin film transistor* (TFT) [4]. Shortly thereafter, Peter Brody of Westinghouse constructed a liquid crystal display using thin film transistors as the addressing switches, and this was the forerunner of the modern mass-produced liquid crystal display.

But Brody's work met the fate of so many others in America's large electronic companies, as his research was not supported by Westinghouse's upper management, and he had to turn to the U.S. Air Force and Naval Research laboratories for support. After almost 10 years, Brody's endeavors were rewarded in 1972, when he demonstrated the world's first thin film transistor active matrix liquid crystal display. Brody's thin film transistor active matrix concept is schematically illustrated by a present-day *thin film transistor active matrix liquid crystal display* (TFT-AMLCD) addressing system in Figure 17.4.

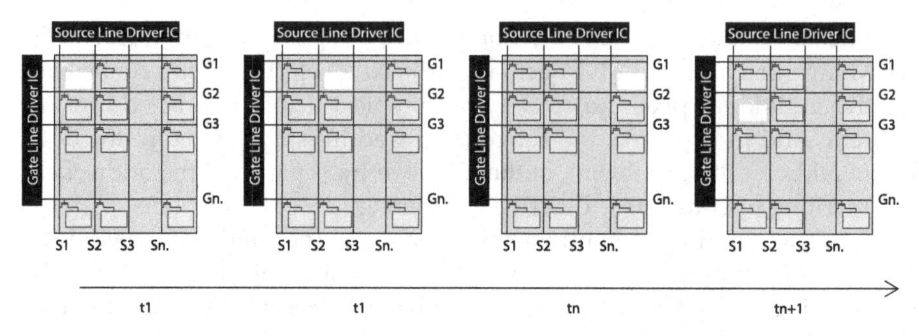

Figure 17.4 Active matrix addressing system.

The gate line and source line of the thin film transistor mounted on the glass substrate are connected respectively to a gate line driver IC and a source line driver IC, as shown in Figure 17.4. In operation, the selected pixels are addressed by the receipt of a signal on the gate of their transistors; then the grayscale image data signal voltages are transmitted through the source line to be written on the selected pixels. In Figure 17.4, the open, nonshaded transistors are those that have been activated for receipt of data; the abscissa in the figure is time. This circuit design addresses and writes each pixel independently, and thus is called *active addressing*.*

Figure 17.5 is a representative column driver schematic. After the addressing signal from the gate driver opens the transistor's gate, the dynamic shift register source driver will sequentially switch on the selected transistors, then upon reception of the video red, green, and blue data signals, they are stored sequentially in the storage capacitors (denoted by ⊣⊢), thereafter the amplifier (denoted by ∇), sends the amplified signal to each pixel's column line. The "COM" in Figure 17.5 is the common voltage, the "OE" denotes the Output Enable signal, and "V_b" is the source voltage [5].

The driver IC circuit's display's gate driver and source line are typically mounted on the TFT glass substrate using a contact pad. Tape automated bonding (TAB) or the more advanced chip-on-glass (COG) are the commonly used methods of mounting the driver ICs on the glass (see Chapter 25).

In order to avoid charge leakage from the transistors, in the early days, the liquid crystal itself would need to have relatively high bulk resistivity. The electrodes for producing the applied electric field also may absorb charge. In contrast to simple passive addressing systems, when there is no

* A gate line is also called an address line or a row and it addresses the pixels; a source line is also called a data line or a data bus line, or a column, and it transmits data to the pixel.

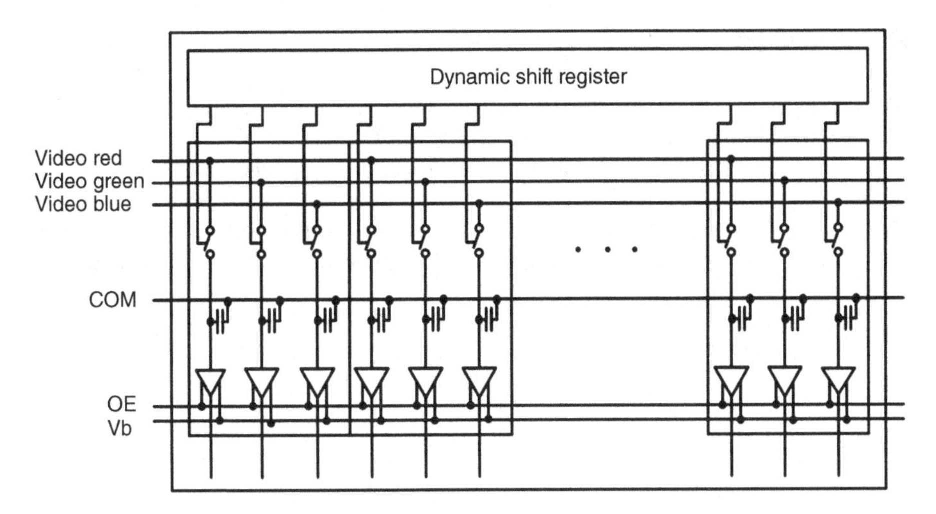

Figure 17.5 Column driver schematic.

signal, the charge will not be leaked by the source electrodes, and when there is a signal, the storage capacitors discharge fully, thus the active matrix addressing system operates under full charge. In this way, a merely adequate voltage can promote a complete and fast turning of the liquid crystal molecules, resulting in fast response times and minimal light leakage all at low-voltage operation. Furthermore, the bulk resistivity requirement of the liquid crystal itself is decreased because the storage capacitors hold the charge. Still further, since the number of rows and columns are only limited by the rise time (see Chapter 24), active matrix addressing does not require the steep t/v curve slope, and therefore is not constrained by the ΔV and V_{th} ratio restrictions, so the transmission to voltage curve and slope requirements of the *active matrix liquid crystal display* (AMLCD) are not as stringent as for the passive matrix LCD (PMLCD).*

The active matrix addressing liquid crystal display has become the standard for high-resolution large screen displays, but its development required the aforementioned low-voltage fast-switch transistor, mountable on a glass surface and capable of driving the millions of pixels in a modern liquid crystal display, particularly television. The development history and technology of the thin film transistor will be described in Chapters 19–21.

---

* The t/v curve slope analysis, however, did serve to instigate the development of the STN liquid crystal display, so it still played a critical role in liquid crystal display development.

References

[1] Schadt, M., and Helfrich, W. 1971. Voltage dependent optical activity of a twisted-nematic liquid crystal. *Appl. Phys. Lett.*, **18**, 4, 127–128.

[2] Kawamoto, H. 2002. The history of liquid-crystal displays. *Proc. IEEE.*, **90**, 4, 487ff.

[3] Lechner, B., et al. 1971. Liquid crystal matrix displays. *Proc. IEEE.*, **59**, 1576–1579.

[4] Weimer, P. 1962. The TFT—a new thin-film transistor. *Proc. IEEE.*, available at http://www.ieee.org.

[5] Lueder, E. 2001. *Liquid Crystal Displays*. Wiley, Chicester, p. 233.

18

New Screens

Although there were fits and starts in the process of development as described above, the ultimate pervasiveness of liquid crystal televisions is testament to its technical brilliance and popular appeal. Which begs the question: in retrospect, why was it that a company like Sharp could aggressively seize new technologies like the liquid crystal display, while the more technologically advanced companies at the time, such as RCA and Sony, demurred? RCA was the pioneer in the creation of most of the technology behind the liquid crystal display (LCD), yet did not develop a product and the company itself ultimately disintegrated. Sony, although the prime consumer electronics purveyor of the time, shunned the LCD, and ultimately never caught up with the liquid crystal tsunami, and was bypassed as the supreme electronics brand by Samsung, who rode the wave for all it was worth (confer Chapter 26).

The obstacle to RCA and Sony's entry into the new realm of liquid crystal television ironically could be just the fact that the two largest television makers at the time were RCA and Sony themselves, that is, the television giants perhaps were not able to avoid the overriding influence of their respective flagship products, the shadow mask and the Trinitron® cathode-ray

---

Liquid Crystal Displays, First Edition. Robert H. Chen.

tube (CRT) televisions. Sometimes the inertia of commercially successful products devolves to a stubborn paranoia; RCA and Sony relentlessly tried to enlarge the CRT screen size, lighten it, and slim it down, all in ultimately futile attempts to combat what proved to be an inexorable rise of the light-weight, huge-screen, wafer-thin liquid crystal television. As prices for the increasingly mass-produced LCD television continued to drop, the market determined the fate of the CRT, and with it the fates of RCA altogether, and the Sony Trinitron® in particular.

Twisted Nematic Television

Sharp, in contrast, had no CRT/shadow mask/Trinitron® legacy burden. Travelling light, Sharp could nimbly venture into the new product unknown, secure in the knowledge that it had little to lose in failure and much to gain in success. And of course, Sharp did not fail, and in 1988, under the dynamic leadership of Washizuka Isamu, Sharp succeeded in producing a hitherto inconceivably large screen 14-inch liquid crystal color television (Figure 18.1). The set weighed 1.8 kg and was only 27-mm thick, about 1/13th the thickness of a 14-inch CRT television. Although Sharp's first

Figure 18.1 Washizuka and Sharp's twisted nematic 14-in liquid crystal color television. From Kawamoto, H. 2002. *Proc. IEEE,* **90**, 4, 475, April.

liquid crystal color television was only what was considered the minimum screen size for a living room television, it was the harbinger of the multibillion dollar liquid crystal television market to come.

That liquid crystal displays could evolve from the display of simple watch and calculator numerals to the complex dynamic images of television can be traced to the development of the twisted nematic liquid crystal and the active matrix addressing system enabled by the amorphous silicon thin film transistor (Chapter 20). All the basic elements were in place, most invented by RCA; the task was to consolidate them into a marketable product.

In a policy push to promote the capture of a new commercial market for flat-panel televisions by Japanese companies, the Ministry of Trade and Industry (MITI) in 1988 brought together the big brand-name electronics firms, Hitachi, Sharp, NEC, Seiko-Epson, Casio, and Sanyo, and several upstream component manufacturers, altogether 12 companies, and ordered them to cooperate to develop a 40-in liquid crystal color television for mass-market production. With a target date of 1995, the "Development of Fundamental Technologies of Giant Electronics Devices" seven-year plan designated Japan's large industrial chemical companies to provide the upstream components: so Toppan Printing and Dai Nippon Printing would make the color filters, the glass would come from Asahi Glass, the liquid crystal from Chisso Chemical, the polymer sealants from Nippon Synthetic Rubber, and Ulvac Equipment would provide the fabrication equipment. It was a grandiose plan for an unprecedented scale of cooperative effort for Japanese companies.

The target, however, was not reached, and some might say that MITI's plan was yet another case of government interference in commercial activities best left to companies themselves. But the later devolution of the participating companies into the pillars of an enormous LCD industry to come could be attributed at least in part to the cooperative lessons learned and technology developed under Japan's 40-inch LCD TV seven-year plan [1].

Before LCD television took off however, the driving force behind the development of liquid crystal displays was the soon-to-be ubiquitous notebook computer.

Notebook Computer Screens

Computers of course now have a well-known and storied history, and their introduction to society is likely one of the great transformational events of mankind. The first "societal" computers were the IBM 360 series

Figure 18.2 The IBM 360/40 mainframe computer. From IBMArchives, IBM.com.

mainframes, the computational workhorses of the Sixties; every university's computer center had IBM "big iron" taking up its entire basement (for example, the IBM 360/40 shown in Figure 18.2).

Students using the computer facilities would take their boxes of computer punch cards and line up for *batch processing* by the mainframe computer. The centralized computing system obviously took up a lot of time, and projects were at the mercy of Computer Center scheduling. Predictably, the situation led science and engineering departments in particular to set up their own computer systems on site in a distributed computing scheme using minicomputers, such as the then-ubiquitous Digital Equipment Corporation's (DEC) PDP-8 (shown in Figure 18.3), Hewlett-Packard's 9830, and the advanced calculators from Wang Labs. Although the minicomputers were much smaller than an IBM mainframe, they generally could still take over a room, but with computer capacity increasing exponentially and the size of the circuits likewise shrinking, there came to be not only increased computing power, but also downsizing of the computer itself.

And so by the 1970s, the Xerox Palo Alto Research Park (PARC) in California was developing a personal computer, complete with mouse and

Figure 18.3 The PDP-8 minicomputer. From http://www.pdp8.net, courtesy of D. Gesswein.

overlapping screens. In those years, Steve Jobs could often be seen kibitzing the PARC researchers, looking for keys to the new personal computer treasure box. As is now well-chronicled, Jobs and Steve Wozniak, an electrical engineering student at Berkeley, founded Apple Computer in Jobs' parents garage and began their great personal computer adventure. Figure 18.4 is a photograph of the first personal computer, designed by Wozniak, with a fancy wooden case and typewriter keyboard; a CRT television screen was used for the display. In 1977, Apple's first commercial computer was sold; and it was not until 1981 that IBM unveiled the 5150 PC with its memorable green and white CRT screen. Within two years, IBM had sold 10 million of the now quaint-looking IBM PC machines.

In contrast to Apple's proprietary closed system, IBM took the open system approach, although such a choice was at least in part a response to the U.S. Department of Justice's antitrust investigation. IBM thus increased its purchases of non-IBM parts and services to avoid further charges of monopolistic behavior, and so for example, in its personal computers, the central processing unit (CPU) came from Intel and the memory from Texas Instruments. As for software, the most commonly used personal computer operating system at that time was Digital Research's *CP/M*, and it was the

Figure 18.4 The first personal computer by Apple used a CRT television for the display. From www.oldcomputers.net, courtesy of S. Stengel.

natural first choice for IBM, but of course Microsoft's DOS came to be the dominant PC operating system.*

The PC open system meant that any company could manufacture IBM PC-compatible computers using Intel's CPU and Microsoft's operating system. In particular, Acer's founder Stan Shih, saw a momentous opportunity, but there was a gnawing and pivotal question of what the Japanese would do. That is, would the vaunted giant electronics companies with all their technological and marketing power also enter into and dominate the personal computer business just as they did in the consumer electronics business? The answer came in a meeting in Taipei in 1984, when Shih

* The oft-recounted legend is that when IBM came to see Digital Research's president Gary Kildall about adopting his CP/M, he was up in the air flying his private plane and so enjoying himself that he was in no particular hurry to come down. When eventually he did land, IBM's men had already left in a huff, feeling insulted that they had not been accorded the respect they felt they deserved, being as they were from IBM. Not long after, IBM made the decision to use DOS from a new company called "Microsoft" for the IBM PC software platform, and the rest as they say is history; Bill Gates of course went on to become the world's richest man, and Kildall, what must he be thinking these days? Refer to Ferguson, C.H. and C.R. Morris, *Computer Wars* (Times Books, Random House, 1993), p. 25, ,as well as asking almost anyone in Silicon Valley.

asked the then Chairman of the Japanese word-processing giant NEC whether they would manufacture and sell IBM-compatible personal computers. To Shih's surprised relief, Chairman Kobayashi disdainfully replied: "What kind of company do you think NEC is, one that would *follow* someone else?" [2].

It was a rare instance of a lack of modesty for which the Japanese are rightfully known and admired, and they subsequently paid dearly for it, ceding the global personal computer market almost wholly to manufacturers in Taiwan. Acer and many other PC manufacturers in Taiwan made the hardware for almost all the big brand name personal computers, including Compaq, Dell, HP, and significantly IBM itself, and even Apple, as well as the local shops, and myriad no-brand computers. The manufacturing of course required paying through the nose for Intel's microprocessors and being licensed and forking over hefty hardware and software royalties to IBM and Microsoft, respectively (and grudgingly respectfully), and paying Sharp, Toshiba, and NEC for the very expensive liquid crystal screens of the computer monitors and notebook computers. The domination of personal computer manufacturing by companies in Taiwan and the high cost and uncertain supply of LCD screens led naturally to Taiwan's entry into the manufacture of LCDs and its co-domination with Korea of that industry, as will be described in Chapter 26.*

Of course, the monitor was a critical personal computer peripheral device, but the CRT was an entrenched and cost-effective, low-priced product that performed fully well, making it very difficult to dislodge. So although the beachhead for the liquid crystal display was the monitor, the conquest would accede from the computer screen for the notebook computer.

Among the first "portable" computers was Compaq's *fourteen kilogram* "minicomputer" (pictured in Figure 18.5). The 9-inch black-and-white (green) CRT screen was the primary culprit for all that heft. It was clear that the key for the realization of the intriguing concept of the portable personal

---

* Accompanying and augmenting the PC, Taiwan also came to dominate the manufacture of IC controllers, logic chips, CPU chipsets, dynamic random access memory (DRAM), Static RAM (SRAM), application specific integrated circuits (ASICs), NAND gate nonvolatile memory (flash), and also light-emitting diodes (LEDs), scanners, keyboards, mouses, printers, and monitors. The chip fabrication foundries, such as Taiwan Semiconductor Manufacturing Company (TSMC) and its rival United Microelectronics Corporation (UMC), also grew to become the world's two largest contract chipmakers, IC design firms such as Mediatek produced components for DVD players and LCD televisions, and HTC produced Android smart phones to rival Apple's iPhone, and eventually CPUs were produced to compete with Intel; the source of all these products could be traced back to the personal computer and its progeny.

Figure 18.5 Compaq's fourteen kilogram "minicomputer." From http://www.oldcomputers.net, courtesy of S. Stengel.

Figure 18.6 Tandy hand-held computer made by Kyocera displaying a full eight rows of text on its liquid crystal display. From http://www.oldcomputers.net, courtesy of S. Stengel.

computer was the development of an alternative thin, light, low power consumption display screen.

The first "computers" to use liquid crystal displays were electronic typewriters developed by Sharp, Matsushita, and Tandy with memory only sufficient to produce a single line of words, which could be duly displayed on a primitive liquid crystal screen much like that of a calculator. Then in 1983, Tandy turned out the popular TRS-80/100, made by Japan's Kyocera (pictured in Figure 18.6), which could display a full eight rows of text on its liquid crystal display. Seiko also developed a full screen 25-line, 80-letter capable display with a fluorescent backlight primarily for computer games use. These displays, however, all used the original dynamic scattering mode (DSM) invented by RCA, with the attendant poor contrast and small viewing angle.

The invention of super twisted nematic displays by Scheffer and Nehring thus was a timely technological advance propelling the development of notebook computer displays. Their company Brown Boveri, like Hoffman-LaRoche, was not interested in manufacturing the STN displays themselves, so in 1985, they licensed the technology to several Japanese manufacturers. Within a year, the licensed companies demonstrated their own high resolution *640 × 400* displays at the All-Japan Electronics Show. The weight and thickness of the displays were further reduced by Kyocera's *chip-on-glass* (COG) technology replacing the printed circuit board drivers, and in 1989, NEC proudly showed off what it called the world's first thin-screen super twisted nematic portable computer; already in the now-familiar notebook computer form factor (Figure 18.7).*

Figure 18.7 The first thin screen super twisted nematic portable computer. Courtesy of obsoletecomputermuseum.org.

* Chip on glass (COG) means that the driver is mounted directly on the glass substrate (see Chapter 25). The first portable computer was the 1981*Osborne I*, but the company shut down in 1983. At the time, many Silicon Valley engineers scornfully called the oft-malfunctioning TRS-80 the "Trash-80."

Not long after, and in spite of Alt and Pleshko's *t/v* curve contribution to the development of fast liquid crystal displays, IBM's upper management decided to withdraw from liquid crystal display research and development. The advent of the rapidly growing notebook computer market caused IBM to reconsider, but because LCD panel manufacture required huge capital outlays for plant construction and production, IBM chose to spread the costs and risks by joining with Toshiba to form the company Display Technologies Incorporated (DTI) in Japan's Yasu Prefecture. DTI began LCD production in 1991, and its first product was a 10-inch super twisted nematic active matrix liquid crystal display panel for use in notebook computers. DTI was the first company to mass produce twisted nematic displays for use in notebook computers, and in 1998, IBM DTI produced a 2560×2048 super high-resolution display for medical applications, proving once and for all that liquid crystal displays could match the quality of cathode ray tube displays.

The development of liquid crystal displays presents an interesting question: if there were no such thing as a notebook computer and its requirement of a thin, light, low-power display, would liquid crystal display products ever have developed? Or conversely, if there were no thin, light, low-power liquid crystal displays, would notebook computers ever have developed? The two were no doubt mutually influential, but did one require the other to prosper? Notwithstanding the lack of a definitive answer, this chicken and this egg both went on to develop into multibillion dollar industries.

The active matrix carried liquid crystals into computers and color televisions, the "active matrix" that made that possible relies on the transistor and the integrated circuit, the subjects of the following Chapters 19–21.

References

[1] Castellano, J.A. 2005. *Liquid Gold*. World Scientific, Singapore, p. 202.
[2] Chen, R.H. 1997. *Made in Taiwan: The Story of Acer Computers*. McGraw-Hill International, Taipei, p. 114.

19

The Transistor and Integrated Circuit

Liquid crystal displays work because the liquid crystal can polarize light; the polarization changes the intensity of a source of light that passes through each individual picture element making up the whole image. The polarization depends on the voltage that turns the liquid crystal molecules, and that voltage depends on semiconductors and integrated circuits to control and deliver the appropriate voltage. The liquid crystal display and almost all the other now all-too-familiar electronic devices of the modern age owe their existence to the transistors and integrated circuits within. Giving life to those devices is the semiconductor, the very soul of the transistor and integrated circuit, but just exactly what is a semiconductor, and how do transistors and integrated circuits control the voltage to the picture elements?

The Bohr Atom

The transistor is the source of our modern electronics era, and the spirit of the transistor is found in its semiconductor soul. But just what is a "semiconductor"? Is it an electrical conductor, is it an insulator, or both; or is it half of each? But how can one be two, and just what does "half a conductor"

Liquid Crystal Displays, First Edition. Robert H. Chen.

mean? The seemingly mystical *yin* and *yang* of conductive-resistive duality in fact will be seen to owe its essence to foreign matter.

Any material thing can be described and distinguished by its resistivity (resistance to electrical conduction); for example, the aluminum wiring used in transistors has a resistivity at room temperature of $10^{-6}\,\Omega$-cm, and the silicon dioxide (SiO_2) used as an insulator has a resistivity of $10^{16}\,\Omega$-cm, the span between conductivity and insulation being an astounding 22 orders of magnitude!

By convention in electronics, for resistivities less than $10^{-2}\,\Omega$-cm, the thing is considered a conductor, and for resistivities greater than $10^{5}\,\Omega$-cm, the thing is called an insulator. Those things that fall in the space between the conductor and insulator are the putative "semiconductors," but this categorical placement does not seem to really explain exactly what they are. The beauty, and indeed the utility, of the semiconductors lies in the fortunate happenstance that their conductivity and resistivity can be changed and controlled by an injection of appropriate neighboring elements in the *Periodic Table*.

The story of semiconductors must go back at least to the year 1789 with the publication just three months before the French Revolution of Lavoisier's *Elements of Chemistry*, wherein the father of modern chemistry identified *elements* and accounted for each element in a "balance sheet" of reactions. Eighty years later in 1869, the Russian chemist Mendeleev completed his great work of categorizing the behavior of matter by a table of elements. As each element was first listed in sequence of its atomic weight, the observable features of the elements displayed some distinct periodic behavior. Later in 1913, Henry Moseley, a student of the discoverer of the nucleus Rutherford, continuing his mentor's experiments on the atomic nucleus, discovered that listing the elements according to their atomic number (nuclear charge, number of protons) rather than atomic weight could produce predictable chemical behavior as manifested by the number of electrons in the atom (that must equal the number of protons to maintain atomic charge neutrality). Now the *Periodic Table of the Elements* revealed an even greater periodicity, and from this order, the Table (Figure 19.1) was used as a basic point of departure for chemical research.*

However, according to the classical electrodynamics of Maxwell, Rutherford's model of electrons orbiting the nucleus in a solar system-like

* Lavoisier was guillotined in 1792 just when he was making his fundamental contributions to the new science of chemistry. Geiger and Marsden, who participated in Rutherford's discovery of the nucleus by alpha particle bombardment, found themselves on opposing sides in World War I, and Moseley at 28, tragically was killed in action at Gallipoli.

1	2	3	4	5	6	7	8	9	10	11	12	13	14	15	16	17	18
1 **H** 1.0079																	2 **He** 4.0026
3 **Li** 6.941	4 **Be** 9.0122											5 **B** 10.811	6 **C** 12.011	7 **N** 14.007	8 **O** 15.999	9 **F** 18.998	10 **Ne** 20.180
11 **Na** 22.990	12 **Mg** 24.305											13 **Al** 26.982	14 **Si** 28.086	15 **P** 30.974	16 **S** 32.065	17 **Cl** 35.453	18 **Ar** 39.948
19 **K** 39.098	20 **Ca** 40.078	21 **Sc** 44.956	22 **Ti** 47.867	23 **V** 50.942	24 **Cr** 51.996	25 **Mn** 54.938	26 **Fe** 55.845	27 **Co** 58.933	28 **Ni** 58.693	29 **Cu** 63.546	30 **Zn** 65.38	31 **Ga** 69.723	32 **Ge** 72.64	33 **As** 74.922	34 **Se** 78.96	35 **Br** 79.904	36 **Kr** 83.798
37 **Rb** 85.468	38 **Sr** 87.62	39 **Y** 88.906	40 **Zr** 91.224	41 **Nb** 92.906	42 **Mo** 95.96	43 **Tc** (98)	44 **Ru** 101.07	45 **Rh** 102.91	46 **Pd** 106.42	47 **Ag** 107.87	48 **Cd** 112.41	49 **In** 114.82	50 **Sn** 118.71	51 **Sb** 121.76	52 **Te** 127.60	53 **I** 126.90	54 **Xe** 131.29
55 **Cs** 132.91	56 **Ba** 137.33	57-71 *	72 **Hf** 178.49	73 **Ta** 180.95	74 **W** 183.84	75 **Re** 186.21	76 **Os** 190.23	77 **Ir** 192.22	78 **Pt** 195.08	79 **Au** 196.97	80 **Hg** 200.59	81 **Tl** 204.38	82 **Pb** 207.2	83 **Bi** 208.98	84 **Po** (209)	85 **At** (210)	86 **Rn** (222)
87 **Fr** (223)	88 **Ra** (226)	89-103 #	104 **Rf** (261)	105 **Db** (262)	106 **Sg** (266)	107 **Bh** (264)	108 **Hs** (270)	109 **Mt** (268)	110 **Ds** (281)	111 **Rg** (272)	112 **Uub** (285)	113 **Uut** (284)	114 **Uuq** (289)	115 **Uup** (288)	116 **Uuh** (291)	117 **Uus**	118 **Uuo** (294)

* Lanthanide series	57 **La** 138.91	58 **Ce** 140.12	59 **Pr** 140.91	60 **Nd** 144.24	61 **Pm** (145)	62 **Sm** 150.36	63 **Eu** 151.96	64 **Gd** 157.25	65 **Tb** 158.93	66 **Dy** 162.50	67 **Ho** 164.93	68 **Er** 167.26	69 **Tm** 168.93	70 **Yb** 173.05	71 **Lu** 174.97
# Actinide series	89 **Ac** (227)	90 **Th** 232.04	91 **Pa** 231.04	92 **U** 238.03	93 **Np** (237)	94 **Pu** (244)	95 **Am** (243)	96 **Cm** (247)	97 **Bk** (247)	98 **Cf** (251)	99 **Es** (252)	100 **Fm** (257)	101 **Md** (258)	102 **No** (259)	103 **Lr** (262)

Figure 19.1 Periodic table of the elements. From Department of Chemistry, Queen Mary University of London.

atomic model should continuously emit radiation, thereby losing energy and quickly collapse into the nucleus under the opposite charge attraction. Why this did not happen portended a momentous change in thinking, requiring a different physics for the subatomic world. The transition to the new atomic physics was begun by Bohr (also in 1913 in Rutherford' laboratory) using the quantum concept of discrete energy levels (as first proposed by Planck for radiation and Einstein for the photoelectric effect) for electron *stable* orbits with *ground state* electrons closest to the nucleus at lowest energy in an orbit that accommodated only two electrons in accordwith the *Pauli exclusion principle*.* Elements with more electrons distributed them in a build-up (*aufbau*) according to their increasing energies away from the nucleus, settling into *orbitals* of *allowed* energies that constituted *subshells* that in turn made up *shells*, including an outermost shell inhabited by the final *valency* electrons.

In the Bohr atom of early quantum mechanics, the conductivity of an element is determined by the number of electrons in the outermost shell of the system. In fact, the horizontal sequence of columns of elements arranged by atomic number in the Periodic Table just follows the number of electrons in the outermost shell. Of course this is not just happy coincidence; the number of electrons in the outermost shell determines the chemical behavior of the element. A few electrons in the outermost shell gives them freedom of movement, giving rise to conductivity tendencies, and conversely full complements of electrons in the outer shell will offer little room for maneuver, resulting in more insulator-like tendencies.

Following Bohr's quantum number rules, the outer shell's full complement is eight electrons. The elements with the highest conductivity are, for example, copper (Cu), silver (Ag), and gold (Au), which all indeed have only one outermost shell electron (refer to column 11 of the Periodic Table in Figure 19.1). And the elements of the highest resistivity (insulators), for example, in column 18, neon (Ne), argon (Ar), and krypton (Kr), indeed have eight electrons in the outer shell (helium [He] does not have sufficient electrons to reach eight). By convention, elements having fewer than three outer shell electrons are designated as conductors, and those having more than five outer shell electrons are deemed insulators.

Common sand, found virtually everywhere on Earth, contains the element silicon (Si), which has four electrons in its outer shell, and another important

* The Pauli exclusion principle states that no two electrons in an atom may have an identical set of quantum numbers, or two electrons with the same spin may not occupy the same point at the same time.

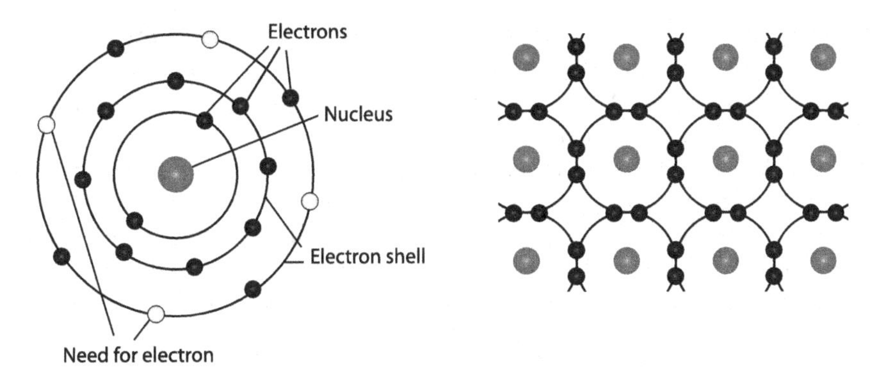

Figure 19.2 Stable insulating silicon lattice.

semiconductor material germanium (Ge) also has four electrons in its outer shell; so Si and Ge are right between the conductor and insulator designations. In the silicon crystal lattice, owing to the tendency of atoms to form the most stable lattice of eight electrons in the outer shell, the four neighboring silicon atoms will naturally aggregate so that they each share an electron, thus each silicon atom will combine with four neighboring silicon atoms to form an interlocking eight-electron full-shell silicon lattice. Since such a lattice structure has no free electrons, the structure is a very stable insulating silicon lattice, as shown schematically in Figure 19.2.

The atoms in the crystal lattice are clearly not isolated from one another, and the chemical bond among them implies that the electrons in neighboring atoms can exchange places (or in quantum-mechanical terms, their precise position and momentum [or energy and time] cannot be simultaneously determined). Because of that, the formerly sharply defined atomic energy states (as determined by the electrons' shell in Bohr's theory) are in actuality broadened to become *energy bands* that describe the energy distribution in the crystal as a whole. The lower energy band is called the *valence band* and electrons in it, owing to thermal lattice vibrations, can jump to a higher energy upper band, called the *conduction band*. Since there are fewer electrons in the conduction band (because the energies are higher), the electrons are freer to flow under an applied field to constitute an electric current. The quantum-mechanically allowed energy states constitute the valence and conduction bands, and there are certain energy states that are not allowed between the allowed states, and these form a *band gap* (also called variously, *energy gap* or *forbidden gap*). This band gap is a critical characteristic of a semiconductor material that determines its electrical functionality.* It turns out that silicon has an ideal band gap (1.94×10^{-19} J, pure Si) for transistor functions, while that of germanium is slightly too narrow, making it less useful for electronics than silicon.

The stable silicon lattice order produced by the sharing of electrons can however be disrupted by the invasion of an appropriate foreign element called a *dopant*. If a dopant having five outer shell electrons, for example, phosphorus (P) or arsenic (As), is added to a silicon atom with four outer shell electrons, the resulting nine-electron outer-orbit combination will result in one electron with no place in the orbit as shown in Figure 19.3. In other words, because the attractive force of the nucleus on the dopant atoms' extra electron is smaller than the lattice strength of the combination, that electron will be free to move about the lattice.

One of the inventors of the transistor, William Shockley, came up with a cogent analogy, likening the free electron in silicon doped with phosphorus or arsenic as a car driving around looking for a space in a parking lot. As Figure 19.3 shows, since the free electron is not bound by the phosphorus or arsenic atoms nor the silicon lattice, it is free to move, and with the application of just a small external electric field, the free electrons can form a sub-

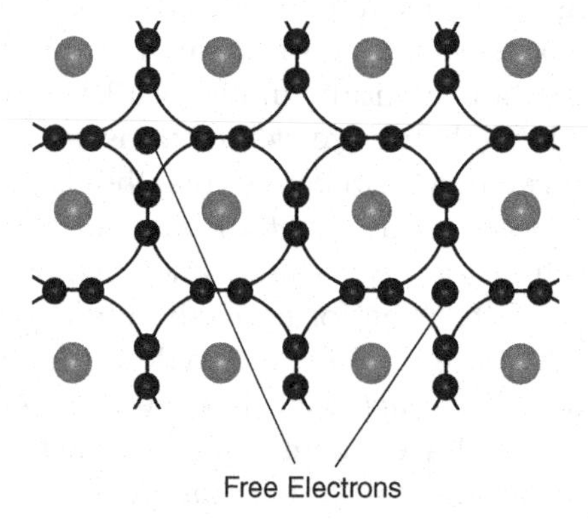

Figure 19.3 Silicon atom and five outer shell electron dopant resulting in a free electron.

* Heisenberg's uncertainty principle states that momentum and position cannot be determined with complete precision simultaneously; the same for the other conjugate pair, energy and time. The *allowed states* can be calculated by the Schrodinger equation of modern quantum mechanics, and they determine the *orbital* energies of the electrons. For many-electron atoms, particularly the transition metals, the *aufbau* gets quite complicated.

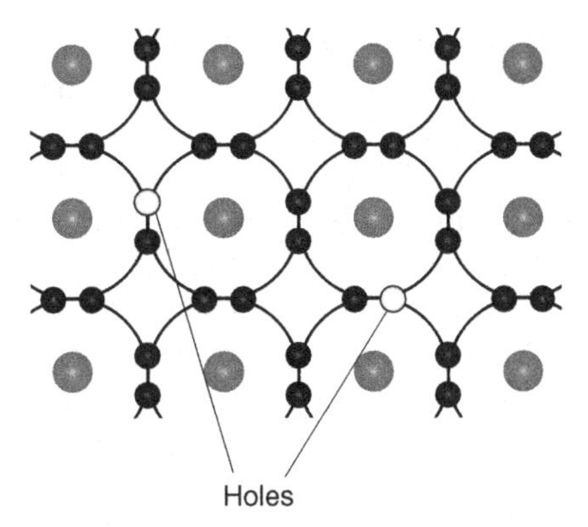

Figure 19.4 Silicon atom and three outer shell electron dopant resulting in a hole where the eighth electron would be.

stantial current of negative electrons rushing to the positive pole of the field. So, the originally nonconducting silicon lattice has become a "semiconductor" with the addition of the dopant, and because the charge carrier is a negatively charged electron, this type of semiconductor is called an *n-type semiconductor.*

Conversely, if the doping element only has three electrons in its outer shell, for example, boron (B), the combination with the silicon atoms will result in only seven electrons in the outer shell, with an empty space where the eighth electron would be, as shown in Figure 19.4. This positive "hole" will attract the negatively charged silicon electrons surrounding it, and once a neighboring electron moves over into the hole, its previous position now becomes a new hole. Continuing in this manner, the holes can be seen as a movement through the boron-doped silicon lattice of positive charge. Actually, the movement of the electrons in this case could also be used as the current carrier, but since there are far more electrons in this configuration, it is easier to calculate the hole movement, so it is the holes that are seen as the charge carriers.

In this case, Shockley's analogy is a full parking lot except for a single space, and the car beside the space moves into it, leaving the space it previously occupied empty. Of course a real parking lot doesn't operate in this way, but in the electron world, because the neighboring car is negatively-charged, it is most strongly attracted by the positive hole next to it, so it moves into the space. So continuing inductively, although it is the negative car that is moving, from a bird's eye view, it can also be the positive space that is moving in the opposite direction, and just as in the case of the electrons above, the hole being neither bound by the boron atom nor the silicon lattice, the hole becomes a "free hole." Applying an electric field will then supply the energy to cause a "hole current" to start to flow towards the negative electrode, producing a substantial positive current (the holes constitute a positive current because the negative electron has left and the charges of the other nearby electrons is not sufficient to counter the positive charge of the nucleus). The result is that once again the originally insulating silicon lattice has become a semiconductor by dint of the doping of boron. Because it is the positive holes that are the current carriers, this is a positive charge carrier semiconductor, and thus is called a *p-type semiconductor*.

Now if two different kinds of dopant are added to the silicon lattice, it is not surprising that there will be both free electrons and free holes, and depending on the relative concentration and effect of the dopants, one of the electrons or holes will be the majority carrier, with the other the minority carrier. Upon the application of an electric field, the electrons and holes will move in opposite directions. Although the minority carriers are always a significant minority, as shall be seen below, they play a major role in the operation of a transistor.

The Point Contact Transistor

The electron and hole jumping and flowing routine can be described theoretically by a quantum mechanical probability function, that is, that the electron (or hole) will jump and move is governed by a probability wave function indicating the probable position of the electron (or hole) and its subsequent traverse through the silicon lattice. The quantum mechanical calculation predicts that just as described above, upon application of an electric field, there should be a substantial negative current flow of electrons (or a positive current flow of holes). However, experiments confoundingly never showed the theoretically predicted current flow.

It was at this time that John Bardeen, freshly arrived at America's Bell Labs, came to study the problem and surmised that since the mobility of the fast, light electrons was much higher than the mobility of the relatively heavy and slow ions that had just lost the electrons, the spatial probability function of the electrons was much greater than that of the ions, and when an electric field is applied, this greater mobility "somehow" caused the electrons to bunch up near the surface of the semiconductor, forming a negative charge layer that impeded the movement of other electrons, thus barring current flow through and along the surface. The external field would perversely cause electrons to rapidly congregate at the layer, producing a huge *negative charge inversion layer* (also more simply called a *surface state* in experimental jargon) that blocked any other current flow. But although Bardeen's theory could explain why there was no current, there remained the practical problem of just how to free the electron current from the surface state, and further, how to control that current flow and hopefully produce a current gain.

The "somehow" of the formation of the inversion layer was deduced by the Russian physicist Igor Tamm. Tamm's hypothesis was that the surface state was created because when the semiconductor material was sliced to form a wafer, this caused the formation of molecular *dangling bonds*. The dangling bonds captured free electrons, thereby presenting a negative repulsive force holding the other free electrons at bay near the sliced surface of the semiconductor, and thus the inversion layer was formed.

Based on the Tamm hypothesis, Bardeen and his experimentalist colleague Walter Brattain proceeded to carefully place a spring-controlled, highly conducting gold foil *point contact* at the surface of the semiconductor, thereby successfully establishing a conducting channel to break the current barrier caused by the dangling bonds' accumulation of negative charges; the electrons could then freely flow through the semiconductor.* Because Bardeen and Brattain's point contact could produce and control current gain, this small point at the instant of contact could be said to be the point in time that ushered in the great new age of microelectronics.

Indeed, the date of Bardeen and Brattain's discovery, December 16, 1947, is now recognized as the date of the invention of the transistor. The world's first transistor, as pictured in Figure 19.5 was the *point-contact transistor*, and it first used a p-type silicon substrate and an n-type germanium inversion layer with a conducting point-contact gate. Later, an n-type germanium and p-type inversion layer, as shown schematically Figure 19.6, was more commonly used. At an operating frequency of one thousand Hertz, the world's first transistor could produce a 40× power gain and a one hundred-fold voltage gain.

The mission of Bell Labs was to research products for use in electronic communications, so the patent application for the transistor took precedence

* The positively biased point contact is actually injecting holes into the semiconductor, and the holes are being swept up by the negatively biased "collector" substrate, resulting in a current. The many different theories and researches leading to the invention of the transistor are not described here; Bardeen and Brattain's invention was like many other seminal scientific events a culmination of the work of many others as well.

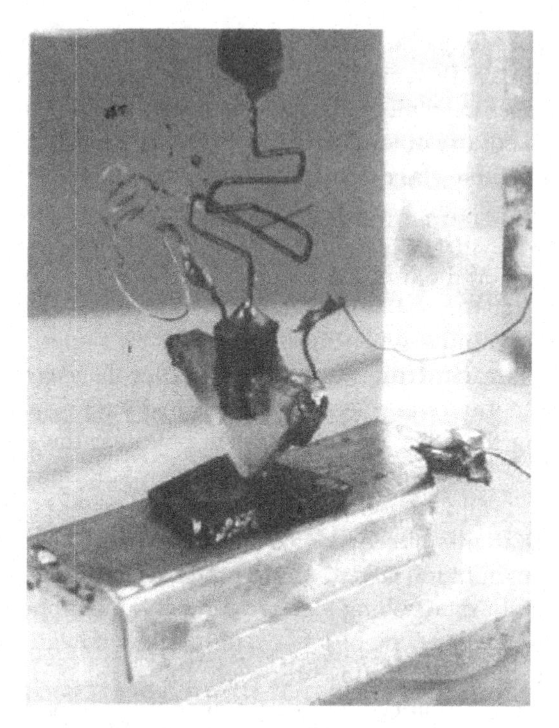

Figure 19.5 Bell Labs photo of the world's first transistor, the Bardeen–Brattain point-contact transistor.

over any scholarly scientific report, and since the transistor was declared "top secret" by Bell management, Shockley, the leader of the small research group, was ordered not to disclose any information about the transistor invention until the patent was filed. But when the irascible Shockley found out that his name was not on the list of inventors in the patent application, his extreme displeasure, manifested in gratuitous insults and belittling, fomented an irredeemable conflict with the gentlemanly Bardeen* and the unaffected Brattain.

But Shockley's anger also fomented his formidable "will to think," a fierce desire that propelled him to invent a more useful transistor that was to become the basis of all of today's microelectronic devices.

---

* John Bardeen is so far the only man to have won two Physics Nobel Prizes (the other is for his theory of superconductivity, which was alluded to in Chapter 12 with regard to de Gennes' smectic liquid crystal theories).

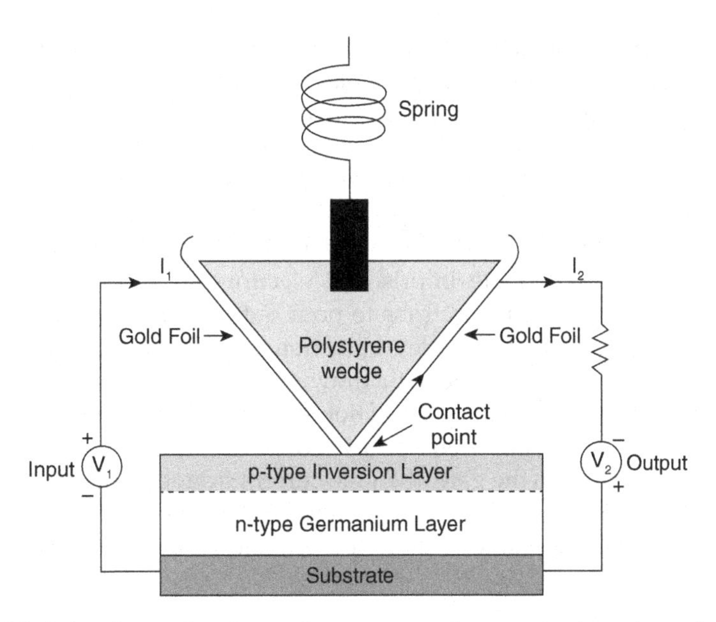

Figure 19.6 A schematic of an n-type germanium and p-type inversion layer point-contact transistor.

The Junction Transistor

Although Bardeen and Brattain's point contact transistor worked, there were practical problems of mass manufacture, primarily due to the point contact's fragility, and when placed in a circuit, it was difficult to control the delicate spring; so before it could obsolesce the vacuum tube, the transistor would have to be redesigned for robustness.

After his perceived humbling by Bardeen and Brattain, the recalcitrant Shockley marshaled his formidable will to think to attack the practical problems. It could be said that his real motivation was to show up Bardeen rather than to advance science and technology, but whatever the driving force, he worked day and night in feverish haste, and after only two weeks on New Year's Eve 1947, he broke through.

It was a sandwich. Specifically a gallium (atomic number 31, p-type) and antimony (atomic number 51, n-type) dopant with a germanium base (atomic number 32, n-type) forming a "p-n-p junction." The sandwich configuration eliminated the problems of manufacture, spring control, and fragility of the point contact transistor. The minority carriers at the junction interface acted as Brattain's point contact; that is, the transistor function was performed by the small number of minority carrier holes from the p-layer, which infiltrated the n-layer and opened the door of the inversion layer, allowing the trapped electrons in the inversion layer to pour out, generating a huge current gain.

A rigorous treatment of the physics of the junction transistor requires solid state quantum mechanics, but in a simplistic analogy, Shockley's minority carriers played the role of Bardeen and Brattain's point contact, and the "shock troops" rescued the imprisoned electrons by breaking down the surface barrier, freeing the electrons to pour out in a liberated current flow. Figure 19.7 is a picture of Shockley's junction transistor experimental apparatus, and Figure 19.8 is a schematic drawing of its operating principle.

The upper layer of Shockley's junction transistor is an n-type source, the bottom layer is an n-type drain and in-between is a p-type gate. If there is no liberating current to the gate, then the current from the source is blocked, and the drain has no (or very little) output current. If there is a liberating current to the gate, the gate will be opened, and a current, amplified by a

Figure 19.7 Bell Labs photo of Shockley's junction transistor experimental apparatus.

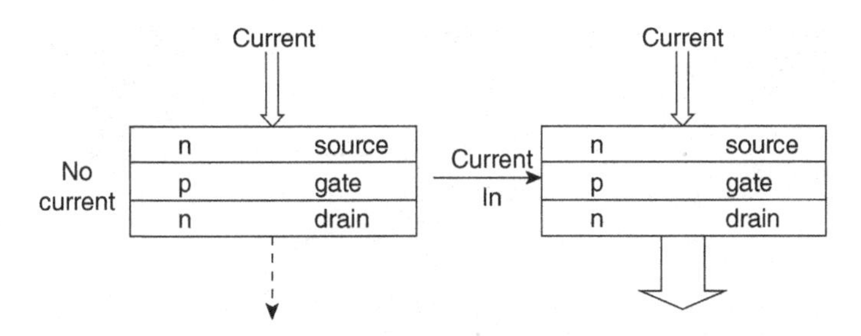

Figure 19.8 Junction transistor operation schematic.

cascade of newly liberated electrons, will flow from the source to the drain; furthermore, the amount of liberating gate current is proportional to the amount of current to the drain, so the degree of amplification can be controlled. Just like that, Shockley's junction transistor could turn a current on and off and control its size by changing a small gate current; that is, his very small, cool transistor could take the place of all the big, fat, hot vacuum tubes in any electronic circuit.

The transistor could not only be a substitute for amplifying vacuum tubes, the tiny transistor's hundred-fold amplification of a signal also can provide high-speed switching that would be essential for new high-performance electronic devices, in particular, the miniature, binary microsecond switching necessary for the rapidly developing electronic computer, and especially the small, fast analog to digital precision voltage control required by the liquid crystal display. Both would depend on the cool, small, fast, and dependable semiconductor transistor for their further development.

Today, all transistors, including the multipurpose integrated circuits and the thin-film transistor, are descendants of Bardeen and Brattain's point-contact and Shockley's junction transistors. The historical invention of the transistor was recognized in their joint award of the Nobel Prize for Physics in 1956, and commemorated in the Bell Labs publicity photo showing Shockley (seated) with Bardeen and Brattain behind him (Figure 19.9). Shockley, as the group leader, arranged the poses and is actually sitting at Brattain's lab table; the modest Bardeen and the affable Brattain acquiesced.

After inventing the junction transistor, Shockley left Bell Labs to return to his family home in Northern California's Santa Clara Valley. He was bent on establishing his own company to develop his semiconductor technology, and true to his self-aggrandizing nature, the company he founded was duly named *Shockley Semiconductors*. Shockley, riding on his personal (and Bell

Figure 19.9 Bell Labs publicity photo showing Shockley (seated) with Bardeen and Brattain behind.

Labs') repute, was able to enlist the best and brightest from universities and companies all over the world, including among them the would-be founders of Intel, Robert Noyce and Gordon Moore. But despite the wealth of talent (or because of it), with a man of Shockley's temperament as boss, the company's fate was predictable. Not long after their arrival, unable to bear his overbearing presence, Noyce, Moore, and six other "traitorous eight" researchers left to form their own company in Santa Clara; funded by the East Coast defense supplier Fairchild Camera, and duly named Fairchild Semiconductor. Shockley Semiconductors struggled a bit and then gave up the ghost, closing its doors for good the next year [1].

Apparently in yet another example of a man of surpassing intelligence failing as a corporate boss, Shockley demonstrated that analytical ability by itself is not sufficient for business success; commercial acumen, people skills, and at least some semblance of an emotional quotient are just as important as a high intelligence quotient. But a Nobel Prize winner, even with the foibles of a William Shockley, should not have to worry about finding a job. Not long after the demise of the eponymous company, Shockley was offered and accepted a professorship in Stanford University's electrical engineering department. However, a tiger could easier change its stripes than Shockley avoid controversy. His days at Stanford were fraught with acrimony and contention, particularly in regard to his racial superiority theories, to the extent of numerous student and faculty protests against him personally and his espoused theories intellectually.

But Shockley's other influence was nevertheless momentous, for he founded the first semiconductor company in what is today Silicon Valley, named as it is for the material of semiconductors. Thus, although himself an abysmal business failure, the companies that followed, including Fairchild, Intel, Hewlett-Packard, and many others mushroomed throughout the Valley, rapidly growing in size and influence, and quickly transforming the fruit orchards into the Silicon Valley that many countries have sought to emulate as both a progenitor and symbol of an advanced society.

The Tyranny of Numbers

The earliest all-transistor radios used germanium transistors, but although germanium is relatively easy to work, because of its small band gap, higher temperatures caused more electrons to jump to the conduction band, which led to more current and even higher temperatures, creating a positive feedback phenomenon called *thermal runaway* that rendered germanium unsuitable for most commercial electronics use. Silicon, on the other hand, has a wider band gap (but not so wide as to inhibit conduction), and so is sufficiently heat-worthy to avoid thermal runaway. But because a pure silicon crystal is hard to grow and brittle by nature; its natural rock-like crystalline structure, shown in Figure 19.10, while stable, can be easily sheared along the crystal lattice boundaries as shown in the silicon lattice cluster of Figure 19.11; the manufacturing of a silicon transistor was fraught with difficulty.

In 1954, Texas Instruments (TI), a company originally engaged in the production of geophysical instrumentation for oil exploration that later transformed itself into a manufacturer of precision instruments, being familiar with minerals and measuring their properties, was perfectly placed to achieve the process breakthrough. After only a few years of process development, TI overcame the crystal growth and brittleness problems, and succeeded in cost-effectively producing silicon-based transistors in quantity.

With the rapid development of Shockley's junction transistor technology and TI's great quantities of inexpensive silicon transistors, many formerly only conceptually possible new products were now suddenly feasible. But just as the display of even simple words required multiplexing connections as described in Chapter 17, the new electronic products with many more

Figure 19.10 Rocks of natural silicon. From cpushack.net, courtesy of J. Culver.

components and concomitantly more complex wiring schemes could not avoid an embrangling web that was the dreaded hot wire "spaghetti" coursing through the apparatus. The prime feature of small size ironically made transistors much harder to connect than vacuum tubes, and the labyrinthine wire connections among the tiny components required thousands of delicate and difficult soldering operations. The upshot was that although many brand new products were now possible, the sheer number and convolutions of the component connections within those new products made assembly long and arduous, subjugating the advance of new products to what engineers at the time called the "tyranny of numbers."

In particular, the rapidly developing new technology of electronic computing machines, yearning with the prospects brought by the transistor, was shackled by this tyranny of numbers. The second generation digital computer, the first to use transistors in place of all the vacuum tubes, owing to the small size and relatively cool operating temperatures of the transistors, allowed the number of components to rise dramatically and greatly increase computing capability; for example, the most advanced computer of the time, the Control Data CD 1604, used 25,000 transistors, and 100,000 diodes, resistors, and capacitors. The number and complexity of the wire connections among these components were the mundane technological barrier to further advance. With tremendous computing capability near on the horizon, the

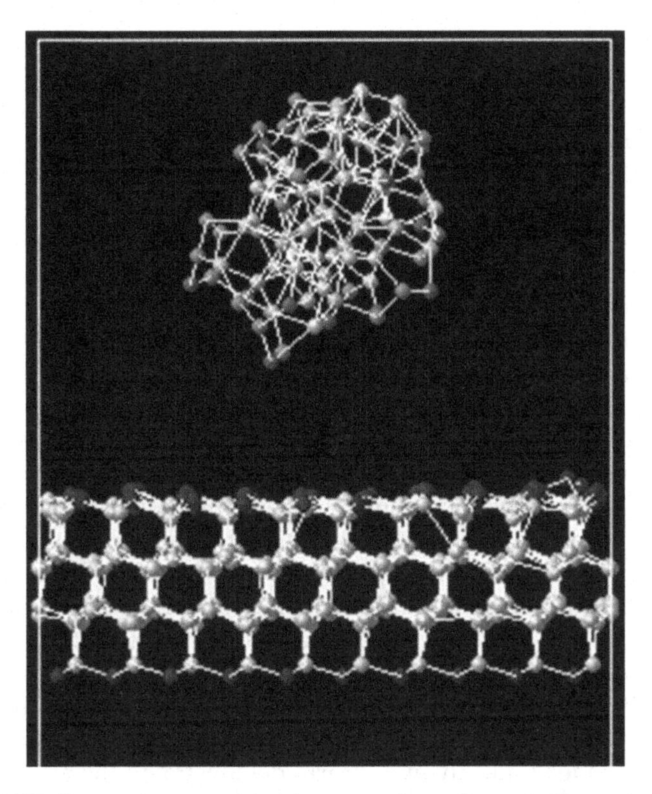

Figure 19.11 Computer graphics image of a silicon lattice cluster. From photon.t.u-tokyo.ac.jp, courtesy of Professor S. Maruyama.

computer industry desperately cried out for a siliconite hero to save the industry from the tyranny of numbers and carry it into the promised land of the new computer age.

Monolithic Component Integration

The silicon crystal lattice described earlier as very stable and electrically insulating should give a hint of the what could be done and the consequent creative thinking to come. That is, why not use the pure silicon as a resistor in a silicon semiconductor? And so in 1958, having recently arrived at TI and eager to make his mark, Jack Kilby began to think: a silicon semiconductor, because it could both conduct and resist, could naturally act as a diode, and if the p–n junction were appropriately doped, it could also act as a capacitor. The premise was, the four basic electronic circuitry components, the transistor, resistor, capacitor, and diode, could be formed on a single piece of silicon substrate. That is, with the proper manipulation of the p-n junction at various places in the silicon, all the required electronic components of a circuit could be formed on (or within) the silicon substrate. Furthermore, this would allow the circuit to be fabricated in a single process, and it could be made very, very small,

The concept of Kilby's monolithic silicon substrate appeared theoretically sound, but to achieve proof of concept, he had to build a model to turn his theory into a company practice. That is, Kilby had to persuade the bosses at TI that they should embark on an entirely new line of products based on his ideas. After hearing Kilby's technical report, his mindfully enthusiastic research group manager chose a phase-shift oscillator as the device to demonstrate Kilby's monolithic circuit. A phase-shift oscillator turns a direct current into an oscillating current by changing the phases of the signal, and when connected to an oscilloscope, the effect is very clearly demonstrated when the straight horizontal dc voltage line suddenly becomes an ac sinusoidal curve; it was an ideal choice for a demonstration circuit.

The world's first monolithic integrated circuit is shown in the photograph of Figure 19.12, and shown in Figure 19.13 is the famous "flying wire" drawing for the patent application. This homely device, hooked up to an oscilloscope, indeed quickly and clearly demonstrated the ac–dc oscillator

Figure 19.12 Texas Instruments Incorporated photo of the world's first monolithic integrated circuit.

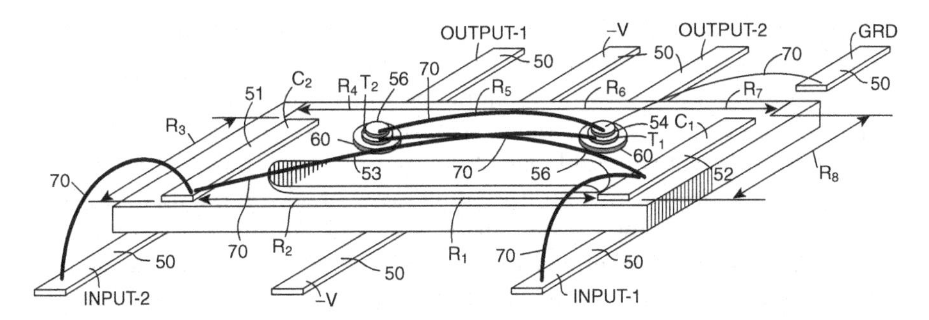

Figure 19.13 The "flying wire" drawing for the monolithic integrated circuit patent application.

effect. The knowledgeable TI bosses in gleeful surprise were easily persuaded, and the primitive flying wire prototype monolithic integrated circuit was the portent of the great new microelectronic age to come. And in that new age, Texas Instruments would become one of the great semiconductor companies of the world.

Kilby's flying wire contraption was indeed the world's first circuit device having basic electronic components integrated into a single monolithic silicon substrate. Although Kilby worked on integrating the wiring into the silicon substrate, his patent application figure did not show the connecting wires integrated into the body of the circuit because in the rush to apply for the patent, that was the only figure available at the time [2]. So the flying wires remained in the patent drawing, and Kilby's patent application thus did not disclose the solution to the problem of the tyranny of numbers.

Monolithic Circuit Integration

After leaving Shockley Semiconductors and founding Fairchild Semiconductor in 1957, while producing rather mundane transistorized devices such as oscillators, Robert Noyce often observed the fabrication process in person. The typical production process began with many small transistors being cut up from a large crystal silicon wafer, then an army of young female assemblers (only the dexterous hands of young women could handle the job) would painstakingly solder the transistors to each other and to the resistors, capacitors, and diodes to form the completed electronic device. However attractive the story would be, Noyce did not claim that he thought of a way to save these thousands of young women from a life of drudgery by observing their labors; overcoming the tyranny of numbers instead would come from the solution to another, different problem.

In the fabrication process of the junction transistor, there was a problem of particle contamination of the crystal silicon; to protect the substrate, Noyce's colleague at Fairchild, one of the traitorous eight, Jean Hoerni from Switzerland, suggested that an insulating layer of silicon dioxide (SiO_2) be deposited to seal off the crystal silicon substrate during fabrication, and further, holes in predetermined places could be cut in the silicon dioxide layer through which dopants could be injected to the silicon layers below. Fairchild implemented Hoerni's sealing step, and its efficacy was such that it was to become the standard for integrated circuit process to this day. Since the layer was formed in a plane over the silicon substrate, Hoerni's SiO_2 deposition method was called a *planar process*.

While applying for a patent for Hoerni's process, in order to write as broad a claim as possible, Fairchild's diligent patent lawyer doggedly peppered the engineers with questions about other uses for the silicon dioxide layer planar process. This led Robert Noyce to think: Why not place the connecting wiring in the silicon dioxide layer; like candles in the icing of a cake, the wires could be stuck into the SiO_2 and be both stable and insulated from each other. Gaining momentum, Noyce then thought, actually those later-added wires could be done away with altogether, the copper wiring connecting the components could be printed in a pattern on a circuit board and placed directly on the silicon dioxide layer. Preparing a printed circuit board would be much easier than soldering the wires to the components one-by-one, and the wiring circuit printed beforehand could be applied as a whole during the fabrication of the device, thus constituting only a single step for all the wiring of the entire device. Taking things further, as with Kilby, the fabrication process did not have to be limited to the silicon transistor, it could encompass all the electronic components, resistors, capacitors, and diodes, in one process, and the printed circuit could connect all the components together.

Thus the wiring among the components was formed on the substrate by a printed circuit, thereby transforming the dish of hot spaghetti into a plate of lasagna. The tyranny of numbers could be overthrown by Italian food.

In his Fairchild engineering notebook, Noyce wrote in 1959 [3],

> In many applications now it would be desirable to make multiple devices on a single piece of silicon in order to be able to make interconnections between devices as part of the manufacturing process, and thus reduce size, weight, etc., as well as cost per active element.

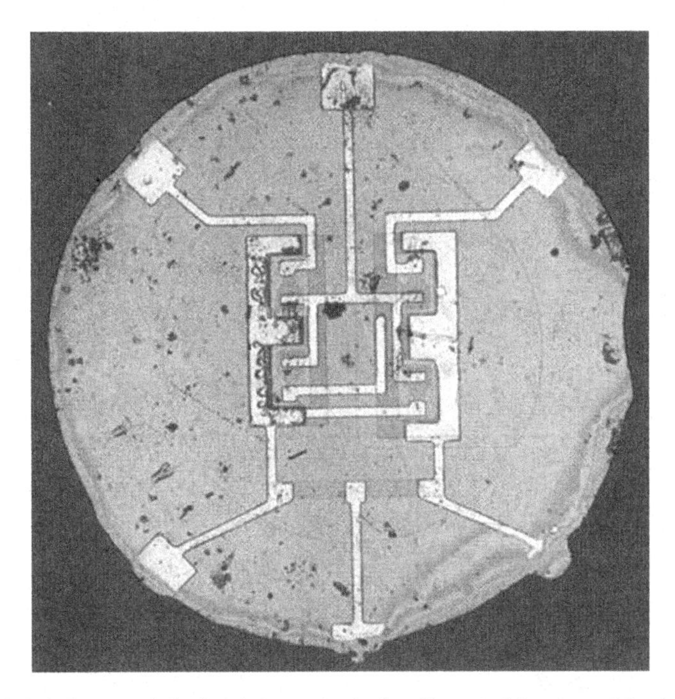

Figure 19.14 The world's first integrated circuit, a resistor-transistor logic (RTL) flip-flop. From Fairchild Semiconductor Corporation.

One could imagine the excitement that Noyce's rapid-fire creative thinking generated both within himself and throughout Fairchild Semiconductor.

Figure 19.14 is a picture of the world's first integrated circuit, a resistor-transistor logic (RTL) *flip-flop* with the embedded wiring connections clearly shown; Figure 19.15 is the schematic drawing from the patent application and the circuit diagram.

Noyce achieved circuit integration from thinking about the wiring problem, Kilby from the formation of electronic components on the silicon substrate; although the conceptual points of departure were different, they reached the same goal: the monolithic integrated circuit.

Noyce and Kilby brought together the transistor and other electronic components and their wire connections into a single silicon substrate, to form what is today the basis for literally all electronic devices and products. Their respective ideas were conceived at virtually the same time, and this precipitated what is called in patent law a *patent interference procedure* to

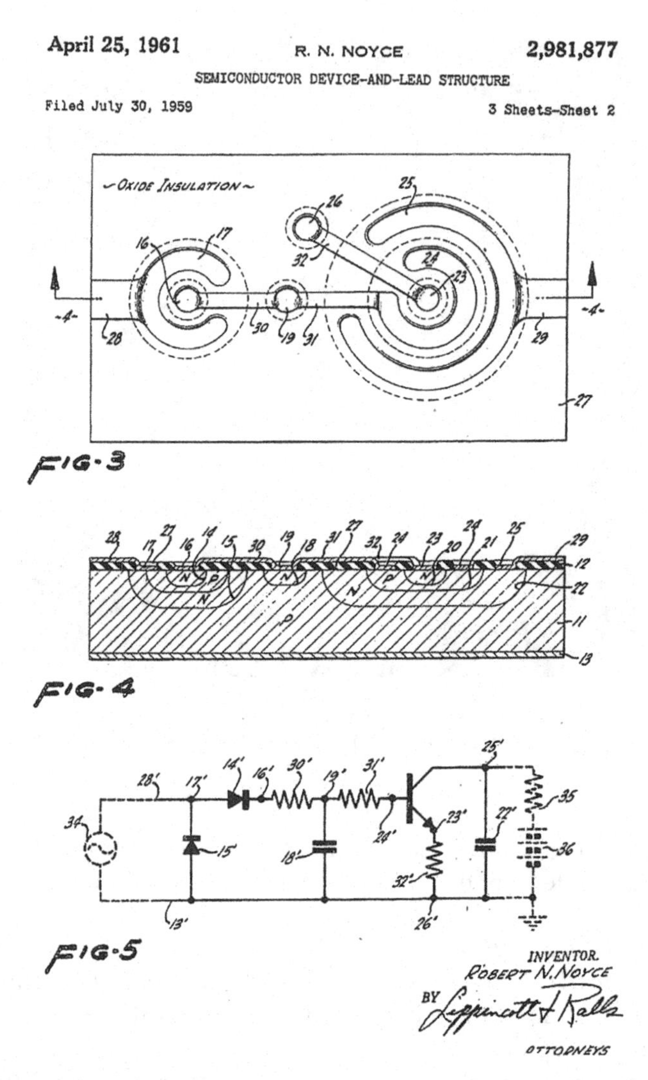

Figure 19.15 The drawing from Noyce's patent application and the accompanying circuit diagram.

determine who was first to conceive of the idea of an integrated circuit. From the purely patent law perspective, Noyce won the interference (primarily because of Kilby's flying wires), but in accord with the mutual respect and scholarly character of the principal engineer/scientists, the two companies amicably resolved the issue of prior conception through a cross-license.

But it was only Kilby who was awarded the Nobel Prize for Physics in 2000 for work on the integrated circuit. The reason is that Noyce unfortunately died in 1990, and thus was ineligible at the time of the award; however, Kilby graciously lauded Noyce and recounted his contributions to the invention of the integrated circuit in his acceptance speech.*

A little reflection about today's microelectronic world, with proper notice paid to the myriad and delightful devices emanating from the invention of the integrated circuit, would lead one to believe that the surpassing value of the integrated circuit would be evident to all. But at the time, in spite of the efforts of companies like Fairchild Semiconductor to promote it, the electronics industry actually did not particularly appreciate this great new invention. The reason was cost; at the time, many an electronic component such as a resistor or diode cost only a fraction of a cent; compared with the costs of the not yet mature integrated circuit process, economically, there was no question as to which to choose, and the integrated circuit languished on the laboratory bench for some time.

The savior of the new invention was once again the American Government. Although having no interest in consumer electronics, the cost-is-no-object Department of Defense wanted small and light electronic devices for its weapons and the Air Force's warplanes. But especially to fulfill President Kennedy's vision, proclaimed in his 1961 inaugural address, that within the decade America would put a man on the moon and bring him back safely. This of course required a spaceship with small and light electronics, and for this the integrated circuit was not only ideal, it was necessary. With the government's active funding, the fabrication process for the integrated circuit gradually matured and costs quickly decreased; mass production soon brought the economies of scale into play, and the integrated circuit finally took off (with a boost from the Saturn V rocket).

Texas Instruments, Fairchild, and its successor Intel rapidly became the pillars of the new technology. But the rich, staid old electronics brand companies, such as RCA, GE, and Westinghouse rested on their vacuum tube laurels, and just as in the case of the liquid crystal display, they all missed the new integrated circuit rocket express; was it yet another case of the paralyzing inertia inherent in hidebound giant companies?

---

* Two requirements for a Nobel Prize award are that the theory must be proven to be true and the recipient must be living. The most famous nonrecipient of a Nobel Prize in spite of a seminal work is of course Albert Einstein and his general theory of relativity, which has not yet been conclusively proven experimentally primarily because gravity has not yet been integrated with the electromagnetic, strong, and weak nuclear forces, and the graviton has so far not been detected.

References

[1] Shurkin, J. 2008. *Broken Genius*. Macmillian, New York.
[2] Reid, T.R. 2001. *The Chip*. Random House, New York.
[3] Berlin, L. 2005. *The Man Behind the Microchip*. Oxford, p. 104.

20

A Transistor for the Active Matrix

With supply burgeoning, the prices for liquid crystal display calculators and watches began a precipitous decline, and in order to make any profit at all, companies raced to upgrade existing products and develop completely new products. The new active matrix liquid crystal display (AMLCD) could provide the resolution and speed necessary for the new products, but although the concept of the active matrix was clear, the engineering problem was to find a suitable transistor to drive the liquid crystal.

Peter Brody's early research on thin-film transistors (TFTs) at Westinghouse used a cadmium selenide (CdSe) transistor, and although workable, it was extremely sensitive to variations in composition and impurities and highly toxic to boot, making it unsuitable for consumer products. Sharp later developed a tellurium (Te) TFT that could be deposited directly on the glass substrate, but its electron mobility—the average speed of electrons in an applied electric field—was so high as to have a serious current leakage problem, which caused the on/off current ratio to be far from the 10^6 requirement for display use (see Chapter 22).

Because silicon semiconductors by this time had been extensively studied for use in transistors, it was the favorite of electronics engineers everywhere, so it was natural for silicon to be a candidate for thin film transistors for

Liquid Crystal Displays, First Edition. Robert H. Chen.

liquid crystal displays. However, the transistors in calculators were made from pure crystalline silicon, and although more than capable of serving as the addressing elements for a liquid crystal display, it was difficult to grow the single crystal in a thin film on a glass substrate. This made commercial mass production of larger screens impracticable, so pure crystal silicon displays were limited to the occasional appearance of expensive miniature artifacts. For example, a Seiko pure crystalline silicon transistor wrist television (shown in Figure 20.1) made its appearance as worn by James Bond in the movie *Octupussy*.

The research effort then proceeded naturally to *polycrystalline* silicon, whose electron mobility, albeit lower than that of single crystalline silicon, still should be sufficient for display addressing use. As early as 1983, Seiko surprisingly produced a 2.13-inch polysilicon flat-panel liquid crystal color television, and in only 4 years, more than three million were sold, indicating a palpable demand. Shortly thereafter, Sharp produced a polycrystalline

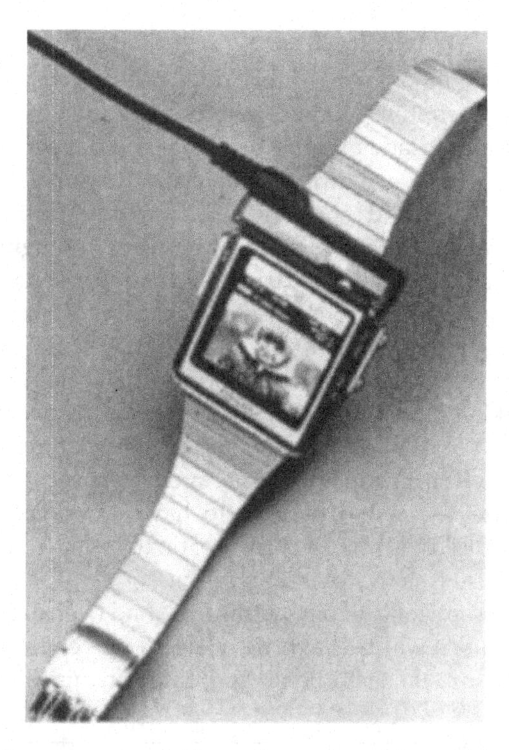

Figure 20.1 Seiko pure crystalline silicon transistor wrist television. From Kawamoto, H. 2002. *Proc. IEEE,* **90**, 4, 494, April.

Figure 20.2 Sharp polycrystalline transistor wall-hanging liquid crystal color televisions. From Kawamoto, H. 2002. *Proc. IEEE,* **90**, 4, 496, April.

transistor liquid crystal color television that could be hung on the wall (shown in Figure 20.2). It appeared that Heilmeier's vision of a large-screen, wall-mounted flat panel television was finally at hand.*

But it was not to be, at least not yet. Polysilicon deposition requires processing temperatures of over 600°C, much too high for deposition on glass, which would melt at those temperatures. Substituting quartz substrates would allow the high-temperature thin-film deposition of polysilicon, but because of the difficulties of maintaining material uniformity over large surfaces made of quartz and its high cost, the polysilicon liquid crystal display was limited to small screen applications.†

So now the engineering problem devolved to finding a semiconductor material that possessed sufficient but not too high electron mobility, and that

* Castellano in his book *Liquid Gold,* believes that James Bond movies actually hindered the development of the liquid crystal display industry because in the movie *Live and Let Die,* Bond wore a fancy light-emitting diode (LED) watch, causing many companies to switch from LCD to LED watch manufacture. In the long run, however, because the LED watch required pressing a button for display to save energy, and even then its batteries only lasted a few months, LCDs became the dominant digital watch display.

† In the 1990s, a low-temperature polysilicon process (LTPS) was developed, and screen sizes could be increased, but uses are still limited to small- and medium-size displays such as mobile phones and navigational aid screens.

further could be deposited in thin films on the glass substrates holding the liquid crystal at temperatures that would not melt the glass.

Hydrogenated Amorphous Silicon

In early 1969 while performing experiments on silicon semiconductors, Britain's Standard Communications Laboratory researchers were in the process of vapor depositing crystalline silicon using a silane gas (SiH_4), when they noted a glassy, brown by-product congealing in globs around the apparatus. Since the globs had little or no crystalline structure but nonetheless were still silicon, they were named *amorphous silicon* (a-Si), "amorphous" meaning "having no determinant shape," and in mineralogy and chemistry, meaning "not composed of crystals in physical structure" [1]. The amorphous character can be analogized to a random pile of matchsticks that compared with the orderly rows of matchsticks in a matchbox, are disordered with little or no structure, but are nevertheless still matchsticks.

The disorderly amorphous structures maintained their molecular silicon structure, but there were voids at grain boundaries within and at the surfaces in lieu of atoms, as shown schematically in Figure 20.3. Just as in the surface

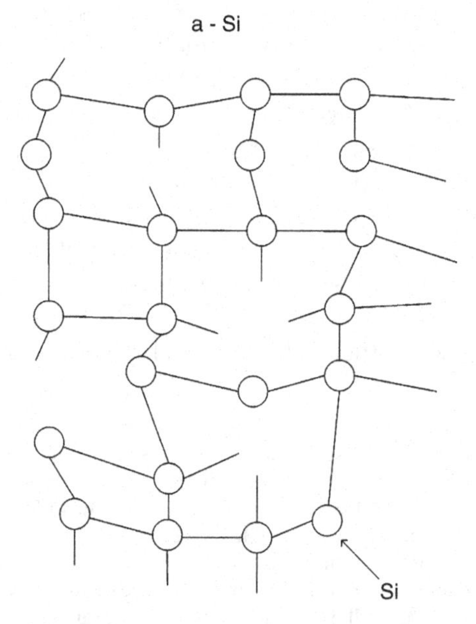

Figure 20.3 Amorphous silicon dangling bonds.

effect of the early transistors, these quaintly named *dangling bonds* would capture free electrons and thereby impede the flow of electrons through the material. This caused the electron mobility of amorphous silicon to be much lower than that for pure crystalline silicon. The low electron mobility of course cast doubt as to the efficacy of using amorphous silicon for transistors, but its ease of preparation and underlying potential were intriguingly attractive.

Albeit disordered, amorphous silicon nonetheless was silicon, so hope for greater electron mobility, while dim, was tangible. The eccentric American inventor Stanford Ovshinski earlier had used amorphous silicon in solar cells, thus demonstrating that while low, the electron mobility was still sufficient to allow a current to be generated by sunlight through the photoelectric effect. Several years later Peter LeComber and Walter Spear at Dundee University in Scotland, known for their work on *slow* electrons, were given a sample of the amorphous silicon from the Standard Communications Laboratory for study as a specimen of slow electron material.

LeComber and Spear found that just as in silicon semiconductors, doping the amorphous silicon would allow it to display semiconductor characteristics. In the hope that doping the a-Si with light atoms would terminate the dangling bonds of the amorphous silicon, they tried hydrogen gas as a doping agent, and as shown schematically in the dangling bonds of Figure 20.3, the black dots of Figure 20.4 representing the hydrogen atoms did just that.

The hydrogen atoms closed off the dangling bonds, rendering them unable to trap free electrons, and those electrons would then flow through the material unimpeded, resulting in a much higher electron mobility. In fact, the electron mobility was high enough to allow the amorphous silicon to be doped to exhibit the all-important *transistor field effect* necessary for liquid crystal display switching.

The Field Effect Transistor

As the key concept behind the invention of the transistor was the point contact or junction injection of minority carriers, so the movement of the charge carriers (electrons or holes) near the point contact or the p-n junction is fundamental to transistor operation.

First off and at the very outset, Shockley believed that at the p–n junction, the injected charge carriers were not confined to the surface between the n- and p-type semiconductors (the *surface effect*), but could diffuse through

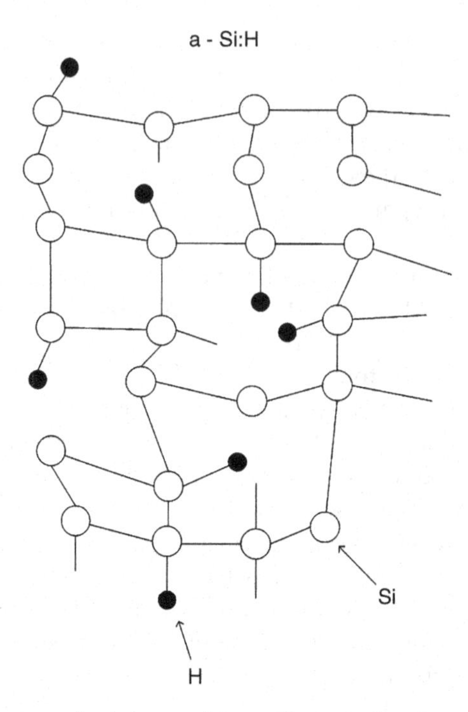

Figure 20.4 Hydrogenated amorphous silicon with dangling bonds terminated by hydrogen atoms.

the body of the semiconductor. This was confirmed experimentally by John Shive at Bell Labs simply by placing wires at the opposite ends of the semiconductor body and measuring the current. So if an n-type and a p-type semiconductor are placed in contact, what happens to the charges in the vicinity of the junction between them, and just how this juxtaposition influences the motions of the electrons in the semiconductor body is the basic question to be posed and answered to explain field-effect transistor action.

Near the n-type surface, there will be an excess of holes from the p-type, and, vice-versa, an excess of electrons from the n-type near the interface in the p-type; so the electrons and holes diffuse through the interface in opposite directions, electrons from the n-side to the p-side and holes from the p-side to the n-side, with the concentration gradient generating the charge carrier diffusion. In the p-side, the invading electrons rapidly recombine with the local majority holes, thereby removing holes in the vicinity of the junction, resulting in a region of negatively charged atoms. Similarly, on the n-side, there will be a region of positively charged atoms owing to the invasion of the holes. This then creates an electric dipole field, which opposes the diffusion of electrons and holes by engendering an electron and hole apposite *drift*, this time with the electric dipole field causing the charged carrier drift [2].

A thermal equilibrium between the diffusion current and the drift current is quickly established, and although individual electrons and holes are still diffusing and drifting, the equilibrium means that there is no net current across the junction. The overall result of the diffusion and drift of charge carriers, however, is that there will be a depletion of free charge carriers (either free electrons or free holes) in the regions near the junction; this is the so-called *depletion region* that forms a positive space charge on the n-side and a negative space charge on the p-side. The space charge separation thus inherently acts just like a capacitor with a voltage across it (and this is why Kilby and Noyce could make capacitors as well as transistors and resistors in their integrated circuits).

Now if a voltage is applied in the *forward* direction (making the n-side negative and the p-side positive), the voltage is reduced across the junction, and because the electric field is thereby rendered weaker, the drift current is reduced, but since there is no effect on the concentration gradient of charge carriers, the diffusion current is unaffected, and now constitutes a net current in the form of a minority carrier current across the junction. The electric field is weaker across the junction, which means that the charge in the depletion region must be reduced, causing a narrowing of the depletion region, and the current consequently flows more freely through the semiconductor; this forward-biased p–n junction is performing a *bipolar* transistor function. If a reverse direction voltage (*reverse-bias*) is applied, it opposes the diffusion current and the electric field across the junction increases, so the depletion region expands, thereby inhibiting the flow of current.

A typical present-day *junction field-effect transistor* (JFET), shown schematically in Figure 20.5, has a lightly doped n-type layer formed on top of a

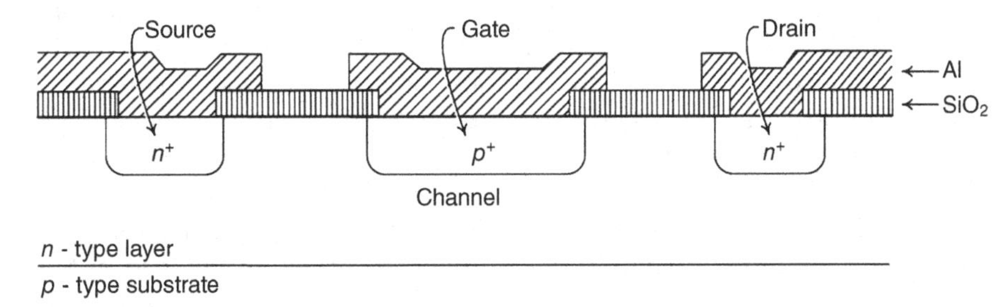

Figure 20.5 Schematic of a junction field-effect transistor (JFET).

p-type substrate. Two heavily doped n-type regions (denoted by "n^{+}") are added by diffusion, which form the source and drain of the transistor, and are in good ohmic contact with a so-called *channel region* between them. "Good ohmic contact" means that "the contact itself offers negligible resistance to current flow when compared to the bulk," and also refers to a linear relation between the voltage and the current (Ohm's law $V = IR$), and that the resistance of the majority charge carriers between the source and gate is essentially zero. This means that the voltage drop across ohmic contacts is negligible compared with voltage drops in nonohmic contacts elsewhere in the device, and thus no power is dissipated in the contacts, and the ohmic contacts are at thermal equilibrium even when currents are flowing. The consequences of all this is that all carrier densities at an ohmic contact are unchanged by current flow, that is, the densities remain at their thermal equilibrium values at good ohmic contacts.

The source provides majority carriers to the channel, so the n-channel JFET (NMOS) conventional current is from drain to source, and in a p-channel JFET (PMOS, replace the *n* with *p* in Figure 20.5), the conventional current is reversed as source to drain.

Referring to Figure 20.5, the borders of the channel region are defined from above as the depletion region below the gate, and from below as the depletion region at the substrate p-n junction. A reverse-biased p^{+} region above the channel then can act as a gate. That is, if the p-substrate is set at ground potential (the usual condition), and if the drain is positively biased, current will flow from the drain through the channel to the source, but if the source is now grounded and a negative potential (the reverse bias) is now applied to the p^{+} gate, the depletion region in the *pn*-junction region will expand and the channel region will narrow, causing its resistance to increase, thereby decreasing the current from the drain to the source. Therefore, a reverse bias signal to the p^{+} gate controls the current flowing through the channel by means of an electric field; this is the fundamental operating principle of a *field-effect transistor* (FET) [3].

The metal-oxide semiconductor field-effect transistor (MOSFET) quickly became the standard in the electronics industry. MOSFETs are particularly favored because of their small size and low power requirements compared with the bipolar transistor, and these attributes were further enhanced by the development in the late 1960s of the *complementary metal-oxide semiconductor* (CMOS).

The CMOS paired an NMOS and a PMOS transistor; the second transistor has its gate connected directly to its drain and is placed in the load of the first transistor. This then forms a switch that uses power only when activated and otherwise consumes no power, ideal for the liquid crystal display addressing system, as described in Chapter 17. Switching speed depends primarily on the length of the channel region, therefore the smaller the transistor, the faster it is, and although the bipolar transistor is capable of faster switching speeds, the CMOS switching speed is about 10^{-11} seconds, which is fast enough for a liquid crystal display.

The a-Si:H Field Effect Thin-Film Transistor

For use in liquid crystal displays, amorphous silicon is derived from the deposition of SiH_4 on a heated glass substrate to form a thin a-Si layer; the dopant, for instance, phosphorous, derived from PH_3, also can be added at the same time. The active layer, where the amplification happens, in the a-Si:H FET is made of *intrinsic hydrogenated amorphous silicon (i-a-Si:H)*, where "intrinsic" means an undoped semiconductor where the number of negative and positive charges are equal, and "hydrogenated amorphous" refers to the addition of hydrogen atoms to promote electron mobility so that the active layer can act as an amplifying field-effect transistor. A transistor fabricated from the doped amorphous silicon furthermore demonstrated a low current-on threshold and minimal current leakage (meaning that its electron mobility was not *too* high). Importantly, it also could be deposited on glass in thin film form at relatively low temperatures.

This was the breakthrough to large-screen liquid crystal television, but in 1979 LeComber and Spear announced the new hydrogenous amorphous silicon thin film transistor for *solar cells*, not for liquid crystal display addressing [4].

Three years later, when Spear was invited to Japan to speak on the amorphous silicon solar cell, near the end of his talk, he mentioned that the hydrogenous amorphous silicon thin film transistor could demonstrate a strong field effect and the current-on/off ratio was 10^6. In the audience was a Sharp engineer who, upon confirming the last sentence of the talk with Spear, hurried back to report to his supervisor and suggest that Sharp should change over to use a-Si:H instead of tellurium for the thin film transistors they were developing for liquid crystal displays. Once again demonstrating their technical acumen, Sharp immediately invited Ovshinski to Japan to talk about amorphous silicon transistors, but the subject again was solar cells, and not liquid crystal display addressing transistors.

The first global Oil Shock in 1974 dealt an inordinately cruel blow to a country that relied entirely on imported oil. In response, some electronics companies in Japan, Sharp and Sanyo among them, embarked on major solar cell research and development projects, with a goal to lessen Japan's dangerous total oil dependence, and also hopefully to exploit a new natural energy resources market. However, by 1986, the shock had long worn off, oil prices had dropped, and production and consumption had already returned to normal levels. The relatively low oil prices, as has happened time and again, dampened the market for solar cells. Following the vagaries of the market, Sharp nimbly turned its amorphous silicon development focus to liquid crystal display use, and Sharp's solar cell research experience would prove invaluable for the development of the active matrix thin film transistor liquid crystal display addressing systems.

Following in Sharp's wake, the major Japanese electronics producers Canon, Fujitsu, Toshiba, Hitachi, Sanyo, and Matsushita all developed hydrogenous amorphous silicon thin film transistor active matrix displays.

The arduous journey from cadmium selenide to tellurium and through a desert of silicon sands to the hydrogenated oasis was finally completed; the semiconductor of choice for the large-screen liquid display active matrix thin film transistors was hydrogenous amorphous silicon. In summary, the voyage was made with the following stops on the itinerary:

Cadmium Selenide → Tellurium → Crystalline Silicon → Polysilicon
→ Amorphous Silicon → Amorphous Silicon:Hydrogen

Also at this time, universities in France, England, and Germany also commenced research into amorphous silicon. In America, the big defense companies like General Electric, Rockwell-Collins, and Honeywell Aerospace also began research on military-use amorphous silicon liquid crystal displays. But in particular, the Japanese consumer electronics companies were poised to take advantage of a huge new commercial market.

Now at long last, the liquid crystal, the addressing scheme, and the thin film transistors were all developed, and the new large screen displays were clear on the technological horizon. The basic science, the product prototype, and the manufacturing process are the indispensables of a successful high-technology product, and it was apparent that considerable progress had been made on the first two fronts, but the mass production of a liquid crystal display still required a standardized mask process and precision control of the glass substrate cell gap; in short, the process, fabricating machines, techniques, and know-how that are essential to the efficient production of precision electronic devices.

References

[1] *Oxford English Dictionary*. 1985. Oxford.
[2] Orton, J. 2004. *The Story of Semiconductors*. Oxford, p. 81.
[3] Muller, R.S., and Kamins, T.I. 1986. *Device Electronics for Integrated Circuits*, 2nd edition. Wiley, New York, p. 202ff.
[4] Le Comber, P.G., Spear, W.E., and Ghaith, A. 1981. Amorphous-silicon field-effect device and possible applications. *Electron Lett.*, **15**, 6, 179–181; LeComber, P.G., and Spear, W.E. 1984. *The Development of the a-Si:H Field-Effect Transistor and Its Possible Applications, Semiconductor & Semimetals, Vol. 21: Hydrogenated Amorphous Silicon*. Academic Press, New York, Chapter 6, p. 89.

21

Semiconductor Fabrication

Silicon certainly possessed the exotic requirements for effective semiconducting, but the principal reason for its widespread use in integrated electronic circuits was more mundane: silicon's own oxide, SiO_2, could be formed easily on the silicon itself to provide a stable metal oxide insulator to control electronic function. The SiO_2 layer also can shield the silicon layer below from the high-energy ion beams or high-temperature gaseous diffusion implantation of dopants in the fabrication process. That is, the superior insulating property of the SiO_2 layer could be used to control the doping of the silicon into specified areas with extraordinary precision, and in the very component-dense environment of integrated circuits, this ability to precisely control the areas of doping is critical. Further, the protection of the silicon semiconductor during the harsh fabrication process was of paramount importance for a high-quality yield. This is why the *metal oxide semiconductor field effect* transistor (the MOSFET) and its progeny, the *complementary metal oxide semiconductor* (CMOS), quickly became the flagship transistors in integrated circuits for modern electronic products.

Liquid Crystal Displays, First Edition. Robert H. Chen.

Growing Crystals

To make silicon semiconductors, high-purity crystalline silicon (less than one part per billion impurity) is formed from the extraction of SiO_2 from common sand through several chemical reactions to first form polysilicon. Heating the polysilicon to form a melt, and then lowering a seed crystal placed on a slowly rotating rod into the melt, the strained outer portions of the seed crystal will dissolve in the heat of the melt, exposing fresh, pure crystalline silicon to the polysilicon melt. Upon slowly pulling the rod out of the melt, the pure silicon crystal will attract the silicon crystals in the polysilicon melt, and as the composition cools, the silicon crystals will adhere to the rod, replicating the pure silicon crystal structure onto a now larger rod of pure crystalline silicon. This *pulling* method to produce pure silicon crystals is called *Czochralski crystal growth*, named after the Polish scientist who devised it way back in 1918. In preparation for use in electronic devices, the pure silicon rod is sliced by a diamond saw into thin wafers to be processed into integrated circuits. The maximum size of pure crystal silicon wafers made by the Czochralski method so far is practically limited to about 300mm diameter, although wafer sizes will likely increase [1].

As described in Chapter 20, because growing pure crystalline silicon transistors on glass was difficult, they were used only in wristwatch displays and very small-screen liquid crystal televisions. The formation of polysilicon transistors required high temperatures that obviated glass substrates, and uniform quartz substrates could only be made in small and medium screen sizes. The thin-film amorphous silicon transistors thus became the mainstream transistor addressing elements in large-screen displays.

Thin-film hydrogenated amorphous transistor fabrication uses the almost standardized photolithographic CMOS process. In the following descriptions, the historically significant principles of the semiconductor fabrication process will be introduced with the particulars of the thin-film transistor fabrication described where appropriate.

The Planar Process

The planar process fabrication of silicon semiconductor integrated circuits proceeds plane layer by plane layer with the active layer, the metal layer, and the insulator/passivator layer deposited and etched in turn to form the transistor elements [2].

The principal processes of semiconductor transistor fabrication are photolithographic patterning, etch, and deposition. At the beginning of the process, contact with air will oxidize silicon, but electronic components require an oxidation layer considerably thicker than the 2-nm layer of SiO_2 that is formed naturally by oxidation in the air, so either thermal oxidation or chemical vapor deposition (CVD) is used for forming the oxidized layer. After the SiO_2 layer is formed, the next step is to pattern the SiO_2 layer using *photolithography* so that dopant can be delivered through preformed apertures to reach selected regions of the Si layer below and be diffused therein.

Photolithography

Although closely associated with modern high-tech industry, *lithography* in fact has a long history in the art world; it was originally a way of reproducing pictures by embossing several flat slabs, each with a different color found in the original painting, and then precisely aligning the slabs to form the picture color layer-by-layer. Pressing the slabs sequentially on a sheet of paper would integrate the different colors into a reproduction of the original picture. In this way a great number of pictures of identical quality can be produced quickly and quite inexpensively.

The same procedure is used to fabricate integrated circuits. In the photolithographic process for semiconductors, the first step is to place the chemical vapor deposited SiO_2 layer on a platform as shown in Figure 21.1, which is then rapidly rotated to evenly spread a *polymer resist* coating dispensed from

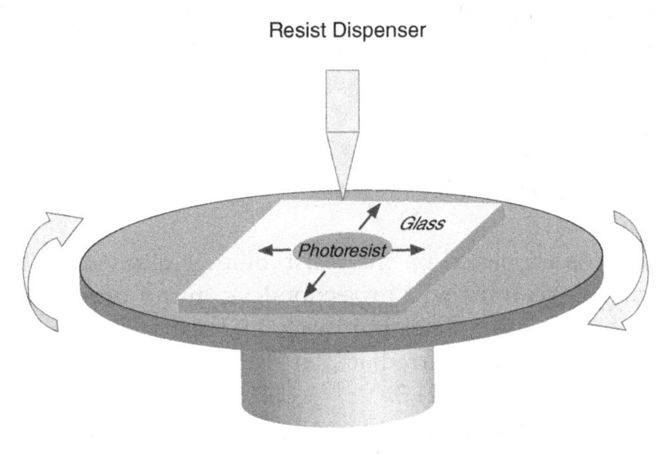

Figure 21.1 Spreading the resist on substrates using the spin coating technique.

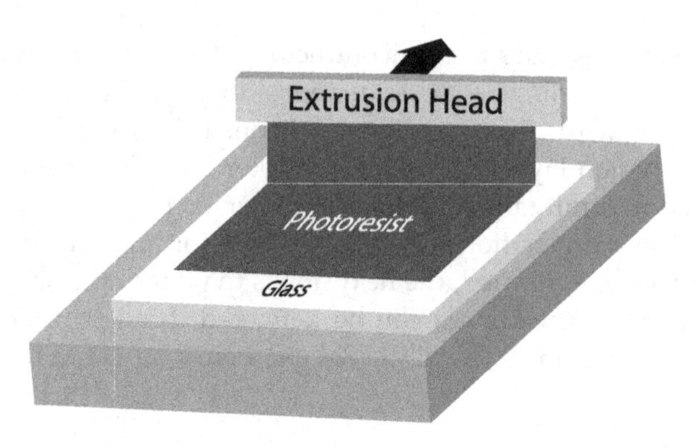

Figure 21.2 Spreading the resist on glass substrates using the extrusion slit technique.

the overhead resist dispenser onto the surface of the SiO_2 layer, in a process called *spin coating*. Spin coating works well for the typically 300-mm diameter or less silicon crystal semiconductor wafers, but for meter-scale large-size liquid crystal display glass substrates, it can be imagined that rapid spin coating can engender significant mechanical and safety problems, so typically the *extrusion slit* technique is employed to spread the resist on large glass substrates, as shown in Figure 21.2 [3].

After the resist is heat dried, the next step is to apply a mask with the doping hole pattern inscribed on it. The precise placement of the mask (called "registration") is a difficult but critical step in semiconductor fabrication; for example, a typical 1-μ gate electrode feature off by one-third of a micron will result in a twofold increase in gate capacitance, so registration requires submicron precision. This exemplifies the process engineers' industry maxim:

> the fewer the masks, the higher the yield.

After the mask is registered on the resist, blue or ultraviolet light from a mercury light source, utilizing a projection lens, mirror projection, or direct (proximity) radiation, irradiates the mask, and the areas of the resist not covered by the mask will be irradiated and photochemically react, as schematically depicted in Figure 21.3.

Resists come in two types, negative and positive, that respectively become less (negative) and more (positive) soluble in a developer solvent after exposure to light radiation. Irradiation polymerizes a negative resist resulting in

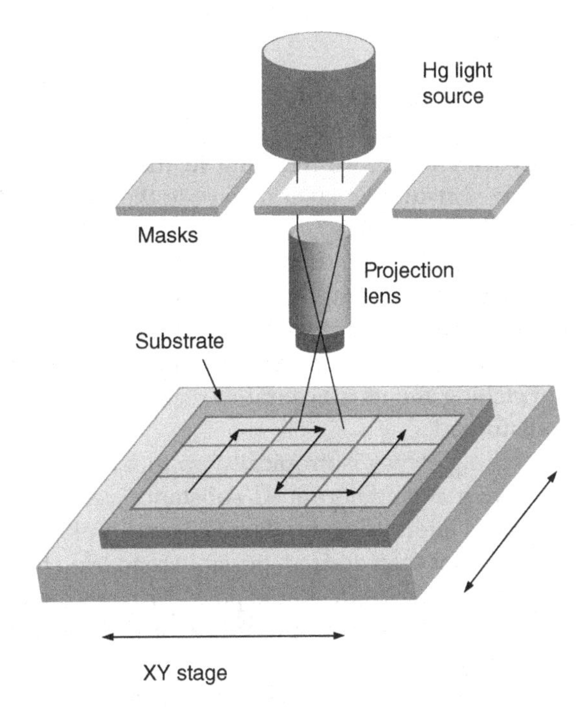

Figure 21.3 Irradiation of masks by an optical stepper.

an acid-resistant protective film; irradiating a positive resist on the other hand will break its molecular bonds, resulting in an acid-soluble resist.

After a negative resist is polymerized by irradiation, those hole regions of the mask so irradiated will become acid resistant, so that when the other regions of the resist are removed by an acidic solvent, the mask pattern will be replicated on the SiO_2 layer. A negative resist is typically used for semiconductor transistors used in many electronic products. For the positive resists commonly used in thin-film transistor fabrication, the photolithography effect is exactly opposite in that the exposed areas of the resist are nonacid resistant and can be removed by an acidic solvent, but the mask pattern is still similarly transferred to the protective insulator layer by the resist [4].

The mask pattern can be radiated all at once, or to improve accuracy, radiated by sectors in a "stepping" motion performed by an *optical stepper*. As shown in Figure 21.3, the substrate is moved by an XY stage to the next region, where it is radiated by the mercury (Hg) light source through the masks [5].

Etch

After the mask pattern is formed by the resist on the SiO_2 layer, those regions not protected by the acid-resistant resist can be removed by chemical etching, for example, by hydrofluoric acid, resulting in the formation of windows through the protective/insulating layer to expose the silicon below through which the doping agents may pass.

The chemical etch is selective in that depending on the chemical used, it can react with the Si-SiO_2 combination, using for example the hydrofluoric acid etch mentioned above, such that only the SiO_2 layer will be etched, and the silicon layer will not be attacked. Such so-called *wet etching* has a wide choice of different etching agents to accurately etch desired regions and leave other regions undisturbed.

However, wet etching etches isotropically, so that the chemical reaction, although only attacking a predetermined substance vertically through the mask, may also proceed laterally under the mask and thus *overetch*, resulting in transistor features larger than desired, possibly rendering them inoperable or at the very least increasing undesirable parasitic capacitance.

To avoid the overetch, *dry etching* procedures using *plasmas*, *reactive ion etch* (RIE), or *sputtering* were developed and are now commonly used. Because it utilizes a spurt of dry particles, a dry etch can be directed and consequently is more sector specific and thus more precisely controllable than a wet etch. Although the dry etch vacuum fabrication equipment is much more expensive, the higher yields and better throughput in modern large-scale integrated circuit fabrication make the investment economically worthwhile, and the dry etch thus has become a standard process for the mass production of thin-film transistors in LCD panels.

Plasma etching is initiated by a plasma comprised of molecules, ions, and electrons in a conglomeration that physically constitutes a fourth state of matter somewhere between a gas and liquid; because of the charge attraction, the state is more morphologic than a gas but its morphology does not reach the level of a fluid.*

A plasma is typically produced by applying a radio signal typically of frequency 10^4 to 3×10^{12} Hz to a gas (such as oxygen, fluorine, and/or chlorine). The free electrons in the gas will be accelerated by the radio waves and collide with the atoms in the gas, and at the extreme, electrons colliding with

* If plasmas are the fourth state of matter, then the mesophase liquid crystals could be considered the "fifth state of matter."

gas molecules can cause the molecules to dissociate, effectively resulting in their disintegration. At a lower energy collision, the electrons can knock off the outer electrons of the atoms constituting the molecules, thereby forming positively charged atoms, that is, positive ions. The free electrons also can excite electrons in the inner orbits of the gas atoms, causing them to jump to higher energy state orbits in the atom, and when those electrons decay back to their original orbit, they give up the energy difference (E) between orbits in the form of a photon of light having a frequency (ν) that is characteristic of the particular atom involved, in accord with the famous Planck equation $E = h\nu$ (where h is the fundamental Planck constant). All the molecular dissociation remnants in the gas, the electrons, ions, excited atoms, and decaying atoms, ultimately will devolve through collisions to an equilibrium state, but because of the constraints imposed by Coulomb forces, the plasma is self-sustaining and stable. Plasma equilibrium means that the number of ion–electron pairs created is the same as the number of pairs lost at the electrode surface and the plasma chamber walls; typical densities are 10^9 to $10^{12}\,cm^{-3}$ [6].

The plasma directed at the exposed areas of the planar layer will chemically attack the layer, causing chemical reactions, and simultaneously the plasma ions also will collide with the atoms of the exposed layer in a combined chemical and physical onslaught. For typical transistor fabrication, the attacking plasmas are carbon tetrafluoride (CF_4), plus oxygen to etch the silicon dioxide layer, and a chlorine-based plasma to etch the conducting metal layers made of aluminum.

Because the plasma is comprised of charged electrons and ions, a suitably oriented low-voltage electric field can further direct the dry etch into an even more precise anisotropic etch. But unlike the wet etch, the dry etch is not chemically selective, so in order to avoid dry overetching of nontarget layers, the etching end point must be carefully controlled; this can be accomplished by precise timing, the placement of physical etch-stops, or by monitoring the emission of characteristic light. With regard to the latter, the extent of a dry etch can be monitored by the light emitted because of the above-mentioned Planck equation, $E = h\nu$; different molecules have different characteristic electron energy levels in its atoms, the decaying electrons falling back into lower orbits will emit light characteristic of the matter being excited, thus measuring the wavelength of emitted light can determine whether the etch has entered a region of different matter. Reactive ion etching (RIE) also can perform an anisotropic etch using the *hexode* invented by Bell Labs, but it is only anisotropic insofar as it differentiates vertical from transverse etching because of the different respective etching efficiencies.

Deposition

For the conductive wiring layers, sputtering deposition is typically used. The positively charged ions in an ionized inert gas (such as argon) are accelerated by an electric field to impinge a negatively biased (cathode) target material composed of the matter to be deposited. The target material must be sufficiently nonvolatile so as to not evaporate upon being attacked by the ions. Upon collision with the ions, momentum transfer causes the target material atoms to be scattered off the target material and driven to be deposited on the semiconductor layer exposed by the mask. The advantage of sputtering deposition is that it is a purely physical process, such that the chemical characteristics of the sputtered material remain unchanged, so that an alloy (e.g., a conducting layer of aluminum plus copper), can be transported chemically unchanged to the desired deposition layer [7].

Again, the apparently modern high-tech sputtering deposition method actually has over 150 years' history of use in coating jewelry and eyeglass lenses; it was invented in England in 1852 by William Robert Grove.

Because of its affinity for SiO_2, early semiconductor devices utilized pure aluminum for the conductivity layers, but because the contact between aluminum and the minority carrier n-type semiconductor layer is nonohmic (resistance is too high), elements having better conductive adhesion to silicon, such as chromium (Cr) and molydenum/tantalum (Mo/Ta), were added to form metal layer alloys. In addition, to avoid buckling of the aluminum layer under high-temperature processing that form the undesirable *hillocks* in the conductive layer, titanium (Ti) has been used to cover the aluminum layer as well.

For thin-film transistors, higher dielectric constant silicon nitride (Si_3N_4) with different adhesive properties is also used for the insulation/protective/passivation layer in place of or in addition to the SiO_2.

The formation of the thin-film transistors used in liquid crystal displays primarily use sputtering deposition (sometimes called physical deposition, PD) and chemical vapor deposition in conjunction with plasmas (called *plasma enhanced chemical vapor deposition*, PECVD).

The scale of the production equipment for fabricating the thin-film transistors used in large-size liquid crystal displays can be appreciated from the 8th Generation PECVD machine from AKT pictured in Figure 21.4.

The PECVD process has been largely automated as shown in the Figure 21.5 schematic of a so-called *cluster tool*, wherein a transfer robot takes cassettes holding the LCD panels (load) from the cassette station and succes-

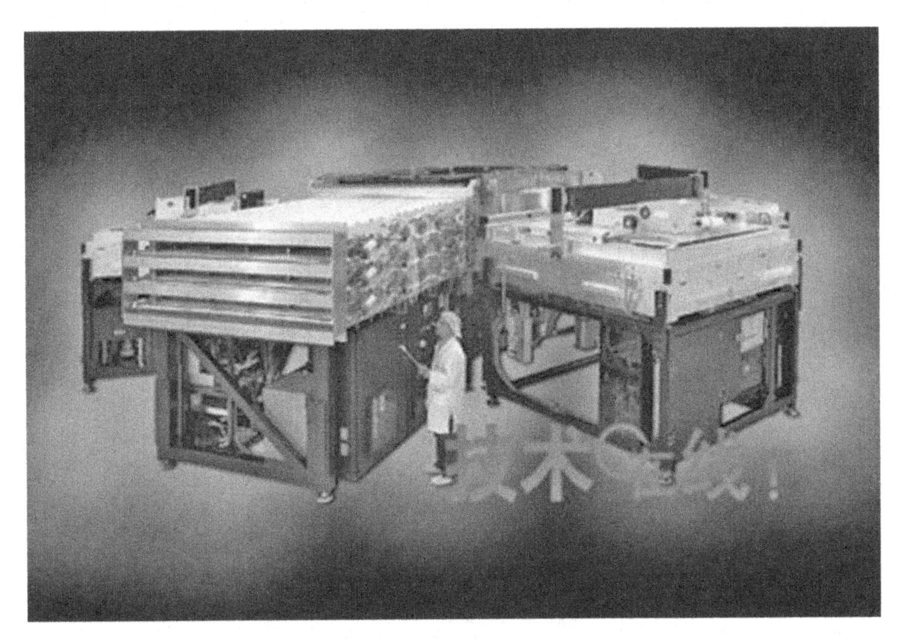

Figure 21.4 AKT eighth-generation PECVD machine courtesy of *Nikkei BP Report.*

sively sends them to the heating chamber, three process chambers, and finally to the unload back to the cassette station [8].

The above description of the fabrication of semiconductor devices above is merely an introduction to the basic principles involved; advanced semiconductor fabrication is an extremely technology-intensive endeavor. But although almost 50 years of evolution and refinement has brought changes, particularly in regard to greater precision and automation of the processes, Jean Hoerni's planar process is still used today as the basic semiconductor fabrication procedure. In the following section, a more detailed description will be given of the fabrication process for the principal thin-film transistor used in a typical liquid crystal display.

The Four-Mask Bottom Gate

The active matrix addressing system described in previous chapters uses integrated circuit transistors to control the voltage on each pixel of the liquid crystal display. The transistors are formed directly on the glass substrate

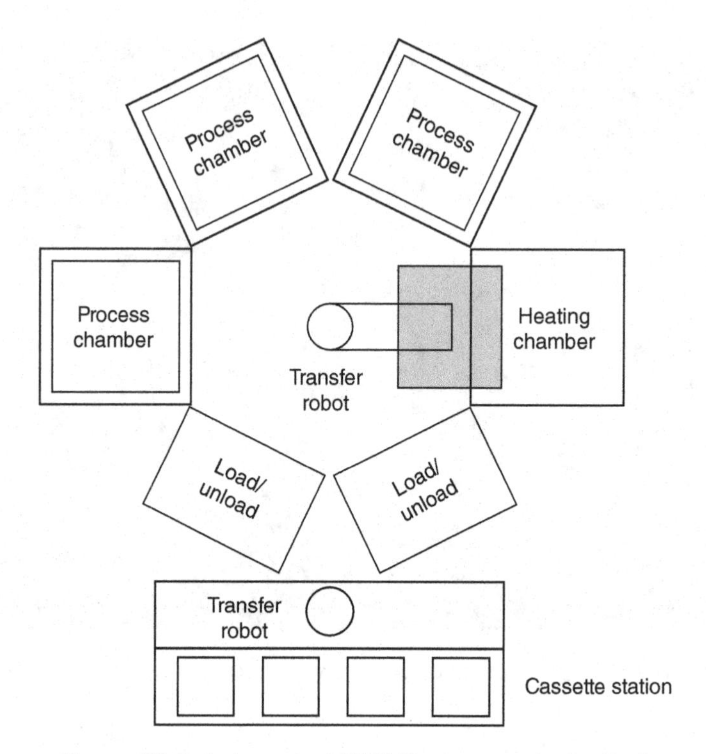

Figure 21.5 Automated PECVD process cluster tool.

holding the liquid crystal using thin-film semiconductor processing technology. An exemplary fabrication process is called the *four mask bottom gate,* and because the registration of the mask on the semiconductor material layer is complex and requires great precision, the few-masks-as-possible maxim drives process engineers to think of new ways to put more processes into a single mask registration step.

A typical thin-film transistor four-mask bottom gate process comprises the exemplary steps listed below and illustrated in the process schematic Figures 21.6I and 21.7II–V [9]:*

1. Clean the glass substrates chemically in an ultrasonic bath in preparation for forming the transistor's active layer, metal layer, and insulating layer (Figure 21.6I(a));

* For a dynamic depiction of the process, refer AUO website www.auo.com.

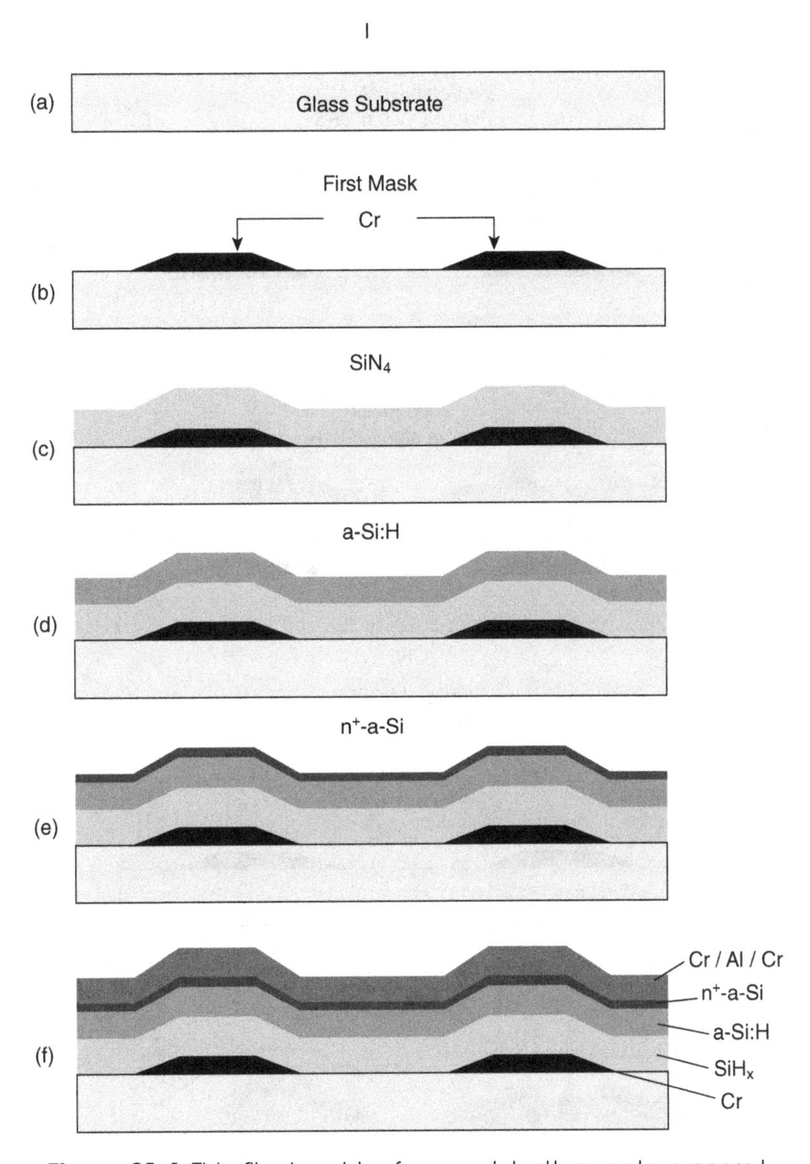

Figure 21.6 Thin-film transistor four mask bottom gate process I.

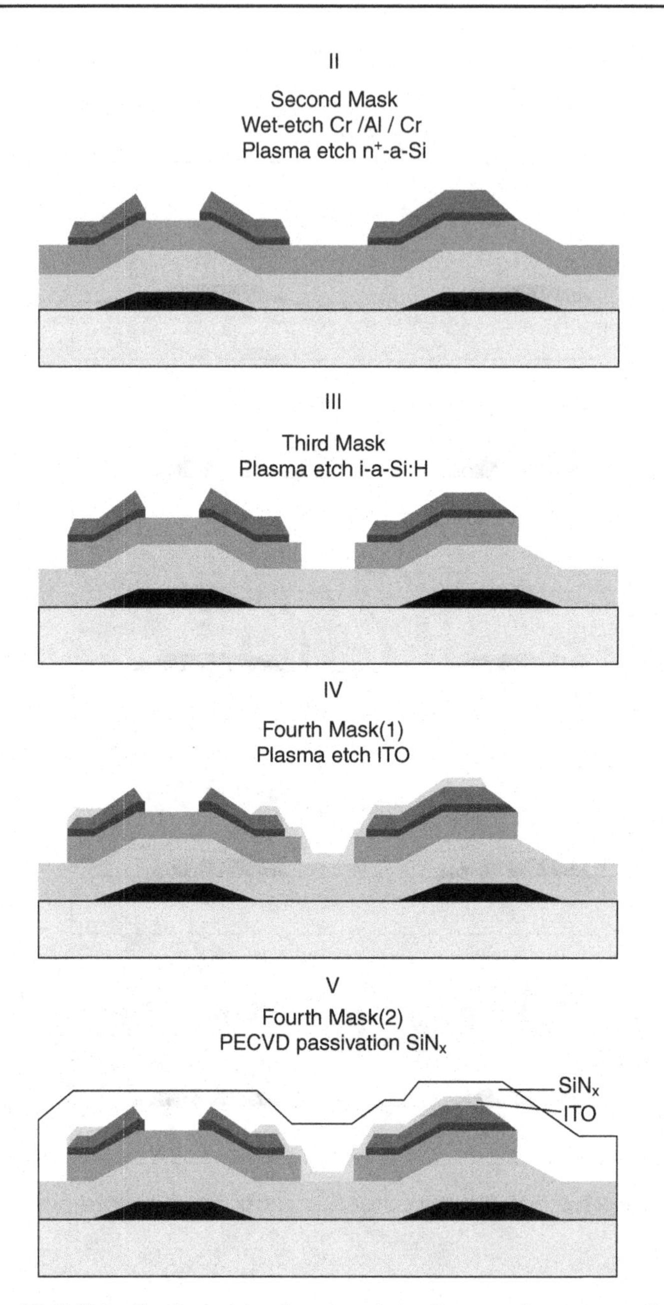

Figure 21.7 Thin-film transistor four mask bottom gate process II to V.

2. Sputter deposit a 200-nm thick chromium (Cr), or molybdenum/tantalum (Mo/Ta) metal layer
3. Wet etch to form the transistor's addressing rows, gate, and bottom electrode of the storage capacitor (Figure 21.6Ib); this is the first metal layer and requires the first mask;
4. Using PECVD,
   - deposit Si_3N_4 to form the insulator for the gate electrode and the storage capacitor (C_s),
   - deposit the *i-a-Si:H* active layer,
   - deposit the heavily doped (P from PH_3) n-type n^+*-a-Si* layer to provide the source and drain electrodes forming a good ohmic contact;
5. Sputter and wet etch to form a Cr/Al/Cr (or Mo/Ta) source, drain,and storage capacitor top electrode, thus forming a pattern of the second metal layer; this step requires a second mask;
6. The Cr/Al/Cr (or Mo/Ta) layer also serves as a mask for the plasma etching of the n^+ *-a-Si* layer (Figure 21.7II), but since the nonselective plasma etch will also etch the *a-Si:H*, an endpoint control is needed in the form of time control, an etch stop, or monitoring light emissions to ensure that all the n^+ *-a-Si* residue is etched away, and at the same time make sure that the 50-nm thick *a-Si:H* layer can provide the extremely low off current required for efficient operation, this so-called *back-channel etch* is an extremely delicate step in the fabrication process;
7. A second plasma etch of the *i-a-Si:H* layer to form the transistor's *i-a-Si:H* islands; this step requires the third mask (Figure 21.7III);
8. Sputter and a third plasma etch to form the ITO pixel electrode: this step requires a fourth mask (Figure 21.7IV);
9. PECVD to deposit a 350 nm thick passivation layer of SiN_x, and in order to open the bond pads (to be discussed below), a fifth mask is required (Figure 21.7V);
10. Deposit a metallic light shield; this requires a sixth mask, but the fifth and sixth masks do not require the precision registration of the other four masks, so they are not considered critical masks that affect the production yield.

The final complete thin-film transistor is schematically shown in Figures 21.8 and 21.9. Figure 21.8 shows the source, drain, and plasma etching endpoint control etch stopper; Figure 21.9 is a schematic of the final product after removal of the etch-stopper showing the gate and passivation layer.

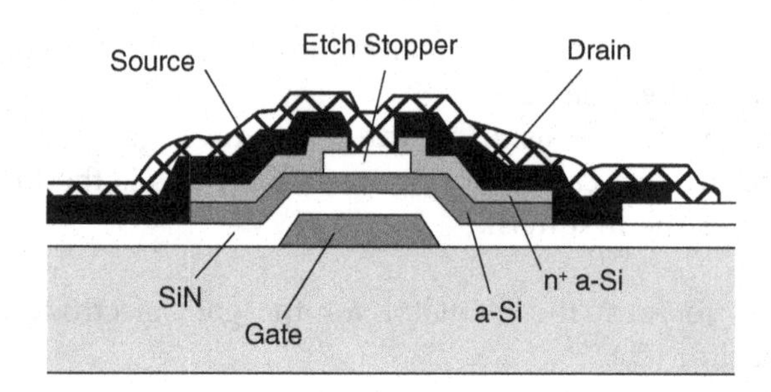

Figure 21.8 Thin-film transistor schematic with etch-stopper.

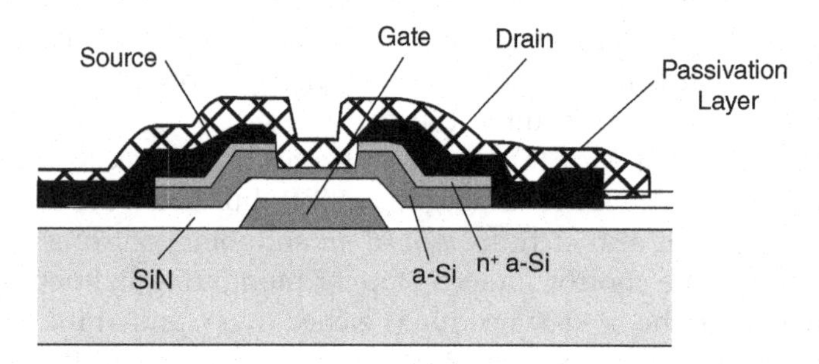

Figure 21.9 Thin-film transistor schematic with etch-stopper removed.

References

[1] Pearce, C.W. 1988. Crystal growth and wafer preparation. In *VLSI Technology* (Sze, S.M., ed.). McGraw Hill International Edition, Taipei, pp. 9–51.
[2] Cheng, H.C. 1996. Dielectric and polysilicon film deposition, Lii, Y.J.T., Etching, and Liu, R., Metallization. In *ULSI Technology* (Chang, C.Y., and Sze, S.M., eds,). McGraw-Hill, Taipei, pp. 205–268, pp. 329–369, pp. 371–468, respectively; Van Zant, P. 1997. *Microchip Fabrication*, 3rd edition. McGraw-Hill, Taipei, p. 351ff.
[3] den Boer, W. 2005. *Active Matrix Liquid Crystal Displays*. Newnes, Oxford, p. 54.
[4] Watts, R.K. 1988. Lithography. In *VLSI Technology*, 2nd edition (Sze, S.M., ed.). McGraw-Hill International Edition, Taipei, pp. 141–181.
[5] den Boer, W. 2005. *Active Matrix Liquid Crystal Displays*. Newnes, Oxford, p. 55.
[6] Lii, Y.J.T. 1996. Etching. In *ULSI Technology* (Chang, C.Y., and Sze, S.M., eds,). McGraw-Hill, Taipei, p. 331.
[7] Liu, R. 1996. Metallization. In *ULSI Technology*. (Chang, C.Y., and Sze, S.M., eds,). McGraw-Hill, Taipei, p. 380ff.
[8] den Boer, W. 2005. *Active Matrix Liquid Crystal Displays*. Newnes, Oxford, p. 53.
[9] Lueder, E. 2001. *Liquid Crystal Displays*. Wiley, Chicester, p. 245ff.

22

Enhancing the Image

With the advent of the new hydrogenated amorphous silicon thin-film transistor, and an efficient process developed for its fabrication as an integrated circuit, the modern active matrix addressing system liquid crystal display was now at hand.

The fundamental cell is shown in an early manifestation in Figure 22.1; each pixel is a *dot*, and contains an a-Si transistor (denoted by its gate, source, and drain), and a storage capacitor C_S. Together with the various insulators, the electronic components take up an area that blocks light passing through, and the ratio of the light transmissive area of pixel to the total area of the pixel is called the *aperture ratio* of a liquid crystal cell, a ratio to maximize in the design for brighter displays.

The active matrix addressing unit mounted on a twisted nematic liquid crystal cell to form a liquid crystal display is shown schematically in Figure 22.2. The liquid crystal layer is bounded by upper and lower transparent ITO plates, which form the electric field; the capacitance of the liquid crystal itself is designated by C_{LC}, the pixel electrode at bottom is connected to the drain of the thin film transistor (TFT) so that when the gate is activated, the image voltages will go through the data bus line from the source through the drain to charge the storage capacitor C_S. When the storage capacitor

Liquid Crystal Displays, First Edition. Robert H. Chen.

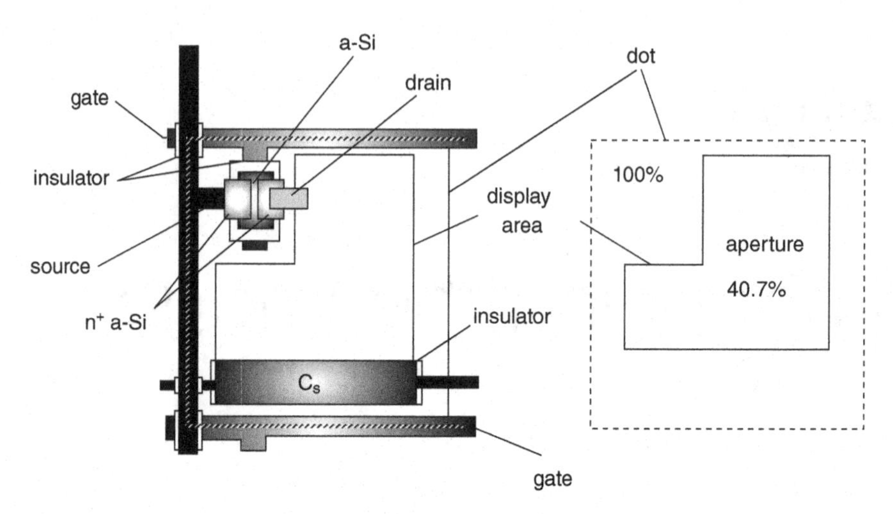

Figure 22.1 Early fundamental TFT cell. Adapted from Yamazaki, T, Kawakami, H., and H. Hori. 1994. *Color TFT Liquid Crystal Displays*. SEMI, Tokyo.

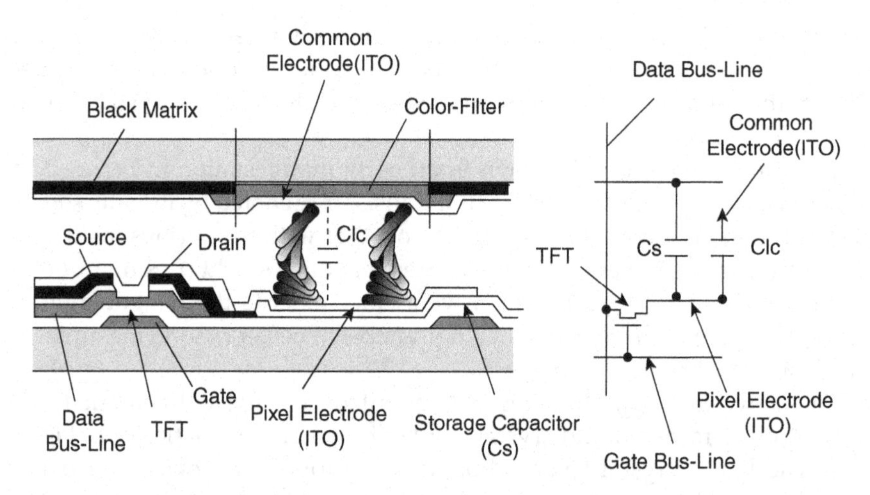

Figure 22.2 Super twisted nematic liquid crystal display cell.

discharges, adding on the charge capacitance of the liquid crystal itself and any parasitic capacitances from the transistor structure, a new voltage is produced, and this voltage is applied to the ITOs.

An electric field is thus generated throughout the liquid crystal layer, and the dipole liquid crystal molecules will respond by turning, and this turning will alter the polarization state of the light passing through, and in conjunc-

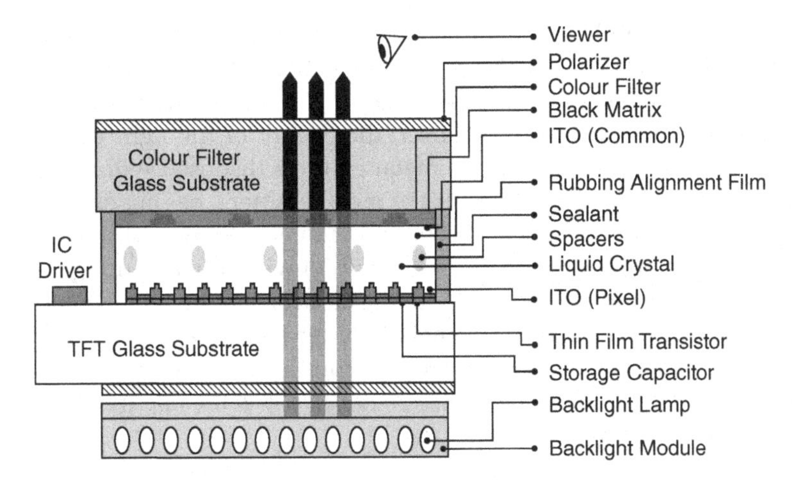

Figure 22.3 Super twisted nematic liquid crystal display system.

tion with a second polarizer, modulate that light to produce a grayscale replicating the light intensity of the image. At the right in Figure 22.2 is the electronic circuit diagram of the display cell.

The twisted nematic liquid crystal display (TN-LCD) cell of Figure 22.2, when mounted on a complete display system, is shown schematically in Figure 22.3. Light from a backlight unit (BLU), made either from cold cathode fluorescent (CCFL) tubes or light-emitting diodes (LED), after being linearly polarized by the first polarizer, is then further polarized by the liquid crystal responding to an applied electric field, and after transmission to a color filter to add the colors to the image, in conjunction with a second polarizer, produces the grayscale. The polarizers, ITO, rubbing alignment layers, liquid crystal, IC drivers, TFTs, and capacitors making up the LCD have all been described previously; in the following, the color filter, black matrix, sealant, and spacers also will be described.

The Grayscale

Perhaps the most difficult part to grasp about a liquid crystal display is just how the light from the backlight module can be processed by the liquid crystal to replicate the original scene, including its colors. The undertaking, while not simple, is a basic electrical engineering signal and image-processing problem with some methods and techniques borrowed from the technology of conventional CRT television.

First, modern video cameras (camcorders) use an array of solid-state charge-coupled devices (CCDs) or complementary metal oxide semiconductor sensors (CMOS sensors) to collect light reflected from the original scene. Then through the photoelectric effect, the energy of the light photons are transferred to the electrons in the metal emitters disposed within the CCD or CMOS sensors. The electrons in the metal emitters, having absorbed the energy from the photons, will thus have an impetus to move, but only those photons with a sufficiently high frequency (ν) will provide enough energy, through Planck's equation ($E = h\nu$), to the electrons in the metal emitters for the electrons to escape from the metal emitter surface and form a current. The kinetic energy (K) of the emitted electrons is given by Einstein's photoelectric effect equation, where the *work function* φ is the energy required to escape from the metal emitter,*

$$K = h\nu - \varphi.$$

Since the photon–electron interaction is based on probability (confer Chapter 4), it is not possible to directly calculate how many electrons an individual photon may photo-emit, but statistically the number of electrons emitted is proportional to the intensity of the light (number of photons with frequency above the work function threshold). As the frequency is related to the wavelength of the light ($\lambda \propto 1/\nu$), different type metal emitters having different escape thresholds in principle can differentially respond to different wavelengths and thus to different colors, and so produce a number of electrons (current) that is proportional to the intensity of the incident light of that color.

Color video cameras typically have three CCD (or CMOS sensor) arrays for capturing three primary colors: red, green, and blue, from which as will be described below, any and all colors can be produced. The electrons form electric currents proportional to the intensity of the different colors of the scene, which currents then generate video signal voltages that analogously represent the light of that color from the image to be replicated.

The analog video signal voltage from the original scene, transmitted to the TFTs through integrated circuit row and column drivers, produces the

* Max Planck won the 1918 Prize for his work on the quantization of energy; the Planck constant $h = 6.626 \times 10^{-34}$ J–s is a fundamental constant of quantum mechanical nature. Albert Einstein received the 1921 Nobel Prize for Physics for his theory of the photoelectric effect. The principal inventors of the CCD, Willard Boyle and George Smith from Bell Labs jointly received the 2009 Nobel Prize for Physics.

electric field on the ITOs to control the orientation of the liquid crystal molecules, thereby producing a controllable phase retardation that alters the polarization state of the light from the backlight module, and in passing through the second linear polarizer, changes the intensity of the light, all in proportion to the different colors of the scene.

The authentic reproduction of the original scene requires exceedingly fine reproduction of the scene's light intensity. *Luminance* is defined as the luminous intensity of a light source per unit area. Dividing the luminance range into levels is what is called in the display industry the *grayscale*, so for example, if the range of voltage representing the luminance is the conventional 2 V, then for the commonly used 8-bit coding producing $2^8 = 256$ levels, each level is separated by only 0.0078 volts, which produces an extremely fine grayscale.

As mentioned above, in addition to the voltage derived from the scene image, the grayscale voltage transmitted by the active matrix addressing system to the transistors of each pixel also has contributions from the charge stored in the storage capacitor (C_S) of the TFT, the liquid crystal's natural capacitance (C_{LC}), and the parasitic capacitance in the TFT itself. This latter capacitance is from the overlap of the gate, source, and drain structures of the TFT, that is, the gate capacitance (C_{GS}), gate-drain capacitance (C_{GD}), and the drain-source capacitance (C_{DS}). The basic capacitance design of LCDs is based on the C_S and C_{LC}, but the parasitic capacitance is an important factor, such that if it is not accounted for in the design, it will seriously erode the quality of the image. In summary, the final grayscale is proportional to the capacitances as follows:

$$\text{Grayscale} \propto \text{Storage Capacitance } (C_S) + \text{Liquid Crystal Capacitance } (C_{LC}) \\ + \text{TFT Parasitic Capacitance } (C_{GS} + C_{GD} + C_{DS}).$$

From this it can be seen that not only must the video signal voltage accurately reflect the light from the original scene, it must then also be adjusted for all the capacitances described above. This adjustment includes many aspects of basic electronics and semiconductor physics, and is a subject of many patents.

Based on the particular design of the display, the process of addressing and writing signals to each pixel is formulated as described in Chapter 17, and after fine-tuning the process considering all the various capacitances, appropriate voltages will be provided for each of the pixels. Because the liquid crystal molecules will re-orient in response to the field established by the voltage, the polarization state of the light from the backlight unit will be

Figure 22.4 Continuum and discrete grayscale. From *Hutcheson Consulting.*

changed, and passing through the second polarizer, the backlight light is modulated, and the pixels will display the appropriate grayscales. The top half of Figure 22.4 shows a continuum of analog grayscale that is broken into levels at the bottom half by scaling the gray using the discrete voltage levels described above. The continuum can provide an infinitesimally small luminance gradation for the red, green, and blue primary colors, and thus an infinite variety of colors, and the more levels of the grayscale, the more the colors that can be displayed.

The entire original scene is reproduced by the composite of voltages to an array of primary color subpixels, each of which has its own grayscale value, and when combined into a pixel, represents its part of the intensity and color of the total scene that is made up of millions of pixels.

The On/Off Ratio

The video voltage signal generated from the original scene is produced by the video processing system from an analog signal through a digital processing procedure to provide a signal suitable for the particular display system. As described in Chapter 17, the signal is transmitted to the gate line driver to address the selected TFT, then the data signal is transmitted through the source line driver.

Liquid crystal molecules are susceptible to structural fatigue and even molecular dissociation at high direct current (dc) voltages, but the modern liquid crystals used in displays are electrochemically stable at the voltages used today's LCDs. However, an electric field force constantly deployed in only one direction will force contaminant ions in the liquid crystal to collect on the alignment layers of the negative and positive electrodes, causing image sticking, and for long-period still images, may result in image burn-in.

Therefore, an alternating current (ac) positive/negative square-wave electric field is used to drive the liquid crystal. With an ac voltage of V_{on}, the on-current from the source line I_{on} would go directly into the capacitance of the liquid crystal C_{LC} were it not for the storage capacitor C_S, which stores the charge at full charge; so when C_S is discharged, the highest possible voltage will be applied, and that voltage accurately represents the scene. Within the time of the turning on of the address line T_{line}, the on-current then must be sufficiently large to discharge the total positive part of the square wave voltage all the way down to the negative voltage part of the square wave in order to adequately transmit an accurate signal; that is [1],

$$I_{\text{on}} > \frac{2(C_S + C_{LC}) \cdot V_{\text{on}}}{T_{\text{line}}},$$

The off-current I_{off} ideally should be zero, but in practice need only be small enough so that the charge on the pixel during the time frame T_{frame} is maintained; that is,

$$I_{\text{off}} < \frac{(C_S + C_{LC}) \cdot \Delta V}{T_{\text{frame}}}.$$

Then considering the total entire capacitance of the pixel, the voltage drop over the pixel cannot be greater than ΔV. Substituting in typical values for the LCD capacitances, voltage, and time, the on/off current ratio is*

$$I_{\text{on}} > 1\mu A \quad \text{and} \quad I_{\text{off}} < 0.5pA \rightarrow \frac{I_{\text{on}}}{I_{\text{off}}} = 2x10^6$$

An on/off current ratio greater than 10^6 is a fundamental requirement for LCD operation. Each pixel must be charged only once within each refresh cycle, which is generally 60 Hz, but fast response time displays can operate at frequencies at 120 Hz, and even 240 Hz. With the low threshold operating voltage of about 2 V and the very small currents (compared to CRT displays), the liquid crystal display addressing systems perform well, and use relatively little power. This makes the LCD ideal for battery-operated portable devices, such as notebook computers, hand-held games, music players, and mobile phones.

* For example, in a typical 15-in liquid crystal XGA monitor, $C_{LC} = 0.3\text{pF}$, $C_S = 0.3\text{pF}$, $T_{\text{frame}} = 16.6\,\text{msec}$, $T_{\text{line}} = 21\,\mu\text{sec}, \Delta V = 20\,\text{mV}$..

The Production of Color

Given that the primary color intensity signals can be accurately transmitted to the display, how then does the liquid crystal system reproduce all the myriad colors of the original scene? Again, starting with the camera's three CCD or CMOS sensor arrays that respectively absorb the red, green, and blue primary colors, the primary color information of the scene is embedded into the video signal. The production of different colors is then based on the principle that any and all colors can be produced by combining appropriate amounts of the primary colors in a mixture.

For example, sunlight that has been diffracted by rain drops produces the rainbow spectrum of colors from red to blue as shown in Figure 22.5, where the colors will have to be imagined as short wavelength blue on the left-hand side, long wavelength red light on the right hand, and intermediate wavelength green light in-between the two. White light contains all the colors represented in the spectrum according to their wavelength.*

Any color light can be similarly divided into the different primary colors, and depending on the color of the light, there will be different proportions of each of the primary colors to constitute the color. The human retina can differentially absorb different wavelengths of light so it can sense the primary colors red, green, and blue, and the mixture of these primary colors by eyes and the brain produces other colors.

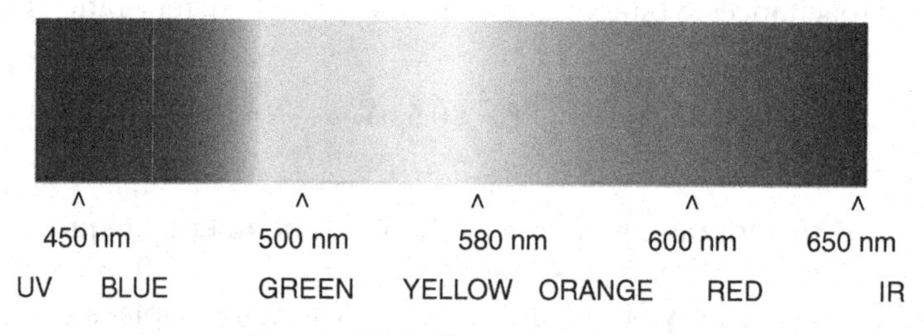

Figure 22.5 Color spectrum.

* Newton first produced the color spectrum using a prism in 1666; in 1814, Fraunhofer using a finer prism and slit found dark lines in the Sun's spectrum. In the 1830s, it was found that substances (e.g., sodium) had their own particular colors lines (yellow) when vaporized. Kirchoff in 1859 showed that the emission (bright line) of sodium was in exactly the same place on the spectrum as the absorption (dark line) in the Sun's spectrum, so proving that there is sodium in the Sun (outer layer as radiated from within the Sun), and so the science of spectroscopy was born.

From the physical optics point of view, as Figure 22.6 illustrates, the RGB primary colors can be mixed to produce other colors, with equal proportions of the RGB producing the white in the overlap of the three circles, so the colors are "additive," such that when the primary colors are mixed in even proportions the result will be white, and the complete absence of color is just the empty blackness of space as indicated by the black background in Figure 22.6.

However, the artists among us will object, saying that the primary colors are really cyan, magenta, and yellow as shown in Figure 22.7, with the white canvas being the absence of color, and that the proper mixtures of the CMY oil paint colors on their palettes will produce all the other colors, with equal amounts of the three primary colors together constituting a black morass, as illustrated by the intersection of the CMY in the center of Figure 22.7.

This science-art dichotomy arises from the different perspective of seeing light as it is physically itself or as reflected off of matter. Physically, light is adding photons to empty space, so colors are "additive" just as energy is additive, so no light is empty black space, and light containing all the colors is white as shown in the center of Figure 22.6.

Looking at the artist's painting, we are really seeing the light reflected from the oil colors applied to the white canvas (that reflects all light). Each newly added color covers up what is underneath, so the process of painting

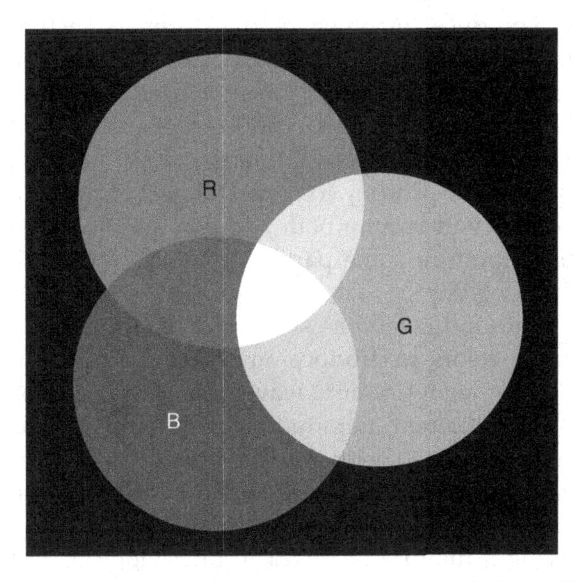

Figure 22.6 Physicists' red, green, blue primary colors additive mode.

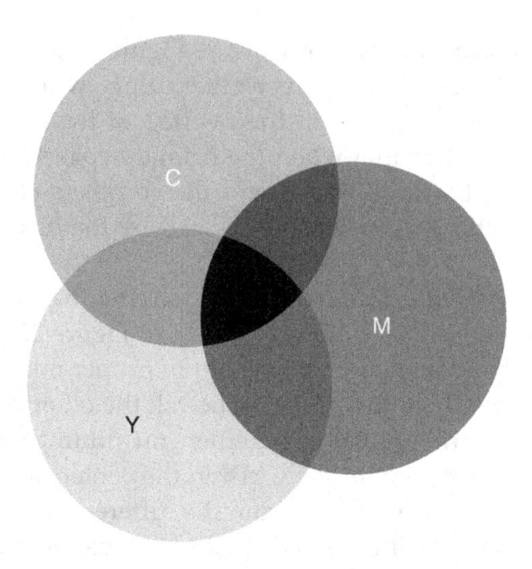

Figure 22.7 Artists' cyan, magenta, yellow primary colors subtractive mode.

is "subtractive." Equal mixtures of the primary colors on top of each other will produce a black composite, which absorbs all light and reflects none, as shown in the center of Figure 22.7.

Actually, not only RGB and CMY, any three separate colors can be appropriately mixed to produce any other color, so "primary" actually just means "separate," which means that no one color of the three can be made up by combinations of the other two. Red, green, and blue are chosen primarily because combinations of RGB can produce the most colors by addition without having to subtract out any primary colors, and so is just a most convenient choice for the primary colors set. However, any choice of a set of three primary colors will require both addition and subtraction of certain colors in order to produce some particular colors, and so the artists and physicists are both right.

In terms of light (and colors) as sinusoidal electromagnetic waves, this mixing of primary colors to produce any other color makes sense, since from Fourier series analysis, any waveform may be produced by an appropriate sum of different amplitude and different frequency sine and cosine functions, and a given color is just a mixture of the primary colors of certain wavelengths (frequencies) in different amounts (amplitudes) of the primary colors. But the phenomenon of color vision as a whole is extremely complex, involving a mixture of physical, physiological, and mental processes [2].

But simply put, only in the physical sense, if it is true that any color can be formed from a combination of R, G, and B, then a simple equation determining any color is just a vector in a three-dimensional coordinate system, with the red, green, and blue serving as the x-, y-, and z-coordinates, and the r, g, and b (they can be negative) being just the component amounts of red, green, and blue in the desired color,

$$\text{Color} = \text{rR} + \text{gG} + \text{bB}.$$

Now if any one color can be represented by a geometrical point in the coordinate space, and another color is another point, the sum of these two colors is the color as a vector sum of the two constituent colors in space. Since increasing the r, g, and b amounts equally (say doubling them) would not change the color but only increase the brightness, the three-dimensional space can be represented by a more convenient two-dimensional plane. So if it is agreed to make all colors the same light intensity so that they can be projected onto a plane, a two-dimensional Cartesian coordinate system *chromaticity diagram* can display any and all of the colors (of equal brightness) as a point in a color coordinate plane.

Figure 22.8 is the 1976 *Commission Internationale d'Eclairage* (CIE) standard chromaticity diagram that is used as the standard for color reproduction for electronic products, and particularly television broadcast; it is also called the *uniform color space*. The original 1931 chromaticity diagram had elliptically shaped regions around each color point within which there is no noticeable difference in color, but these ellipses had different sizes and shapes depending on their location in the diagram. To make the ellipses uniform for easier color analysis, a linear coordinate transformation to new u-, v-coordinates rendered the so-called "just noticeable differences" (JNDs) into circles with equal radii at any point in the diagram [3].

How different colors are composed of the three primary colors and the myriad array of possible colors is clearly demonstrated in the chromaticity diagram. A color may be defined by the r, g, and b *color coefficients* as a point in space, and from the infinite number of points possible, theoretically any color may be constructed by setting the appropriate color coefficients, and a table may be then constructed showing the value of the coefficients for producing any desired color.* The color filter (to be described following) for

* Maxwell constructed such a table, and since different wavelengths of light have different colors, this certainly was in accord with his equations showing that light is an electromagnetic wave.

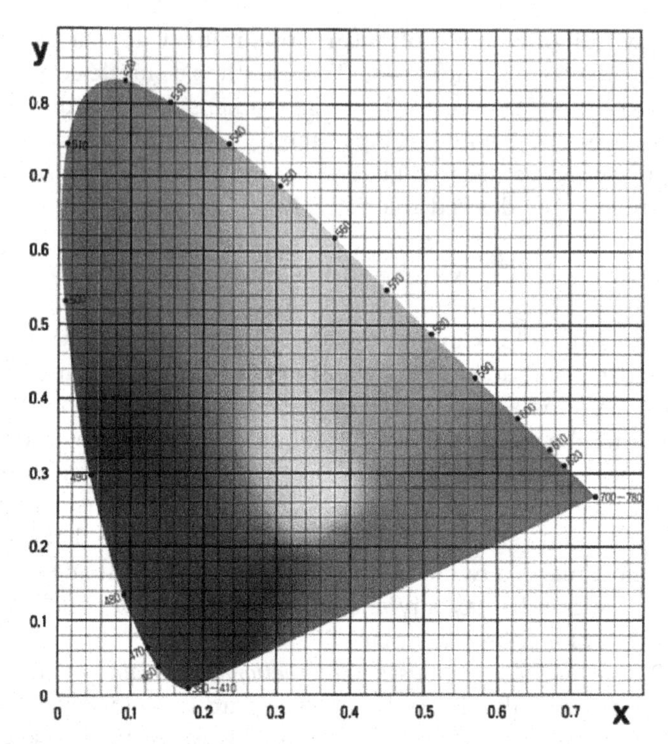

Figure 22.8 1976 Commission Internationale d'Eclairage (CIE) standard chromaticity diagram. From MIT.edu.

liquid crystal displays uses the additive and subtractive primary color principles to reproduce colors from the grayscale derived from the luminance of the original scene.

The triangle in the chromaticity diagram in Figure 22.9 is the *National Television Systems Committee* (NTSC) mandated range of colors that should be available for standard television broadcast; the dashed-line triangle is a sample color range from an exemplary display; the ratio of the area of the dashed-line triangle of a given display to the area of the NTSC standard triangle is called the *color gamut* of the display in question. Thus, the color gamut is a measure of the color quality of a display.

Viewing a screen from an oblique angle will cause a change in the color because the path of the light beam is not the same as from a head-on view (more about this in the following Chapter 23), this so-called *color shift* is measured by the difference of the displayed color from the chromaticity diagram standard.

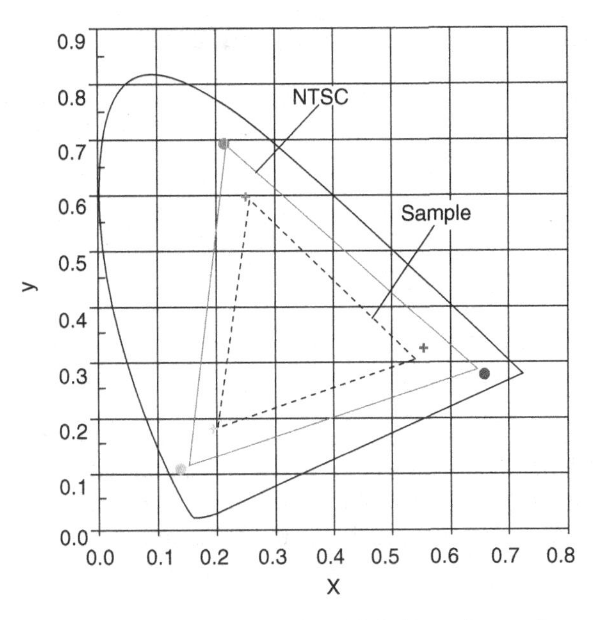

Figure 22.9 Chromaticity diagram with NTSC triangle and a sample color range triangle.

In addition to the artists and physicists, the physiologist can also weigh in on the color perception argument. A human eye senses different primary colors with different degrees of sensitivity, that is, typically, red 30%, green 59%, and blue 11%. So our eyes are most sensitive and most easily absorb the color green.*

The CCFL Backlight and Color Filter

In contrast to a CRT display's direct response to the light reflected from a scene, as described above, LCDs utilize a source of light placed behind the liquid crystal to provide the light. That a separate light source is able to replicate the light from the original scene is a culmination of a complicated signal processing, involving transformation of the light to electric signals, transfer of those signals, including color intensity information, to the active

* This might explain the attraction of a lush green golf course; it certainly can't be the game and all its attendant frustrations; but further elucidation may require input from yet other expert sources, the psychologist and the psychiatrist.

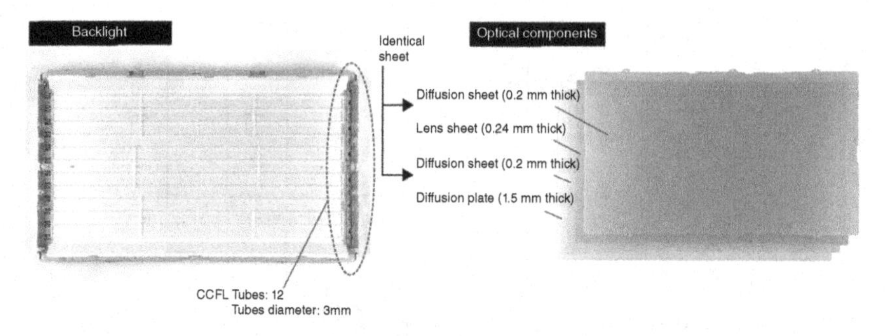

Figure 22.10 A Samsung 12-tube CCFL backlight. From Nikkei Electronics.

matrix addressing transistors to control an electric field to change the orientation of the liquid crystal molecules to alter the polarization state of the light from the light source, and in conjunction with a second linear polarizer, that produces a light intensity that mimics the luminance of the original scene.

In the cold cathode fluorescent (CCFL) tube that was the mainstay backlight source in early LCDs, the inside glass wall of the tube is typically coated with red, green, and blue fluorescent powder to match and enhance the color filter RGB peak transmissivities, with the goal of thereby providing deeper, truer colors; that is, promoting the *saturation* of each color. A typical 12-tube CCFL backlight is shown in Figure 22.10. The backlight includes diffusion sheets and a plate to diffuse the light evenly over the first polarizer and thence to the liquid crystal.

A color filter is employed to provide color by means of the additive combination of primary colors described above. Each separate pixel in the LCD is divided into subpixels that are aligned with the red, green, and blue primary color subpixels of the color filter that is placed over the pixel array. These red, green, and blue subpixels in the color filter transmit only the light of their respective colors and absorb all the other colors; the spatial composite of the transmitted subpixel colors then forms the desired color. The composite perception is possible because the human eye cannot distinguish the subpixels because of their minute size, and the different intensities of the primary colors from each subpixel are perceived as a mixed whole one pixel. This is how the color filter produces the desired color pixel from the different intensities of light from the subpixels.

The subpixels of a typical color filter can be arranged in the simple vertical, square, or so-called *delta* (Δ) configurations as shown in Figure 22.11, but there are many other configurations possible, each promoting particular features, such as intensity enhancement, true color reproduction, and color

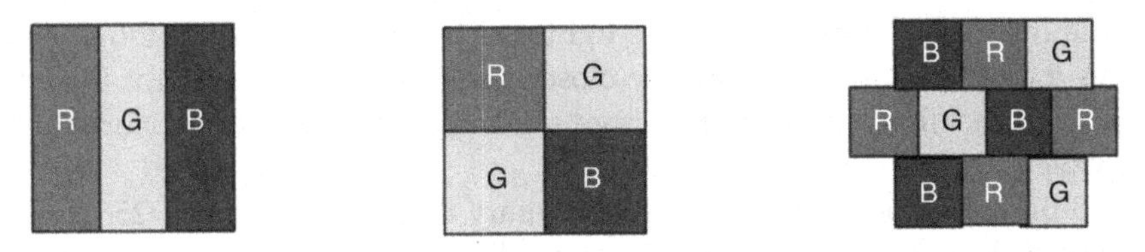

Figure 22.11 Subpixel arrays of the vertical, square, and delta configurations.

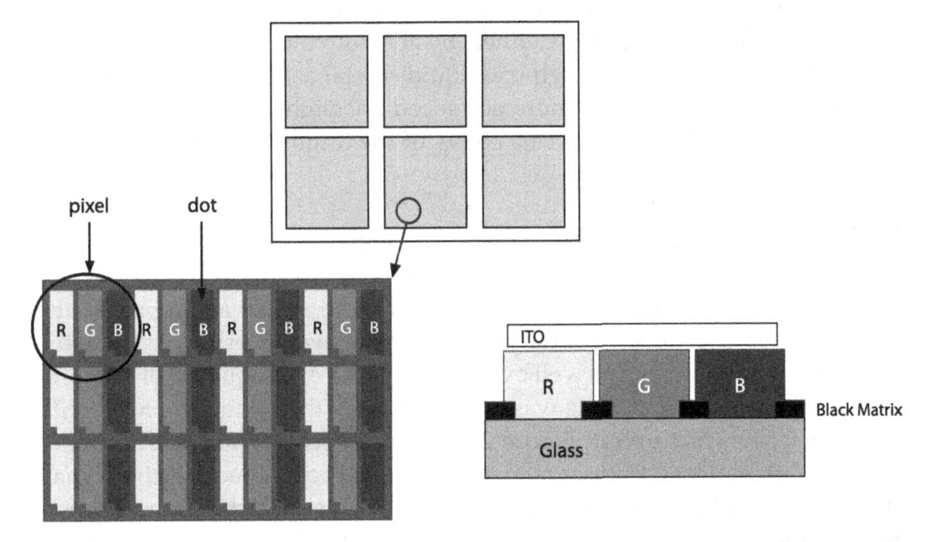

Figure 22.12 Three subpixels cover the pixel and in-between is the black matrix.

saturation. As such, there are also many patents covering the various subpixel configurations. Also, in addition to the RGB subpixels, other primary color subpixels can be added to increase the range of and quality of colors, that is, the color gamut and saturation can be improved.

As shown in Figure 22.12, the three subpixels making up the color resist cover the pixel of the liquid crystal display pixel array, and in-between the subpixels is the so-called black matrix for preventing light leakage.

The color spectrum of the color filter $S(\lambda)$ and the color spectrum of the backlight source ($R(\lambda)$, $G(\lambda)$, and $B(\lambda)$) are combined by convolution as follows,

$$red = \int_{350}^{750} R(\lambda)S(\lambda)d\lambda \quad green = \int_{350}^{750} G(\lambda)S(\lambda)d\lambda \quad blue = \int_{350}^{750} B(\lambda)S(\lambda)d\lambda.$$

The integration over the visible light spectrum (350–750 nm) above produces a convolution spectrum; the closer the spectrum of the color filter to the CCFL backlight coatings, the more saturated and true is the color that is displayed.

If 8-bit color coding is used, each primary color can have $2^8 = 256$ levels of grayscale, so three subpixels can produce $(256)^3 = 16{,}777{,}216$ different colors, enough to practically reproduce the colors of the scene. Higher-bit color coding can produce ever more grayscale levels and thus more colors within the gamut limitations defined by the convolution of the backlight source and the color filters. High-end liquid crystal color television sets with higher bit color coding and more advanced backlight sources can thereby more faithfully reproduce the true colors of the original scene.

Field Sequential Color

The colors from a color filter are obtained by the *spatial* averaging of the different intensities of the primary colors by the eye; in principle then, the colors could be produced as well by *time* averaging of a sequence of different intensity primary colors. This was done in fact by the young Maxwell with his color wheel (pictured in Figure 22.13), which when spun with primary colors of different brightness would produce other colors because the human brain would coalesce the amounts of primary colors into the desired subsequent color.

The field sequential color (FSC) scheme was developed for television in the 1940s, but was eclipsed by RCA's shadow mask for television use in the 1950s. FSC, however, has made a comeback for use in the LCD projector. In a typical FSC, a frame is divided into three subfields, one for each primary color; the backlight flashes red, green, and blue in synchronization with the display addressing; this can be done with a white backlight and a rotating color filter wheel, or a switchable liquid crystal filter. If the time between primary color flashing is sufficiently short, a time average of the primary colors will produce the visual perception of the color desired through a natural retention of the light intensity by the eyes and brain.

The field sequential scheme requires frame memory and image processing for proper subfield operation, and the backlight must be synchronized with the display scanning. If the refresh rate of the display is 60 Hz, then the primary color subfield operates at 180 Hz; this obviously requires the LCD to have a very fast response time of less than 4 ms. Recalling that response time is directly proportional to the square of the cell gap (liquid crystal layer

Figure 22.13 Maxwell holding his color wheel from Cambridge University.

thickness), for example, in a twisted nematic (TN) display, a cell gap of about 2 μ is required, and this has presented manufacturing difficulties. The smaller cell gaps possible in vertical alignment (VA) display modes might make the use of the field sequential technique more widespread, and so far it has been used primarily for LCOS projection displays, near-eye, and mini-projectors (see Chapter 27).

Because of the timing requirements involved, fast-moving objects may produce so-called *color breakup*, which can be demonstrated simply by waving one's hand in the projection of a field sequential color display and observing flashes of different colors between fingers. Because of the rapid, high-intensity switching required, the LED backlight is well-suited for field sequential color display.*

* The digital mirror device (DMD), a mechanical construct using tilting mirrors, uses field sequential color in Texas Instruments' digital light processing (DLP) projectors.

The LED Backlight

Since the color filters described above can absorb up to two-thirds of the backlight spectrum, there is a significant loss of transmittance and the luminance of the display suffers. The major LCD manufacturing companies have developed and are increasingly utilizing light-emitting diodes (LEDs) for use in the backlight sources to take the place of the CCFLs. The LED backlight, through its greater light-producing efficiency, lower heat, lower power consumption, and smaller size, is superior in almost every way to the CCFL tubes. The LED backlight-sourced LCD screen image will be brighter, the screen thinner, and the colors more vivid, but providing all the colors of the spectrum using LEDs is not a simple matter.

The low frequency/long wavelength emissions are easier to produce from semiconductors, so the early LEDs were all red light devices, such as those used in the status indicator lights on almost every electronic device. The emission of high-frequency visible light (e.g., blue) was much more difficult to achieve, and attempts had been ongoing for many years before final breakthrough.

In addition to the technical difficulties, in the beginning there were significant patent royalties to add to the cost, so much so that the technology was in danger of being stifled by the patent holders. Eventually seeing the light, the developing companies formed a complex patent cross-licensing system to avoid legal obstacles to the development of this new lighting technology; their commercial dreams were of course not confined to LCD backlights, but extended to the ultimate replacement of the incandescent light bulb and fluorescent tube by LEDs for all lighting functions.

A diode is a very simple electronic element; it allows current to flow in only one direction. A diode is constructed of a junction of n-type (electrons) and p-type (holes) semiconductors, as shown schematically in Figure 22.14 (top). When there is no applied electric field, the n-type semiconductor electrons fill in the holes in the p-type semiconductor near the junction, as shown in Figure 22.14 (bottom), forming a depletion zone that is insulating since neither the electrons nor the holes can move in it.

When a field is applied with the positive pole on the p-type semiconductor (as shown in Figure 22.15 at top), a forward bias current will flow with both electrons and holes flowing but in opposite directions to produce a strong current. But if the field polarity is reversed (as shown in Figure 22.15 at bottom), there will be no current since the holes will be attracted to the now negative pole of the field and the electrons to the opposite positive pole of the field, and there will be no current flow. This one-way current flow restriction is just the diode function.

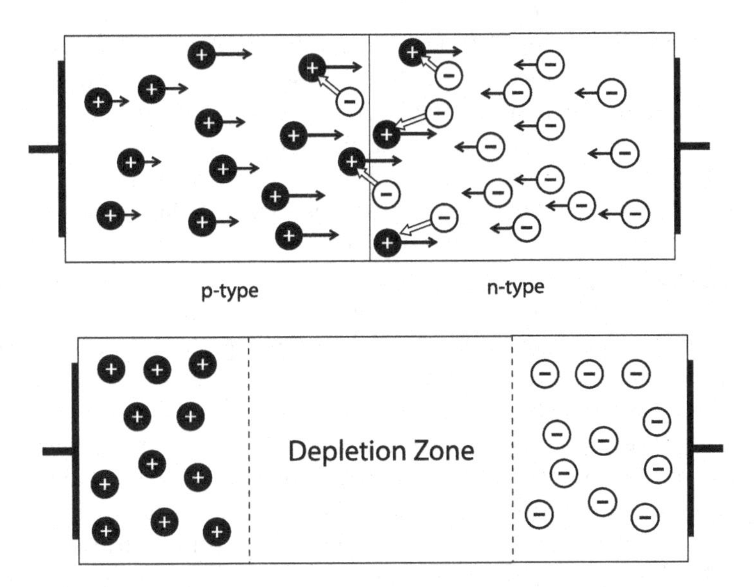

Figure 22.14 Recombination near the junction (top) and depletion zone formation (bottom) in a diode.

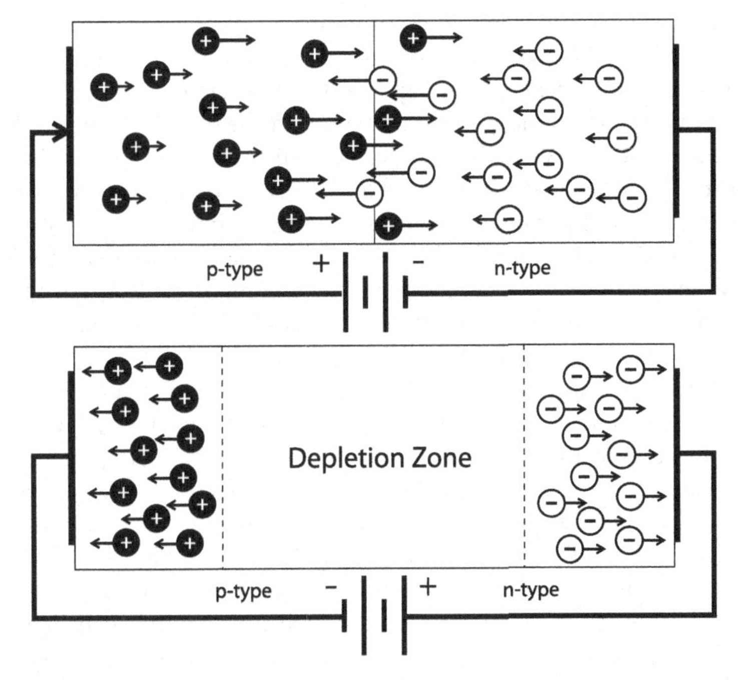

Figure 22.15 Forward biased current (top), reverse biased diode has no current (bottom).

As described in Chapter 19, electrons can be thought of as being in orbits revolving about the nucleus of an atom. These electrons can jump to higher orbits in the atom when stimulated, and depending on the character of the atom, they can emit light when they drop back (decay) to a lower orbit. For semiconductor LEDs, excess electron–hole pairs will be created by the direct injection of minority charge carriers from the forward bias current shown in Figure 22.15 top; in fact, the quantity of light emitted in a forward biased p-n junction diode is directly proportional to the number of excess minority carriers. In good LEDs, the charge carriers are easily pumped into the conduction band, and after being localized near the depletion zone, they recombine with oppositely charged counterparts in the valence band and emit light in a process called *radiative recombination.*

The energy lost by a conduction electron in a band-to-band recombination is emitted as a photon of radiation. The wavelength of the emitted light depends on the energy band gap of the particular semiconductor, for example, doped silicon semiconductors will emit light only in the infrared while the typical LED backlight aluminum gallium arsenide (AlGaAs) and gallium indium phosphorus (GaInP) semiconductors emit in the longer wavelength visible (like red). Thus differently constituted semiconductor materials can be designed to emit different colors of light.

The first blue LED was made of gallium nitride (GaN) with an indium gallium nitride (InGaN) emitting layer in 1993 by Nichia's Nakamura Shuji. GaN has a very wide band gap that can emit in the blue and ultraviolet, but growing pure GaN crystals was extremely difficult, and at the time only Nakamura's re-designed two-flow MOCVD reactor could produce GaN crystals of sufficient purity. Nakamura of course went on to invent the first blue laser as well, also based on GaN, but his patent disputes with Nichia and how much he should be compensated for his invention led to rather nasty fights, and Nakamura ultimately left Nichia for a university post in America.

In products, the light from the LED is directed and concentrated by the shape and internal reflection of the LED housing lamp to primarily exit through the top of the lamp, as shown schematically in Figure 22.16. An array of red, green, and blue LEDs in the additive color scheme described above can provide a 100% NTSC color gamut by varying individual RGB intensities. LEDs emitting white light are actually composed of a high-energy blue LED with a yellow or other visible color phosphor to convert the blue (or UV) color to white much like a fluorescent lamp; this predictably does not provide as good red and green colors as the RGB LED array.

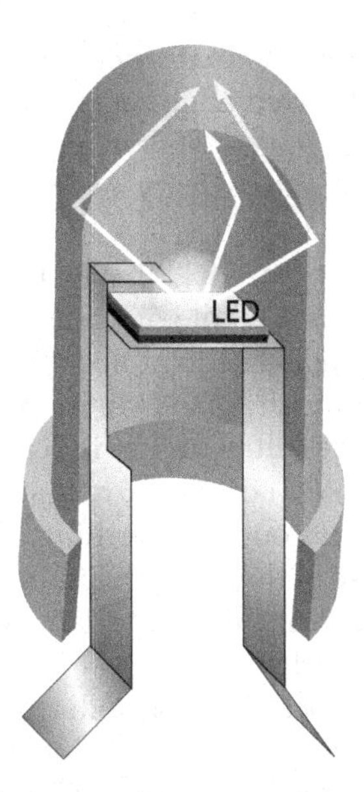

Figure 22.16 LED lamp construction.

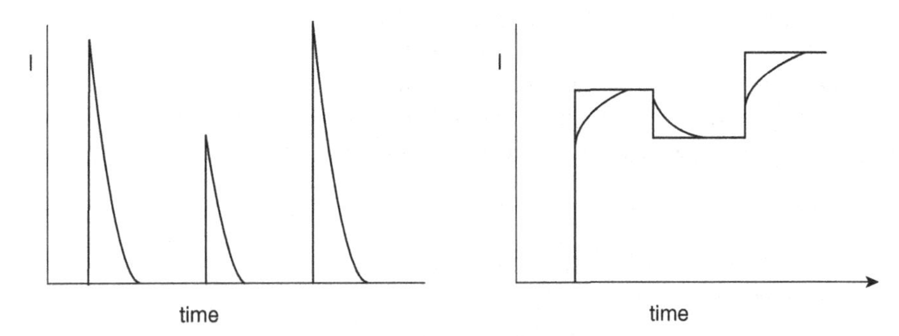

Figure 22.17 Pulse-type (right) and hold-type (left) video display signals.

The traditional CRT display is an *impulse-type* display, meaning that light pulses from the phosphor coating the screen are brief in time and discrete in space when excited by individual electrons impinging the phosphor; these discrete pulses comprising the image can sharply display the edges of a fast-moving object because of the steep slope of the pulses, as shown in the Figure 22.17 at left; it can be seen that there is little difference between the ideal signal (straight vertical line) and the actual signal (the curved line). In contrast, the liquid crystal display is a *hold-type* display, meaning that its luminance varies continuously, so that when integrated over a frame, a fast-moving object will have a blurred edge because of the time lags in the ideal and actual intensity signals; this is shown in Figure 22.17 at right by the curved lines of the actual signal intensity as opposed to the straight lines of the ideal signal intensity.

To emulate the impulse-type CRT, the LCD backlight can be rapidly switched on and off, either as a whole or preferably by scanning the lamps in a predetermined timing sequence to mimic the pulses of the impulse-type display. LEDs with high luminance and dimming with shorter duty cycles for the on/off pulses are well-suited for use in scanning backlights, although pulse width modulation (PWM) of the current supply for the dimming can result in flicker from the rapid on/off switching, but since the dimming can be controlled locally because of the small individual LEDs (as compared with the CCFL global dimming from the large lamps), the LED backlight can provide higher contrast than the CCFL.

A schematic of an LED backlight controller is shown in Figure 22.18; a color controller can take feedback information from the input of a color sensor and adjust the intermediate-range RGB primary color (sometimes plus white) signals by providing intermediate signal power between extremes using a square wave whose pulse width can be varied to change the average value of the waveform (this is called *pulse width modulation*, PWM), and then scanning the lamps to provide a brighter image.

Because the LEDs can themselves emit light in each of the primary colors, red, green, and blue, there is no need for a color filter in LCDs that use LED-

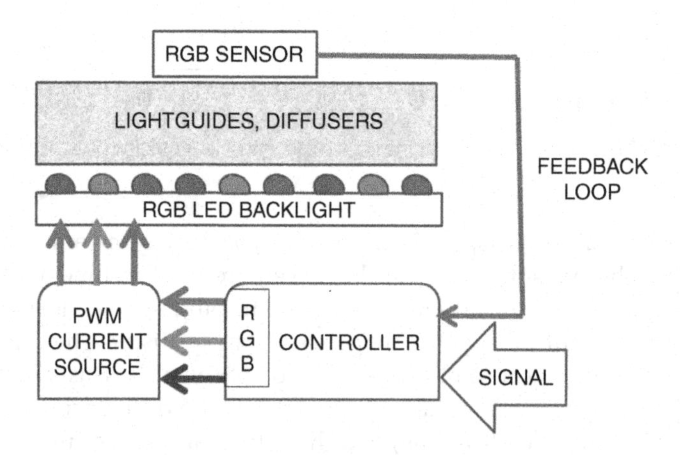

Figure 22.18 LED backlight controller.

backlights, and since light is not attenuated by passage through the color filter, the display can be brighter. Further, since the basic atomic processes involved in LED operation have higher light production efficiencies than filament and fluorescence-based CCFLs, the LED backlight is fundamentally brighter and relatively smaller, and the increased luminance improves the contrast ratio.

In a backlight module, an array of LED lamps placed directly behind the screen is called *dynamic RGB*; there are various combinations of diffusers, reflectors, lightguides, and polarizers employed to reduce light loss, enhance brightness, and provide uniform luminosity. The smaller-size LEDs allow the lamps to be disposed on the periphery of the screens (so-called *edge RGB*) rather than directly behind, and with only lightguides, diffusers, and reflectors behind the screen to guide and diffuse the light, the edge RGB panel can be made much thinner. As with the LEDs themselves, there are many patents covering the many and various dispositions of the lamps and supplementing schemes.

The LED LCD television and notebook computer screens are brighter and sharper, consume less energy, and have a significantly thinner form factor, and are already in common use.* But the laser diode is superior even to the LED, and soon may replace the LED as the mainstream LCD backlight.

Signal Processing

The accurate reproduction of the original scene by a display's signal processing system is not a simple matter, and understanding requires knowledge of display methods and techniques of video signal processing electrical engineering. The following is a simplified outline to give an idea of the signal processing concepts involved in the production of the video image in a liquid crystal display.

The video image signal is produced by a television camera, digital disk, or computer. This signal is fed into a video adapter, which includes a video decoder, an amplifier, and an analog-to-digital converter (ADC), which operate to convert the video signal into the digital signals compatible with the specifications of the display system.

* Interestingly, consumers have confused the newer LED backlight televisions as being something different from liquid crystal display televisions, at times demanding the new LED television and not the "old LCD" television; of course, the LED backlight television set still uses liquid crystals as the display medium, it is only the backlight that has changed.

The video adapter also processes other types of digital image signals, such as text and images from a computer, utilizing other adaptations of the signal. The video adapter also takes the red, green, blue (RGB) color information output signal (or digital color signals from a computer) and color-codes them. The color-coded RGB signals then are transmitted to the multiplexer in the digital signal processing box, which transforms the sequential data stream into three parallel RGB bit streams.

The human eye however does not sense light in a linear fashion; it has been found by experiment based on *just noticeable differences* (JNDs) that the human vision response (l^*) to light can be described empirically as

$$1^* = 116 \cdot (\text{relative luminance})^{1/3} - 16,$$

where the relative luminance is greater than or equal to 0.008856. The so-called *γ-correction* for the grayscale considers the empirical human vision response, or imposes a color standard such as the television standard for CRTs, and the transmittance voltage curve of the display to adjust the RGB signals for truer color perception. The display luminance (l_{display}) for a 256-level grayscale as a function of the maximum luminance (l_{max}) for a given grayscale number (g_x) can then be calculated from

$$l_{\text{display}} = l_{\text{max}} \left(\frac{g_x}{256} \right)^{\gamma}.$$

If the color standard is taken from NTSC, then a conventional CRT-based color television gamma correction look-up table is used. In display industry terminology, the *required luminance* (l_{LCreq}) of a display with 256 gray levels is calculated from the *input luminance* (l_{in}) of the picture source from

$$l_{LCreq} = \left(\frac{l_{in}}{256} \right)^{\gamma_{CRT}},$$

where the gamma is taken from look-up tables. Look-up tables also store the luminance-pixel voltage curves which are transposed with the l_{LCreq} to produce the final applied pixel voltage, which will produce colors in accord with conventional CRT television. A typical γ-correction curve is shown in Figure 22.19 [4].

At the same time, a *feature box* in the video adapter can provide special features, such as subscreens. Finally, the luminance, color signals, and frame

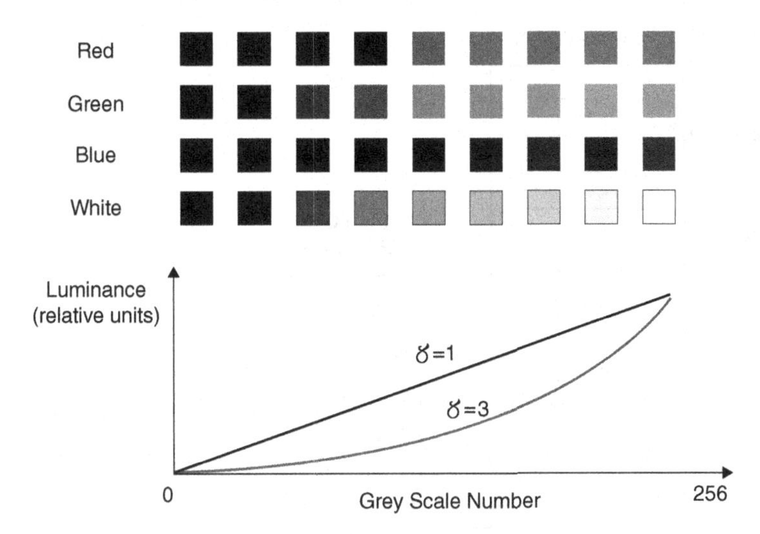

Figure 22.19 γ-Correction curve.

features are all transmitted to a multiplexer so that the *frame generator* can form the final image in half-frame frequency by means of the interlace or progressive scans that are used for television display. The signals are then sent to the pixel driver to generate the appropriate voltages for the transistors attached to the individual pixels. The signal processing system is summarized in the block diagram of Figure 22.20.

Now when the grayscale luminance voltage, carrying the individual red, green, blue digital information, travels through the thin film transistors attached to the ITO, a corresponding electric field is produced enveloping the liquid crystal in the selected subpixel. The liquid crystal molecules in the subpixel then turn in response to the electric field, and the light from the backlight, linearly polarized by the first polarizer, in traversing the liquid crystal has its polarization state altered by the change in the orientation of the molecules, and upon passing through the second linear polarizer presents an intensity of light reflecting the color-coded luminance of the original scene; that is, the grayscale designated for each primary color. Then passing through the color filter's red, green, and blue subpixels, each primary color is presented in its own intensity in accordance with the intensity determined by the original scene's color luminance. The eye now perceives the combination of primary colors that reproduce the color and luminance of the original scene.

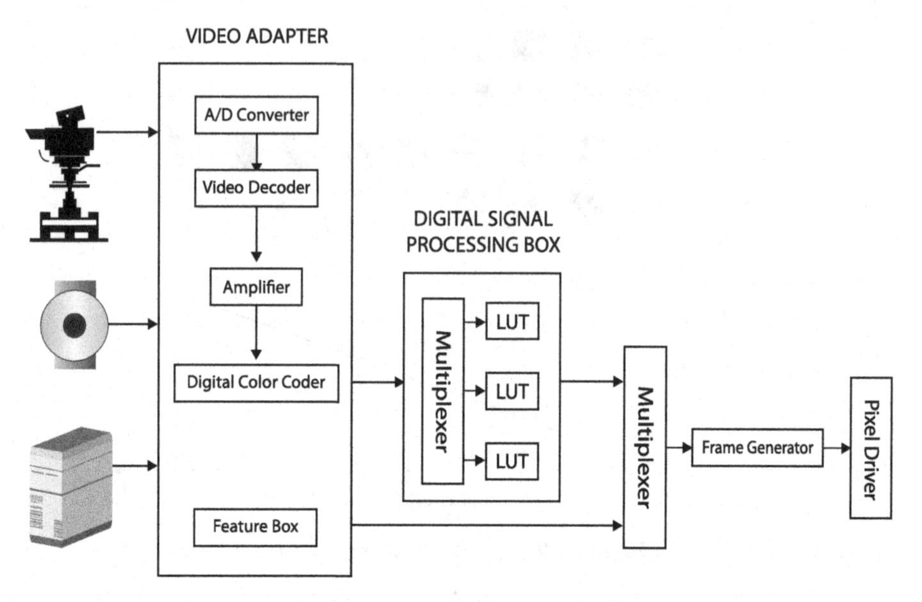

Figure 22.20 Signal processing system schematic.

And so when Lehmann in Europe more than 120 years ago took Reinitzer's carrot sample of cholesteryl benzoate, placed it between two crossed linear polarizers, and saw to his surprise that light from below would pass through the second crossed polarizer, the liquid crystal display industry was conceived. But then some 80 years would pass when Williams in America would control the alteration of light and dark through a liquid crystal with an electric field to create a display prototype; and after more suitable liquid crystal compounds were synthesized by Gray in Britain, the twisted nematic developed by Helfrich and Fergason was turned into commercial displays by the Japanese company Sharp under the leadership of Wada, Sasaki, and Washizuka. Dynamic, high-resolution new products were made possible by the active matrix addressing scheme of RCA's Lechner and Marlowe, and Weimer's thin film transistor was adapted for use in liquid crystal displays by Westinghouse's Brody. Those critical technical breakthroughs, together with many incremental improvements, combined with the application of television-standard signal processing, altogether evolved into today's million-pixel, million-color high-resolution, extraordinarily sharp, clear and bright thin-liquid crystal color television (that doesn't cost a million dollars).

References

[1] den Boer, W. 2005. *Active Matrix Liquid Crystal Displays*. Newnes, Oxford, p. 28.

[2] Feynman, R.P., Leighton, R.B., and Sands, M. 1963. *Lectures on Physics*, Vol. I. Addison-Wesley, Reading, p. 35ff.

[3] Tannas, L.E. 1983. *Flat Panel Displays and CRTs*. Van Nostrand Reinhold, New York.

[4] Lueder, E. 2001. *Liquid Crystal Displays: Addressing Schemes and Electro-Optical Effects*. Wiley, p. 238ff; den Boer, W. 2005. *Active Matrix Liquid Crystal Displays*, Newnes, Oxford, p. 127ff.

23

The Wider View

The bright and beautiful image from the liquid crystal display however, was not equally accessible to all. For the viewer looking from an angle to the display, the contrast, grayscale, and color presented by the twisted nematic screen was not at all the same as when viewed head on; and the more oblique the viewing angle, the greater the discrepancies. For computer monitors and notebook screens, perhaps this viewing angle problem is not all that important, but for the family watching television in the living room, a viewing angle of at least 80° is a commercial necessity, and for the crew and passengers on commercial aircraft, the instruments readings of the pilot and co-pilot should be the same, otherwise the consequences could be dire indeed.

The imperfect wide viewing angle image is basically due to the different optical path taken by the light to the oblique viewer's eye, which will cause a different phase retardation from that of the head-on view. The direction of vibration of the electric vector (E) of a light beam is perpendicular to its direction of propagation; linearly polarized light, upon entering an anisotropic medium having different dielectric constants in different directions, will be phase retarded depending on the angle that the transverse vibrations of the E-vector makes with the directions of different indices of refraction in

Liquid Crystal Displays, First Edition. Robert H. Chen.

the medium. For the different incident angles and subsequent longer optical paths of oblique angle viewing, the phase retardation will increase.*

From the perspective of the liquid crystal, because of the long-rod shape of the molecules, a nematic liquid crystal is uniaxially symmetric, so there is an axis of symmetry that is called the *principal molecular axis* (also sometimes called the *preferred axis*). The dielectric constant in a liquid crystal is different for directions along the molecular principal axis and directions perpendicular to it. The dielectric anisotropy ($\Delta\varepsilon$) of the liquid crystal then is just the difference between its dielectric constant parallel to the principal molecular axis ($\varepsilon_{\Updownarrow}$) and its dielectric constant perpendicular to the principal molecular axis ($\varepsilon_{\perp}$), that is,

$$\text{Dielectric Anisotropy} = \Delta\varepsilon = \varepsilon_{\Updownarrow} - \varepsilon_{\perp}.$$

When $\Delta\varepsilon > 0$, the liquid crystal has positive anisotropy, then $\varepsilon_{\Updownarrow} > \varepsilon_{\perp}$, and the electric dipole moment (denoted by p in Figure 23.1) in the liquid crystal is deemed parallel to the principal molecular axis in positive anisotropy liquid crystals. When $\Delta\varepsilon < 0$, then $\varepsilon_{\perp} > \varepsilon_{\Updownarrow}$, and the electric dipole moment is perpendicular to the principal molecular axis in negative anisotropy liquid crystals.

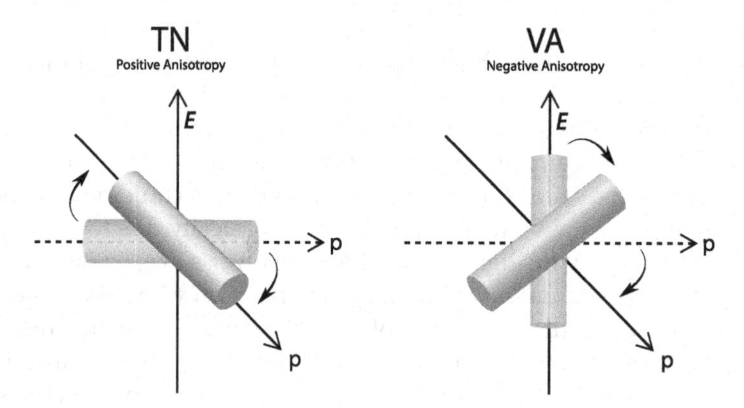

Figure 23.1 Positive anisotropy twisted nematic (TN) and negative anisotropy vertical alignment (VA) liquid crystal molecules turning in the applied electric field (E).

* Recall from Chapter 16 that the phase retardation parameter is given by $\Gamma = 2\pi\Delta n \cdot z/\lambda$), where z is the distance that light is traveling through the liquid crystal layer.

In terms of the polarization of the molecule and the polarization of light, the distinction between parallel and perpendicular dielectric constants is used practically to indicate a *preferred* direction of dipole charge oscillation that is determined by the structure of the material, so it can be said that the linearly polarized light's E-vector will be resolved into components determined by that preferred direction, and from this the phase retardation and polarization of the light is determined.

When an external electric field (E) is applied, because of its intrinsic dipole moment, the nematic liquid crystal will become polar with an induced dipole moment from the re-orientation of the molecules. The dipole moment will turn to align with the applied electric field, thus causing the liquid crystal molecules to turn as depicted schematically in Figure 23.1. For ease of manufacture, most twisted nematic (TN) and vertically-aligned (VA) liquid crystal displays are designed with an electric field perpendicular to the glass substrate surface, so the ITO plate electrodes forming the electric field can be placed directly on the glass surface. In the TN mode, positive anisotropy liquid crystal is primarily used, so the external electric field will cause a positive anisotropy liquid crystal's molecules to turn from their originally homogeneous (lying parallel with the glass substrate) orientation to align with the electric field, as shown schematically in Figure 23.1 on the left. In the VA mode, negative anisotropy liquid crystal is used, and the electric field will cause the originally homeotropic (perpendicular to the glass substrate) negative anisotropy liquid crystal's molecules to tilt away from the electric field, as shown schematically in Figure 23.1 on the right.

As mentioned in Chapter 22, an alternating square-wave electric field drives the liquid crystal. Since there is no distinction between molecular head and tail (there are as many molecules with positive charge up as there are with positive charge down, or left and right for negative anisotropy liquid crystals), it would appear that the tilt orientation of the molecules would all average out under an ac field. But because the response time of the liquid crystal molecules to the external field is much slower than the period of a typical driving ac field, the molecules cannot react directly to the positive-to-negative square waveform of the applied field voltage. Instead, the liquid crystal reacts to the *energy* of the applied field, which is the square of the *root-mean-square* (rms) voltage averaged over time. Since the rms voltage (also called the *effective voltage*) for a sinusoidal wave is 0.707 of the maximum voltage, there is a net effect on the molecular orientation that is consistent with the anisotropy of the liquid crystal and the strength of the electric field.

c-axis a-plate c-plate

The improvement of viewing angle has become a well-developed technical discipline, and as such has naturally engendered its own technical terminology. In order to investigate further oblique viewing angles and the analysis of wide viewing angle quality, it is necessary to first introduce some of the terminology used in crystallography that is applied to liquid crystals.

If light of a particular incident angle passes through an optical element (including liquid crystals themselves) and suffers no phase retardation, that direction of the light through the element defines an *optic axis* of the element, which is called the element's c-*axis* in crystallography (for the "crystal axis"). In a positive anisotropy (also called positively birefringent) uniaxial liquid crystal, the director is nominally the *c*-axis of the liquid crystal. It should be noted, however, that since the director may change directions within the body of the liquid crystal, the phase retardation (and thus the polarization state) of incident light may also change within the liquid crystal itself, and so the *c*-axis of the liquid crystal may also change direction within the liquid crystal.

If the direction of propagation of incident light is aligned with the *c*-axis of the liquid crystal, it will pass through with no phase retardation. But if the light beam direction is not aligned with the *c*-axis, then there can be phase retardation and subsequent changes in polarization. If, owing to the oblique angle of the light beam or because of the change in orientation of the liquid crystal director, the light beam may undergo phase retardation and produce different states of polarization, depending on the amount of phase retardation.

Recall that the anisotropy of a liquid crystal can be characterized by its influence on light as the difference between its extraordinary and ordinary index of refraction ($\Delta n = n_e - n_o$), and if $n_e > n_o$, the extraordinary wave is the *slow wave* (and the ordinary wave is the *fast wave*), and if $n_e < n_o$, then the ordinary wave is the slow wave (and the extraordinary wave is the fast wave). So the positive or negative value of Δn determines which wave is retarded (slower), and denotes a plus or minus sign for the phase difference between the ordinary and extraordinary waves, which is denoted by $\delta = \delta_e - \delta_o$ in the Maxwell equations solutions in Chapter 2, and by the phase retardation parameter Γ of Chapter 16; and so the sign of $\pm\Delta n$ can be said to determine the "direction" ($\pm\delta$ or $\pm\Gamma$) of the change in phase difference or retardation.

Based on the positive or negative phase retardation regime outlined above, the positive or negative dielectric anisotropy ($\pm\Delta\varepsilon$) causing the positive or negative index of refraction ($\pm\Delta n$) of a *phase retardation plate* (also called a *birefringent plate* or *compensation film*) can be used to negate (or "positate") unwanted phase retardation. That is, the effect of the positive or negative character of a phase retardation plate can reverse the "direction" of the phase retardation, so that an oppositely directed dielectric anisotropy phase retardation plate can reverse the undesirable effect of an unwanted phase retardation resulting from viewing at an oblique angle. This is the basic principle underlying viewing angle compensation.

These phase retardation plates have their own technical names, for example, a uniaxial phase retardation plate having an optic axis (*c*-axis) perpendicular to its own plate plane is called appropriately a *c-plate*. For a liquid crystal example, if its molecules are perpendicular to its surface (e.g., as in a vertically aligned (VA) liquid crystal cell), normally incident light will be aligned with the directors, and the electric vector of linearly polarized light will be perpendicular to the directors; in the absence of an external applied field, there will be no phase retardation, and the light will pass through with its polarization state unaltered, like a c-plate.

A uniaxial phase retardation plate having its optic axis parallel to its surface is called an *a-plate*. For a liquid crystal example (as in the field-off, in-plane switching mode), the molecules are laying flat in the plane of the glass substrate with directors rubbed parallel (E-mode) or perpendicular (O-mode) to the linearly polarized incident light electric vector; the direction of propagation of a normally incident light beam will be perpendicular to the plane of the directors, and in the absence of an external applied field, the light electric vector remains parallel (or perpendicular) to the directors, and thus there will be no phase retardation (the light electric vector is not resolved into components that are parallel and perpendicular to the directors); the light will pass through a field-off IPS mode with its polarization state unaltered, like an a-plate.

A positively birefringent c-plate is called a *positive c-plate*, and similarly a negatively birefringent c-plate is called a *negative c-plate*, and similarly for *positive a-plates* and *negative a-plates*.

Oblique angle viewing is just the egress perspective of the light incident on the other side of the liquid crystal layer, and the phase retardation effects are thus conceptually the same as those for ingress non-normal light incidence described above.

The implication of the *c*-axis is that for oblique viewing angles, the paths of light to the viewer that were originally along the *c*-axis now will no longer be through the *c*-axis and there will be undesirable phase retardation. And indeed, the different longer optical path of oblique incidence (or viewing) compared with the *c*-axis path will cause the light electric vector to make an angle with the fast and slow axes (or with the directors in a liquid crystal), thereby engendering phase retardation, causing different polarization states that will be affected by an analyzer differently, so the light transmission will change (just as described in Chapter 4). This departure from the desired intensity is generally called "light leakage" in respect of the analyzer undesirably passing light in a black state. The greater the oblique angle of incidence, and the longer the optical path, the larger the phase retardation causing more departure from the original linear polarization state and subsequent light leakage.

Mid-Layer Tilt

When an intermediate electric field is applied to a liquid crystal, the molecules near the glass substrate, owing to the strong anchoring of the alignment layer, will remain held down parallel to the glass substrate (or with a small pre-tilt angle), but in positive anisotropy displays, the molecules in the middle layers of a homogeneous liquid crystal will tilt towards the direction of the applied field, and in negative anisotropy displays, the molecules in the middle layers of a homeotropic liquid crystal will tilt away from the field direction; in both cases, there will be a so-called *mid-layer tilt* that of the molecules to produce the desired phase retardation.

When viewed from an oblique angle, however, the display's mid-layer tilt will not be seen at the same angle as the head-on view, and so there will be a different phase retardation. The oblique view of the mid-layer tilt was the chief culprit fomenting poor wide-angle viewing in the early liquid crystal displays. Compensating for its effects is critical for improving viewing angle image quality in LCDs requiring wide-angle viewing.

In line with the oblique angle-fomented unwanted phase retardation, an even more severe problem results when an oblique viewing angle just happens to line up with the tilt angle of the liquid crystal molecules in the mid-layer tilt. For example, in the mainstream negative anisotropy vertically aligned (VA) mode for televisions, when an electric field is applied, the molecules in the middle layers of the liquid crystal will turn to align with the field, causing the directors to tilt away from the vertical. When the oblique viewing line of sight just happens to align exactly with the mid-layer tilt angle, what should be phase-retarded light producing a desired grayscale (because the electric field is at an intermediate value) now is not phase-retarded because the light is parallel to the directors and passes through unaltered (like going through the *c*-axis). So for a normally white VA display, there is considerably more transmitted light than desired, resulting in severe light leakage (conversely for a normally black display, there will less than the desired light transmitted for that grayscale because more light will be blocked by the analyzer). The light leakage is severe enough to cause a reverse effect; the intensity not only does not properly follow the variation of the applied field strength, it can be reversed from what it should be. This is called *grayscale inversion*.

From this, it is clear that whatever display has a mid-layer tilt, there will be viewing quality problems at oblique angles and line-of-sight molecular tilt angles. The oblique angle viewing problems affect not only luminance, but also contrast and color quality.

Twisted Nematic Display Oblique Viewing

Recall that for the TN mode, if normally incident linearly polarized light from the first polarizer has its electric vector either parallel to (E-mode) or perpendicular to (O-mode) the rubbing-aligned directors at the surface of the liquid crystal, as the light progresses through the twisting liquid crystal directors, there will be phase retardation, and the plane of the electric vector will rotate to follow the twist of the directors (optical activity), and by appropriately setting the liquid crystal layer thickness, when the light reaches the analyzer its polarization state will be exactly aligned with the analyzer's optic axis and pass through, producing a bright field-off normally white (NW) state.

However, when viewed from an oblique angle, that light's electric vector will not be waveguided to precisely match the transmission axis of the analyzer, and some light will be blocked, resulting in a less bright white state.

Applying a full-charge electric field on a positively birefringent liquid crystal will cause the directors to align with the vertical field (in the mid-layer) and be parallel with the direction of propagation and perpendicular to the *E*-vector of the linearly polarized incident light. Thus normally incident light will just follow the vertical optic axis and suffer no phase retardation in its journey through the liquid crystal, so that when it reaches the crossed analyzer, its polarization state will continue to be exactly perpendicular to the optic axis of the crossed analyzer, and the light will be completely blocked, thereby producing the dark state. The liquid crystal mid-layer in this dark state can be described as a *homeotropic c-plate*.

But when viewing the TN dark state from oblique angles, the egress light will not be completely aligned with the vertically aligned molecules in the mid-layer, and this will cause the polarization state of the light at those oblique angles to depart from exact linear perpendicularity with the transmission axis of the analyzer, and this results in leakage of light through the analyzer and a less than complete dark state. Since the definition of contrast is just the ratio of the brightest white to the darkest dark, the first casualty at oblique viewing is the display's contrast.

In the intermediate field-on state, the mid-layer tilt is set for a desired gray level, but oblique-angle viewing means that the perceived tilt angle is not the same as for normal angle viewing, so the phase retardation at oblique angles is not that which is desired, and the polarization state will not be the one designed for interaction with the analyzer to produce the desired grayscale, resulting in unwanted changes in intensities and colors.

Negative and Positive Compensation

The twisted nematic liquid crystal displays can use compensation films to correct for oblique viewing-angle light leakage as illustrated in the simple example following. In the strong field-on dark state, because the TN mid-layer molecules are oriented vertically (forming a homeotropic c-plate), and since a positively birefringent liquid crystal ($\Delta n > 0$) is used, in the mid-layer region, a TN device is like a VA mode but of positive anisotropy, so it can be deemed a positive c-plate. Therefore, a simple compensation method would be to mount an equally birefringent *negative* c-plate ($\Delta n < 0$), with the same optical thickness (Δnd) on top of this positive c-plate to "negate" the undesirable phase retardation arising from the light traveling through an oblique viewing angle, that is, cancel the unwanted $+\delta$ with a $-\delta$ to reduce light leakage in the dark state.

Of course, such a compensation would have an undesirable polarization effect on the normally incident light, so the optic axis of the negative c-plate compensator must be aligned with the direction of propagation of light (in the conventional coordinate system, the compensator c-axis must be aligned with the z-axis), so that normally incident light will not be phase compensated by the compensator. Further, in order to maintain the left–right symmetry, the negative c-plate can be split into two halves and placed on the left and right sides of the liquid crystal cell. Computer simulations have shown that at a 60° viewing angle, the negative c-plate reduces light leakage from 30% to about 8%, with a concomitant improvement in contrast [1].

But the negative c-plate compensation method for TN displays can only correct the oblique angle viewing of the homeotropic-like positive c-plate molecules in the mid-layer. Those molecules near the glass plate are still lying flat or at a small pre-tilt angle, held by the strong anchoring of the alignment layer, so the negative c-plate will not compensate for any oblique-viewing angle-induced light leakage arising from those near-glass molecules. As described above, the homogenously oriented molecules near the glass substrate technically form a positive a-plate, so naturally, a *negative* a-plate could do the compensating. So adding negative a-plates near the glass substrates should provide compensation for the light leakage from those near-glass, flat-lying molecules. Further, just as in the case for the negative c-plate, to preserve left–right symmetry, the negative a-plate can be divided into two halves, with the optic axis aligned with the rubbing direction of the anchoring compound and placed in front of and behind the liquid crystal layer.

The result of these compensation layers is that for a 60° vertical-viewing angle, adding the negative a-plate reduces the original 8% light leakage down to 1%. So the negative c-plate and negative a-plate combination indeed reduces oblique-angle viewing light leakage in the TN display, and consequently significantly improves contrast.

In summary, the TN display mode dark-state light leakage can be compensated for by using reverse-phase retardation c-plates, but the flat-lying homogeneously anchored molecules in the TN must be consequently compensated for by reverse signage a-plates. From this, it can be seen that the homogeneously aligned a-plate-like display, such as the IPS, in principle will have less oblique light leakage compared with a homeotropic (standing and tilting molecules) c-plate-like display, such as the VA and the mid-layer TN. Indeed, the contrast of an IPS liquid crystal display is inherently superior to that of VA and TN displays.

The Discotic Solution

From the reverse-phase retardation, different-layer, compensating plate principles just described, it becomes apparent that compensation for oblique angle light leakage from different layers of the liquid crystal could be achieved by using many layers of different plates to exactly match the layers of the liquid crystal. That is, in principle, a multiple-layer combination of negative c-plates and a-plates could individually compensate for the unwanted positive (or negative) phase retardations at oblique viewing angles of each specific layer of the liquid crystal. Constructing such a

Figure 23.2 Discotic liquid crystal micrograph. Courtesy of J. Goodby.

compensation system, however, would require a large number of very thin compensation plates all judiciously stacked at appropriate angles in order to properly mimic the liquid crystal layer.

The discotic nematic liquid crystal introduced in Chapter 5 as shown in the micrograph of Figure 23.2 have naturally rigid spines and flexible tails plus wings, and behaves similarly to the calamitic nematic liquid crystal. The central portion of the discotic molecules is composed of firm biphenyls, and the flexible arms move in a plane, so the discotic liquid crystal molecule can be physically modeled as a flat disk. The formula, the structural model, and the schematic flat disk physical model are shown in the top, middle, and bottom of Figure 23.3, respectively.

The negatively birefringent Fuji Film's *Wide View Discotic Film*™ is a nematic liquid crystal (triphenylene derivatives) with disc-like molecules. The film structure is designed to simulate a stack of phase-retardation plates using a "hybrid alignment" to produce the changing molecular orientation. An alignment layer on a tri-acetyl cellulose (TAC) substrate aligns the discotic molecules at the film substrate utilizing appropriate rubbing directions; the orientation angles of the discotic molecules above gradually increase from the anchoring alignment at the top surface of the film. As can be seen in Figure 23.4, discotic molecules near the alignment layer are oriented almost horizontally at an angle by the anchoring layer at a rubbing direction as shown; this layer can simulate the birefringence of a negative a-plate placed near the glass substrates. The molecules in the middle layers of the

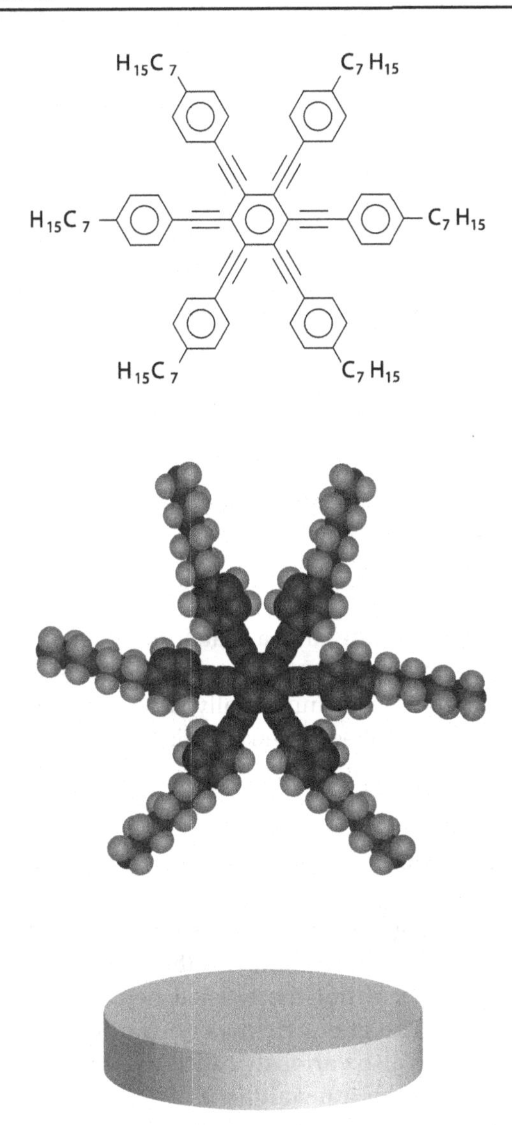

Figure 23.3 Chemical formula, structural model, and flat disk physical model of a discotic liquid crystal. From Yang, D.K., and Wu, S.T. *Fundamentals of Liquid Crystal Displays* (Wiley, 2006), p. 2.

film (the discotic layer) are rotating toward a vertical orientation at the mid-layer angle, and so the discotic layer can simulate the birefringence of a negative c-plate in the mid-layers [2].

In this arrangement then, an appropriately oriented discotic compensation film can appositely mimic the positive liquid crystal and thereby provide

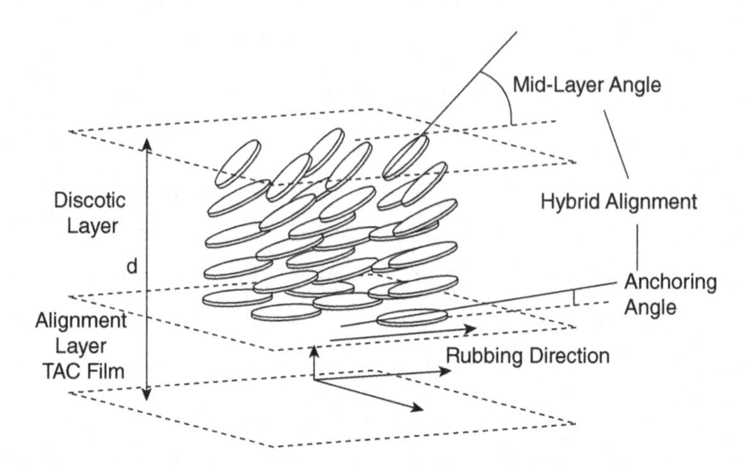

Figure 23.4 Fuji Film's Wide View Film discotic compensation film. Adapted from Fuji Film's WVF data sheet.

negatively birefringent c-plate and a-plate-type compensation in a single film. In addition to effectively decreasing light leakage in TN displays, the discotic compensation film is commercially successful because of its ease of use, needing only to be applied to the polarizer without any other changes to the panel manufacturing process.

Grayscale Inversion

Some of the wide-viewing angle problems were alleviated by multiple compensation films and by the clever design of the hybrid discotic liquid crystal compensator just described. But the twisted nematic display had another more serious problem: The grayscale depends on how much light is transmitted (the *transmittance,* also variously called the *transmission* and the *transmissivity*), and of course the transmittance (T) should be controlled by the applied voltage and be the same for a given voltage (V) at different viewing angles, that is, the relation between applied voltage and transmittance should be perfectly linear regardless of viewing angle, otherwise the display suffers a serious defect called *gray level instability.*

As shown in Figure 23.5, for a normally white O-mode TN display, in the middle regions of the graph, when an increasing voltage (V = 1.56–3.35 V, the parameter in the graph) is applied, at all viewing angles (the abscissa in the graph), the transmittance decreases, thereby providing a proportionally

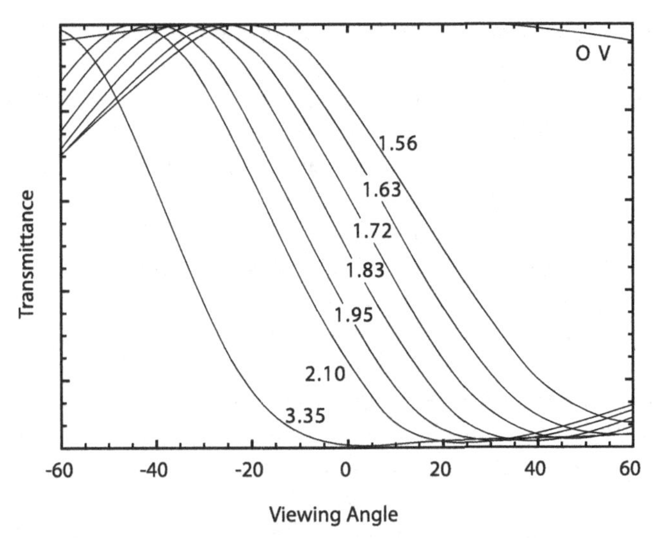

Figure 23.5 Grayscale inversion graph. From Yeh, P., and Gu, C. *Optics of Liquid Crystal Displays* (Wiley, 1999), p. 362.

variable grayscale as desired. In order to produce and control a fine grayscale, the transmittance curves with applied voltage parameters in the transmittance/viewing angle graph must be separated and distinct, that is, if the curves are not separated, different applied voltages will produce the same transmittance, resulting in loss of control of grayscale.

From Figure 23.5, it can be seen that at the voltage extremes, applying no voltage (0 V, top of graph), the transmittance is the highest (normally white state), and there is condignly little variation with viewing angle. Applying the highest voltage in the graph (3.35 V, dark state), the transmittance at the right of the graph is properly lowest (dark state) and at 0° viewing angle (head-on), the transmittance is the lowest, and so the head-on view provides the best contrast (ratio of white to black) as expected. The more separated are the transmittance curves, the more controllable and fine the grayscale can be; meaning no curve overlap, and more area in-between the curves to fit more finitely different voltages.

In the middle of the graph, the transmittance curves are indeed well separated, but with increase in Viewing Angle (abscissa), the separation deteriorates (note particularly the upper left and lower right corners of the graph). And at vertical-viewing angle greater than +20° (viewing from above), at different applied voltages, the transmittance curves near the bottom right of the graph begin to merge and even actually cross! Further, when the viewing angle grows negatively greater than –20° (viewing from below), the transmittance for the same 3.35 V markedly increases, meaning that there is serious light leakage in the dark state at those angles. This was so much so that in the LCD's early days, the quality of a screen could easily be discerned from looking at the screen from below and observing the distinct loss of contrast.

For greater negative vertical-viewing angles, the 3.35 V curve in the upper left of the graph also *crosses* the other voltage curves, even the 0 V curve! This is the dreaded *grayscale inversion*, whereby a lower voltage inversely (conversely and perversely) produces the lower transmittance that a higher voltage should produce, meaning a reversal at that angle where the scene that should be more bright perversely becomes more dark. Needless to say, such a dreadful state of affairs required immediate and careful attention.

The grayscale nonlinearity and inversion problems at oblique angle viewing arise from the relation between the mid-layer tilt angle and the oblique viewing angle. For the normally white TN, positive anisotropy LCD, in the field-off state, the mid-layer tilt of the molecules is zero (the molecules are twisting in the plane horizontal with the glass substrate only). Normally incident light is thus perpendicular to the directors of the flat-lying molecules, and thus the electric field vector of the linearly polarized light is in the same plane as the directors, and for the E- and O-modes, the electric field vector is respectively parallel and perpendicular to the rubbing aligned directors at the incident face. As the liquid crystal molecules rotate helically within the liquid crystal, there is phase retardation producing waveguiding through a 90° twist that aligns the plane of polarization with the transmission axis of the crossed analyzer, and the light thereby passes through unimpeded to produce the normally white state.

When an electric field is applied, the positive anisotropy liquid crystal directors will turn toward the direction of the applied field producing mid-layer tilt, and because of the change in orientation of the molecules, the phase retardation will be less, and the transmittance will decrease. If changes in the voltage cause proportional changes in transmittance, there is the desirable orderly linear relationship between voltage and brightness called *grayscale linearity*.

With the display disposed in the usual manner facing the viewer, for oblique angle viewing, first when viewing at a greater angle above the screen normal (called the *positive vertical*), as the voltage is increased, the brightness decreases as desired because there is less phase retardation (more light blocked by the analyzer), but at a certain applied voltage, the directors of the mid-layer tilting molecules will be turned to be exactly in line with the line-of-sight viewing direction, and suddenly there will be absolutely no phase retardation whatsoever, just as if light were traveling along a *c*-axis, and the transmittance will drop to zero (the plane polarized light has not been twisted and the linearly polarized light is totally blocked by the crossed analyzer). Further increasing the voltage will result in the transmittance perversely increasing instead of decreasing because the molecules will turn more past the line of sight, and there will be phase retardation with the plane of polarization once more twisting; this is the grayscale inversion.

When viewing at a negative angle toward the bottom half of the screen (*negative vertical*), since the mid-layer directors will be tilted away from the viewing direction, there will be greater average phase retardation and more waveguiding resulting in greater transmittance, which explains the greater light leakage at negative vertical viewing angles, and the grayscale linearity is better than at positive vertical viewing angles.

Grayscale inversion can be technically analogized as a *positively* birefringent liquid crystal, due to the mid-layer tilt being aligned exactly within the line of sight, behaving at those viewing angles and greater like a *negatively* birefringent liquid crystal at normal incidence. That is, since the transmittance increases rather than decreases with increasing applied electric field at the inversion viewing angles, it is as if the molecules were turned in the opposite direction by the field, as in a negatively birefringent liquid crystal. And herein lies the solution of the problem. As previously described, compensation in principle should be to add a positively birefringent plate to compensate for the undesirable "negative behavior," but again positioned so that it does not affect the normally incident light; this just describes a positive c-plate.

But as also mentioned above, doesn't a positive c-plate behave just like a homeotropic liquid crystal that itself has severe light leakage at oblique viewing angles; can its addition really help matters? To avoid the positive c-plate's own light leakage problems, it must be oriented such that its optic axis is not aligned with the light propagation direction (the *z*-axis), but rather at some angle with the *z*-axis. In principle, that angle should be perpendicular to the average orientation of the directors in the mid-layer molecules to negate the unwanted phase retardation caused by that orientation. Since the mid-layer tilt angle average is 45° pointing toward the upper half of the screen, the optic axis of the positive c-plate compensator then should be oriented at 45° with the lower half of the screen. This type of orientation for a c-plate turns it into what is called an *o-plate* because its optic axis makes an *o*blique angle with the *xy*-plane. In order to avoid adversely affecting the normally incident light, the o-plate must be combined with a compensation plate, which has an optic axis that is perpendicular to the o-plate, and that is just an *a-plate*.

Fortunately, the o-plate and a-plate combination described above actually can be provided by a single biaxial compensation film. One of the optic axes of the biaxial film must be aligned with the direction of propagation of a normally incident beam to avoid phase retarding the normally incident light, and the orientation of the other optic axis can be adjusted as described above.

Biaxial compensation films are generally constructed by unidirectionally stretching an isotropic polymer material. The stretching will not produce any *stretched* orientational or positional molecular order, but the stretch direction will produce what is called a *compensation film axis of symmetry* while retaining the two optic axes of the biaxial medium. In practice, a negative c-plate and negative a-plate are usually added to the biaxial compensation film to further reduce light leakage.*

Compensation Overview

As an example of the calculations involved for oblique angle view compensation, general expressions for the phase retardations for a c-plate and an a-plate will be derived. For light of wavevector k passing through a uniaxial medium having n_e and n_o extraordinary and ordinary indices of refraction, the phase retardation is

$$\Gamma = (k_{ez} - k_{oz})d.$$

where the wavevector components k_{ez} and k_{oz} of the extraordinary and ordinary waves are with respect to the z-axis (normal to plate surface), and d is the layer thickness. The light beam makes an oblique incidence angle of θ_o with the z-axis; the incident plane makes an angle ϕ_o with respect to the x-axis in the standard coordinate system, and the optic axis of the medium is oriented with a tilt angle of θ_n from the xy-plane and an azimuthal angle of ϕ_n; the wavevectors can be derived from Maxwell's equations as [3],

$$k_{ez} = \frac{2\pi}{\lambda}\left[\frac{n_e n_o}{\varepsilon_{zz}}\sqrt{\left(\varepsilon_{zz} - 1 - \frac{n_e^2 - n_o^2}{n_e^2}\cos^2\theta\sin^2(\phi_n - \phi_o)\right)\sin^2\theta_o} - \frac{\varepsilon_{xz}}{\varepsilon_{zz}}\sin\theta_o\right].$$

* As can be discerned from the discussion, although complicated, thinking about and devising viewing angle compensation schemes can enhance understanding of liquid crystal optics.

and

$$k_{oz} = \frac{2\pi}{\lambda} \frac{n_e n_o}{\varepsilon_{zz}} \sqrt{n_o^2 - \sin^2 \theta_o},$$

where $\varepsilon_{xz} = (n_e^2 - n_o^2)\sin\theta_n \cos\theta_n \cos(\phi_n - \phi_o)$ and $\varepsilon_{zz} = n_o^2 + (n_e^2 - n_o^2)\sin^2\theta_n$. A general expression for the phase retardation of a uniaxial compensation plate can be derived from the equations immediately above as

$$\Gamma = \frac{2\pi}{\lambda} d \left[\frac{n_e n_o}{\varepsilon_{zz}} \sqrt{\left(\varepsilon_{zz} - 1 - \frac{n_e^2 - n_o^2}{n_e^2} \cos^2\theta \sin^2(\phi_n - \phi_o) \right) \sin^2\theta_o} \right.$$
$$\left. - \frac{\varepsilon_{xz}}{\varepsilon_{zz}} \sin\theta_o - \sqrt{(n_o^2 - \sin^2\theta_o)} \right].$$

As can be seen from the oblique incidence equation above, the phase retardation is a function of the optic axes of the medium and the angle of incidence of the light, just as physically explained above. For the c-plate, the optic axis is perpendicular to the surface of the plate ($\theta_n = 90$) and so the phase retardation is independent of the azimuthal angle ϕ_n,

$$\Gamma_c = \frac{2\pi}{\lambda} d n_o \left[\left(\sqrt{\left(1 - \frac{\sin^2\theta_o}{n_e^2}\right)} - \sqrt{\left(1 - \frac{\sin^2\theta_o}{n_o^2} - \right)} \right) \right].$$

For an a-plate, the optic axis is parallel to the surface of the plate ($\theta_n = 0°$), the phase retardation is then

$$\Gamma_a = \frac{2\pi}{\lambda} d \left[n_e \sqrt{\left(1 - \frac{\sin^2\theta_o \sin^2(\phi_n - \phi_o)}{n_e^2} - \frac{\sin^2\theta_o \cos^2(\phi_n - \phi_o)}{n_o^2}\right)} \right.$$
$$\left. - n_o \sqrt{1 - \frac{\sin^2\theta_o}{n_o^2}} \right].$$

So the phase retardations for viewing angle compensation can be calculated, and thus wide-angle view compensation schemes can be theoretically designed and implemented. But from the rather complicated examples given, it becomes clear that these multiple-plate compensating schemes and their system calculations could easily get out of hand.

Furthermore, the many and varied positive/negative compensation combinations and their interdependent optic axes angular adjustments must also take into consideration the reflection of light, as well as its refraction. This back-and-forth reflection added onto the already complicated refraction, together with off-axis transmission effects, makes for a rather complicated optical analysis. And in spite of all the compensation plates and schemes, the basic crossed polarizer and analyzer will always leak some light at oblique viewing angles because their absorption axes cannot be always exactly orthogonal.

These complications can be handled by the *extended Jones matrix* methods, which considers off-axis transmission and reflection, as well as refraction. The transmission characteristics of a birefringent optical system, including multiple surface reflections, as well as refraction, can be calculated using a 4×4 matrix, but upon reflection the calculations are quite complicated. Considering only refractions and single reflections, the optical elements can return to 2×2 matrices in an extended Jones matrix formulation, which is utilized extensively in optical system simulation calculations for LCDs, including the TN, MVA, and IPS, and the reflective and transflective displays (described in Chapters 24 and 27 respectively) [4].

Further fine-tuning for improving oblique angle viewing of the liquid crystal display by even more compensators became ever more complicated, but all the compensation schemes are based on just those "accentuate the positive to eliminate the negative" and vice-versa principles just described, and the contrast and subsequent grayscale and color saturation at oblique viewing angles indeed have been greatly improved.

Finally, the natural molecular vibrations induced by heat will cause particularly the mid-level molecules in any liquid crystal to randomly vibrate and thus change orientation, thereby engendering different polarization effects, so no compensation scheme, regardless of complexity and optimization, can completely eliminate light leakage. From this, one can perhaps better understand the role that heat has to play in any material and electronic device, and why temperature can influence the wide-angle viewing quality of a liquid crystal display.

The twisted nematic display's low threshold voltage requirement and good contrast are among its most attractive attributes, and have made the TN display most suitable for use in monitors, notebook computers, small-screen television, and the small- and medium-size displays of mobile phones and other personal electronics devices. The vertical alignment display's superior contrast and response speed make it ideal for dynamic image reproduction and use in projection displays. But to attain the wide-viewing angle quality required for direct-view television, both the TN and the VA display modes required the complicated compensation film schemes just described.

The compensation schemes further complicated the manufacture of the display system. For instance, the overall combinations require very precise process registration of the compensation plates and films, and all those added compensators can significantly decrease display luminance. Furthermore, there are increases in weight, panel thickness, and cost, all of which must be considered in the final mix.

After all that compensation work, it became clear that the twisted nematic and the vertical alignment display modes were not the best choices for large-screen televisions requiring good wide-angle viewing, and now the mainstream liquid crystal television sets primarily use the in-plane switching (IPS) and the multi-domain vertical alignment (MVA) liquid crystal display modes.

In the IPS mode, since the liquid crystal molecules under an applied electric field only turn in a plane parallel to the glass substrates, ideally the molecules do not tilt vertically out of the plane, so the attendant problems with view angle are obviated; there is no need for c-plate or o-plate type compensation for light leakage and grayscale inversion. The wide-angle viewing in IPS displays is inherently the best of the LCD modes, but there are other issues of fringing field transmittance and manufacturing tolerances (to be described in Chapter 24).

In the other mainstream liquid crystal display mode, an anchoring alignment layer causes the molecules near the glass substrates to stand up, so a negative anisotropy liquid crystal in this vertically aligned (VA) mode can be described as a negative c-plate. Thus a positive c-plate should improve the wide-view angle. But when an electric field is applied, the mid-layer molecules will almost all be tilted away from the field direction, so light leakage caused by oblique-view mid-layer tilt is particularly acute.

To solve this problem, Fujitsu developed a display having "protuberances" on the glass substrate surfaces to promote multiple domains of different director orientation molecules that would effectively cancel out the undesirable phase retardations at wide-viewing angles. This so-called *multiple-domain vertical alignment* (MVA) quickly became one of the two liquid crystal display modes of choice for large-screen televisions. The IPS and MVA technologies will be described further in Chapter 24.

References

[1] Yeh, P., and Gu, C. 1999. *Optics of Liquid Crystal Displays*. Wiley, New York, p. 357ff.

[2] Mori, H., et al. 2001. Necessity of the hybrid alignment structure of the wide view film, *Asia Display 2001*. IDW, p. 593.

[3] Lien, A. 1997. *Liq. Cryst.*, **22**, 171; Yang, D.K., and Wu, S.T. 2006. *Fundamentals of Liquid Crystal Displays*. Wiley, Chicester, p. 209ff.

[4] Jones, R.C. 1941. *J. Opt. Soc. Am.*, **31**, 488; Yeh, P. 1982. *J. Opt. Soc. Am.*, **72**, 507–513.

24

Liquid Crystal Television

A computer monitor or notebook computer is usually viewed by one person at a time. so their wide-viewing angle quality perhaps is not that important, but for a living room television set, wide-viewing angle quality is an objective that must be attained and a technical problem that must be overcome.

The compensation films developed for the twisted nematic (TN) liquid crystal displays described in Chapter 23, albeit improving the wide-view quality, also brought process complexity, loss of transmittance, and undesirable gains in weight and thickness. The addition of more and more increasingly complicated compensation schemes led the Japanese companies Fujitsu and Hitachi to instead attack the problem at its base; that is, at the fundamental liquid crystal molecular orientation level. And so Fujitsu developed the multiple-domain vertical alignment (MVA), and Hitachi the in-plane switching (IPS) display modes to directly address the wide-viewing angle problem.

Vertical Alignment

A transparent silane coupling agent, or a deposition of a silicon polyoxide layer (SiO_x), applied as an anchoring agent on the ITOs and glass substrates

Liquid Crystal Displays, First Edition. Robert H. Chen.

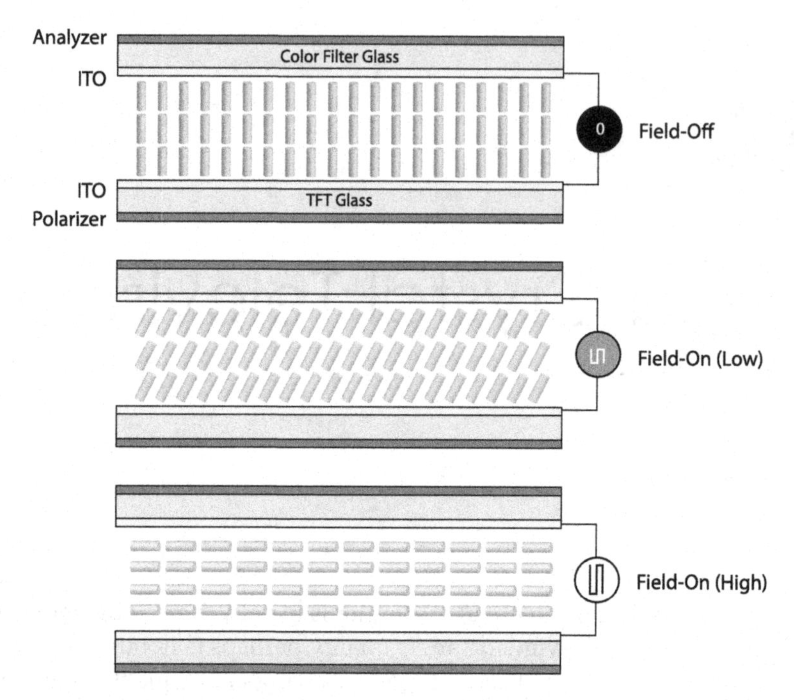

Figure 24.1 Vertical alignment with field-off (top), low field-on (middle) and high field-on (bottom).

can cause the molecules near the glass to attach their feet to the anchoring agent and stand up, as shown schematically in Figure 24.1. Because of the charge polarization structure of the anchoring molecules, they can effectively attract the oppositely charged polarization extremity of the liquid crystal molecules near the glass substrates, and the molecules will stand vertically; that is, perpendicularly to the glass substrate; and as described in Chapters 11 and 23, the field-off homeotropic VA mode can be described as a c-plate.

Incident linearly polarized light propagating in a direction parallel to the *c*-axis (normal to the c-plate) will traverse the VA liquid crystal cell with no phase retardation and consequently without any changes to its polarization state. Thus a crossed analyzer will block light, regardless of wavelength, and there will be no light leakage from different colors that are phase-retarded differently. Further, since there is no polarization required to turn the plane of polarization so the light will be blocked by the analyzer, the VA black state does not depend on the thickness of the liquid crystal layer (the cell gap, d), and since there is no structure-induced phase retardation, structure-altering thermal motion will not have any effect on the polarization state, so the VA black state also is insensitive to temperature changes. Therefore, in the field-off state, the VA black state is "unphased" by wavelength, cell gap, or temperature, so there is very little leakage, producing a very dark black state that contributes to a very good contrast ratio; in fact, the VA panel field-off head-on view contrast is the best of all the different liquid display modes.

When an electric field is applied, it confers a preferred direction for the molecules' intrinsic dipole moments, and they will tend to align with the direction of the applied field.In the negative anisotropy liquid crystal commonly used in the VA display mode, the molecules will turn away (tilt) from a vertical electric field direction. The tilted molecules will engender phase retardation in light passing through and the polarization state of that light is changed resulting in some light passing through the analyzer. This operation is called generally *electrically controlled birefringence* (ECB), and is depicted schematically in a VA cell, comprised of a polarizer, TFT glass, ITO, liquid crystal, color filter glass, and analyzer in Figure 24.1.

The VA c-plate field-off black state in Figure 24.1 (top) shows the vertically aligned molecules forming a *c*-axis through which normally incident light will pass unretarded. When a low power vertical electric field is applied as shown in Figure 24.1 (middle), the negative anisotropy liquid crystal molecules will tilt slightly away from the vertical and the linearly polarized light from the polarizer will be slightly phase-retarded because the angle between the light *E*-vector and the liquid crystal induced molecular dipole moment changes slightly, and the change in polarization state will result in some light being transmitted through the analyzer, thereby departing from the black state to a dark gray state.

When a strong electric field is applied as shown in Figure 24.1 (bottom), the molecules in the mid-layer will turn to lie in a horizontal plane, and the state of polarization depends on the phase difference, which is provided by the birefringence (Δn) and thickness (d) of the liquid crystal layer for light of wavelength (λ). From the phase retardation parameter derived in Chapter 16, the phase difference is given by

$$\delta = \frac{2\pi}{\lambda}\Delta n \cdot d.$$

In order to effect a linear polarization of the exiting light but turned 90° at egress, recall that the phase difference must be an integer multiple of π, so the thickness d of the liquid crystal layer (and the choice of liquid crystal birefringence) must be set such that

$$\delta = \frac{2\pi}{\lambda}\Delta n \cdot d = \pi.$$

This phase difference will turn the linear polarization to exactly align with the transmission axis of the crossed analyzer and the light will pass through, producing the white state (since $d = (\lambda/2)\cdot(1/\Delta n)$, this is sometimes called a *half-wavelength plate*). Unfortunately, this white state is dependent on wavelength, cell gap, and temperature, but that does not negate the good VA black state (which is wavelength, cell gap, and temperature independent, it merely does not improve the contrast ratio any further.

Recall that in Chapters 5 and 12, the viscosity coefficient γ and the liquid crystal layer thickness d is related to the response time as

$$t_s \sim \gamma \cdot d^2.$$

and since the white state transmissive display requirement for the VA is $\Delta n \cdot d = \lambda/2$ from the phase difference equation above, but the requirement for TN is $\Delta n \cdot d = \lambda/1.15$, a smaller cell gap is possible in the VA mode displays, and this can appreciably reduce the response time as seen from the equation above. Further, as will be described below, the VA mode can employ liquid crystals having higher elastic coefficients that snap back faster, so the response time of the VA mode is the fastest of the three mainstream display modes. The superior contrast ratio and fast response times are impressive attributes, but the VA display mode suffers from poor oblique viewing angle quality that required considerable improvement, as described in Chapter 23.

Multiple-Domain Vertical Alignment

In a negative anisotropy VA liquid crystal mode, when an electric field is applied, the liquid crystal directors will turn away from the vertical field and this turning will engender phase retardation and subsequent change in the polarization state of the light passing through, which in conjunction with the analyzer, modulates the intensity of the light resulting in a controllable grayscale. However, as described in Chapter 23, there can be severe light leakage at oblique viewing angles from unwanted phase retardation and line-of-sight co-linearity with the mid-layer tilt. If compensation films are not used, the VA panels are only suitable for applications where wide-viewing angle is not a requirement, such as in nondirect-view liquid crystal on silicon (LCOS) projectors where the illumination and exit angles are fixed. For direct-view

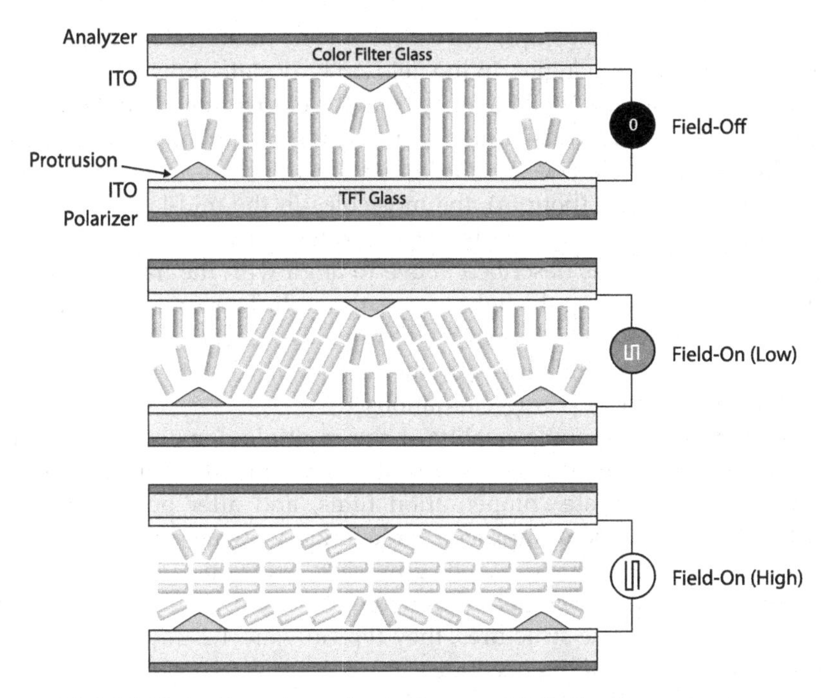

Figure 24.2 Multiple-domain vertical alignment (MVA) with protrusions on the ITO.

television requiring wide-viewing angle quality, the VA display mode used rather complicated compensation plate schemes, such as described Chapter 23, or the problem could be attacked at a more fundamental level.

Following the principle of phase-retardation compensation described in Chapter 23, one solution is to use an opposite mid-layer tilt phase retardation to cancel out the undesirable mid-layer tilt phase retardation causing the light leakage. As shown in Figure 24.2, protrusions are constructed on the ITO surface facing the liquid crystal, and since the VA alignment layer causes the proximate liquid crystal molecules to stand perpendicular to the surface, protrusions having surfaces at an angle with the glass substrate will cause the molecules to stand obliquely in antisymmetric directions owing to the symmetrically-opposite sides of the protrusions. The protrusions thus produce domains of opposite phase retardation, which will cancel out the unwanted phase retardations, which lead to light leakage and thus provide an improved wide-angle view [1].

In the field-off black state shown schematically in Figure 24.2 (top), the different tilt angles in different domains, arising from their antisymmetric orientation on the protrusions, will cancel out any undesirable phase retardations at wide-viewing angles. In the field-on gray state shown in Figure 24.2 (middle), the molecules will tilt away from the electric field, but the opposite tilt molecules in the antisymmetric domains will tilt oppositely, thereby canceling undesirable phase retardations. In the field-on full white state shown in Figure 24.2 (bottom), the molecules in the mid-layer will almost all lay perpendicular to the electric field and the plane of polarization of the light will be rotated as described above to align with the transmission axis of the analyzer, allowing the light to pass through. Those molecules near the glass substrates will be anchored by the alignment layer and tilt in response to the strong electric field, but again the antisymmetric tilting molecules will cancel the undesirable phase retardations.

The wide-viewing angle quality of the multiple-domain vertical alignment (MVA) display developed by Fujitsu can be further enhanced by the addition of appropriate compensation films, and after optimization, the MVA not only produces improved wide-angle viewing, but also achieves good grayscale stability and avoids grayscale inversion.

Figure 24.3 is a scanning electron micrograph showing the MVA protrusions as gentle hill-like structures; they typically are 1.2–1.5 μ in height and cover about 16–30% of the pixel area.

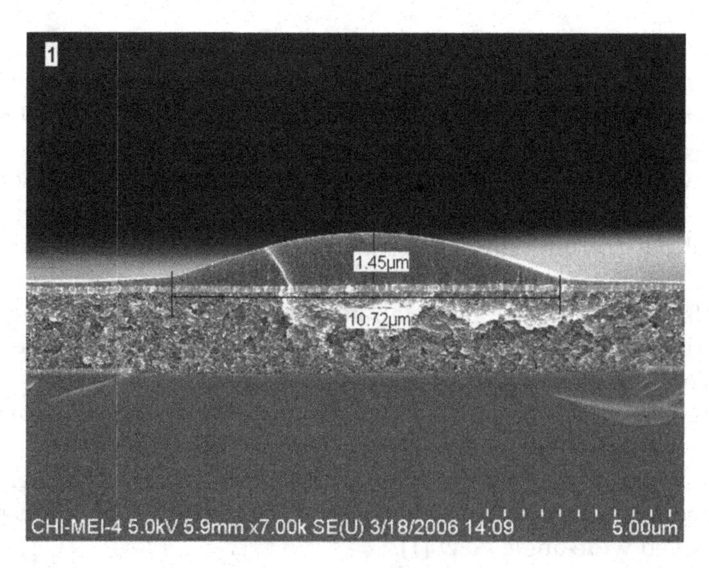

Figure 24.3 Scanning electron micrograph of an MVA protrusion. Courtesy of Chimei Optoelectronics.

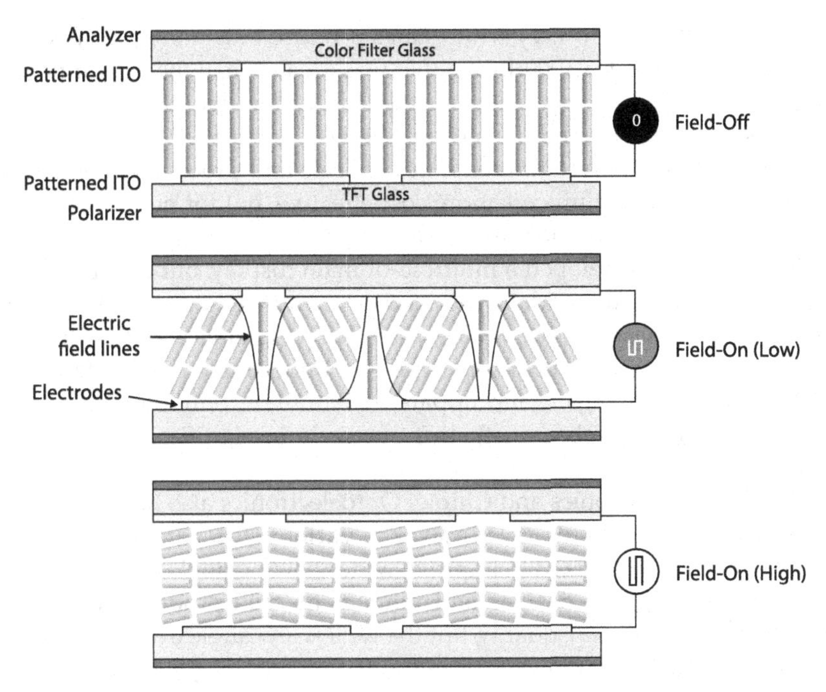

Figure 24.4 Patterned vertical alignment (PVA) with antisymmetric slits.

In the wake of the success of the MVA display mode, Samsung developed a multiple-domain scheme that used saw-tooth patterned slits etched on the ITO and the color filter in place of the protrusions. In the so-called *patterned vertical alignment* (PVA), anti-symmetric slits will produce a slightly bent electric field that will form the domains of different tilt angle molecules.

As shown schematically in Figure 24.4, since there are no protrusions, the molecules near the substrates are all standing straight up (homeotropic alignment) which, as discussed, produces no phase retardation and thus a very dark normally black state as shown in Figure 24.4 (top). When an intermediate electric field is applied as shown in Figure 24.4 (middle), the molecules tilt in opposite directions in different domains, thereby compensating for unwanted phase retardations just as in the MVA, producing good wide-angle viewing grayscale. When the maximum field is applied as shown in Figure 24.4 (bottom), almost all the molecules lay almost flat (homogeneous alignment), resulting in almost no undesirable phase retardation at oblique viewing angles [2].

These very black and very white states produce outstanding display contrast, which can reach a 1000:1 ratio that was greater than the contemporaneous MVA and IPS displays. But since the PVA system requires an additional etching step for producing the slits on the ITO, fabrication costs are higher and yields are lower. Again, economies of scale and process refinement can make the PVA display process more cost-effective, but for now, PVA is used primarily for high-end, more expensive displays.

Sharp also has developed a multiple-domain display, but uses an axially-symmetric field with randomly oriented surrounding domains. This so-called *axially symmetric vertical alignment* (ASV) scheme is predicated on a random orientation of domains to cancel out undesirable phase retardations at oblique-viewing angles, and apparently counting only on statistics and geometry to do the job, nonetheless succeeds in significantly improving viewing angle quality.

Taiwan's AU Optronics and Chimei Optoelectronics also have developed improvements on Fujitsu's MVA, as well as various clever amalgams of MVA and PVA that are optimized to improve wide-viewing angle quality. It is safe to say that future new designs and refinements will produce ever greater improvements in display quality, but the wide-view angle developed to date is already more than sufficient for high-quality television viewing.

In-Plane Switching

In addition to the various multiple domain schemes just described to avoid the mid-layer tilt-caused viewing angle problem, another clever way to improve viewing angle quality is to employ a transverse (instead of vertical) electric field that constrains the molecules to twist solely in a plane parallel to the glass substrate (and thusly avoid the mid-layer tilt altogether). This so-called *in-plane switching* drive mode was first proposed by the French physicist R.A. Soref [3] in the 1970s, developed by Hitachi in the 1980s, and implemented in LCDs in the 1990s [4].

As shown schematically in Figure 24.5, the liquid crystal molecules are in homogeneous mode (lying horizontally with the glass substrate), and anchored at a specific rubbing angle (e.g., 45°) with the electrodes. The polarizer and analyzer are crossed, and the polarizer can be set for E-mode (incident light plane-polarized parallel to the liquid crystal directors at the surface) or O-mode (perpendicular to the directors at the surface) operation.

In the field-off normally black (NB) state shown in Figure 24.5 (top), the light electric field vector is parallel (E-mode) or perpendicular (O-mode) to

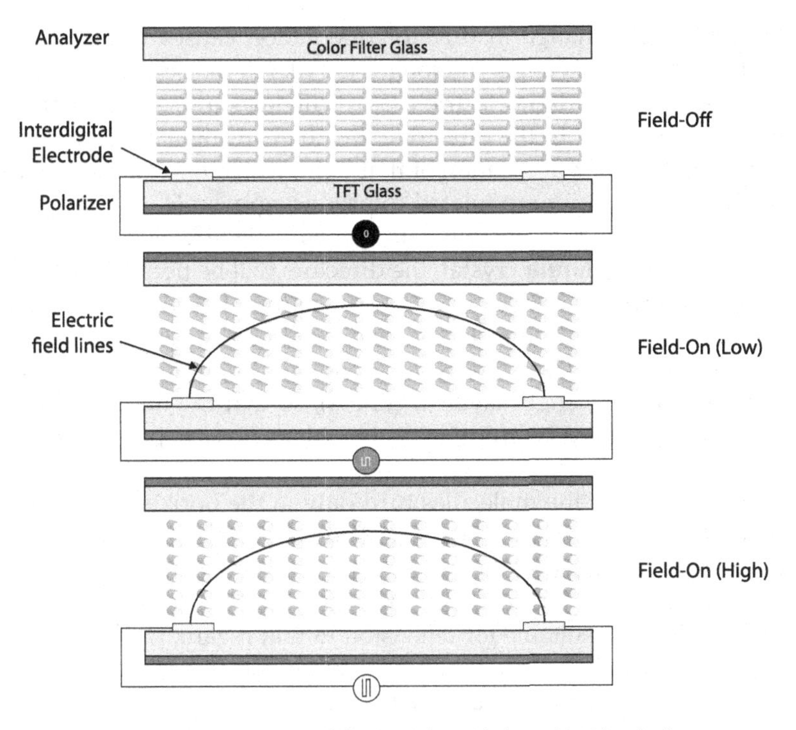

Figure 24.5 In-plane switching (IPS) liquid crystal molecules in homogeneous mode.

the rubbing direction and so also to the molecules' directors, and the light therefore undergoes no phase retardation, since the electric field vector is not decomposed into components. Since the orientation of the molecules does not change throughout the liquid crystal layer, the light passing through suffers no phase retardation and no subsequent change in polarization state, and as in the VA display mode, there is no wavelength, cell gap, or temperature dependence, and the linearly polarized light is ideally totally blocked by the crossed analyzer to produce a good black state.

As shown in Figure 24.5 (top), the in-plane switching mode utilizes *interdigital* common and grid electrodes (separated like fingers on the hand) fabricated in the TFT glass substrate to generate a horizontal (transverse) *fringing* electric field between the electrodes and extending up into the liquid crystal in planes parallel to the glass substrate.

In the field-on state shown in Figure 24.5 (middle), the in-plane horizontal electric fields produced by the interdigitated electrodes cause the liquid crystal molecules to turn within the horizontal plane toward the applied field direction, and the change in director orientation causes the light electric vector to now make an angle (no longer parallel or perpendicular) with the directors, and thus the light electric vector will break into components parallel and perpendicular to the directors, thereby causing a phase retardation and general elliptical polarization of that light, which upon passing through the analyzer, is intensity modulated to produce grayscale.

In the maximum field-on full state shown in Figure 24.5 (bottom), in a negative anisotropy liquid crystal, the directors will be perpendicular to the applied field direction, and in a positive anisotropy liquid crystal, the directors will be aligned with the electric field. To produce the white state upon egress, the phase retardation must be an integer multiple of π as in the VA mode (the half-wavelength layer thickness), so that the linear polarization will be turned 90° to pass through the crossed analyzer to produce the white state.

In principle, since the molecules turn only in the horizontal plane, there are no mid-layer tilt molecules causing oblique viewing angle light leakage (as described in Chapter 23), and so lying down and turning provides a wider view. The IPS display mode provides the best wide-viewing angle quality, and is most suitable for television in that regard.

In practice however, since the interdigitated electrodes are formed only on the TFT glass substrate (and have no counter electrode on the color filter glass substrate), in order for the electric field to reach the molecules in the liquid crystal mid-layer, there must be a strong vertical component of the electric field acting above the electrodes. In the region of this vertical electric field, the molecules may preferably tilt rather than twist, and this upward angle of turning results in an unwanted phase retardation that changes the polarization state, and this can significantly decrease light transmission in regions directly over the electrodes; so that, for example, the transmission of a positive anisotropy, crossed-polarizers IPS liquid crystal display is about 75% of that of a typical twisted nematic LCD.

Since the twist in an IPS display is not a simple function of the light path length through the liquid crystal, there is no simple expression for the transmittance, but a Jones vector analysis of the system can provide a simplified engineering equation for the transmittance (T), assuming a homogeneous parallel alignment,

$$T = T_O \sin^2[2\theta(V)] \cdot \sin^2\left(\frac{\pi \cdot \Delta n \cdot d}{\lambda}\right),$$

where T_o is a composite engineering factor combining the transmittance of the polarizer and the liquid crystal cell aperture ratio, θ is the angle between the changing optic axis (initially parallel to the director in E-mode, perpendicular to director in O-mode) of the liquid crystal and the direction of polarization of the incident light (the light electric field vector), and is thus a function of the applied voltage V as $\theta(V)$, Δn is the birefringence of the liquid crystal, d is the cell gap, λ is the wavelength of the incident light, and the argument of the sine function $(\pi \cdot \Delta n \cdot d / \lambda)$ is just half the phase retardation parameter Γ derived in Chapter 16.

If the typical values of the birefringence and cell gap are taken to make $\Delta nd \sim 0.3\,\mu\text{m}$, and if the wavelength interval is restricted to visible light (350–750 nm), then the far right-hand factor in the equation is approximately equal to unity, and the equation simplifies to

$$T = T_O \sin^2[2\theta(V)].$$

To give a feeling for the veracity of this equation, a simple limiting values analysis can be performed: For a positive anisotropy liquid crystal under E-mode operation, in the field-off state ($V = 0$), θ is still zero, and the transmittance T is also zero, which is just the normally black state. Applying the strongest field (about 7 V, which is generally higher than that for TN and MVA), the positive anisotropy liquid crystal molecules will align with the electric field, and in this case, if the alignment layer rubbing direction is the typical 45°, then the $2\theta(V)$ term is 90°, and then $\sin^2[2\theta(V)] = 1$, so the transmittance T is equal to T_o, and this is just the highest transmittance. Thus the equation at least provides expected results for the extreme field on/off cases for the assumption of a completely homogeneously parallel alignment.

The threshold electric field strength required to cause the mid-layer molecules to begin to turn is a function of the liquid crystal layer thickness d, the twist coefficient of elasticity K_{22}, the permittivity of free space ε_o, and the dielectric anisotropy $\Delta\varepsilon$, and was derived from the Ericksen–Leslie theory for a homogeneous in-plane switching Freedericksz cell in Chapter 12 as

$$E_{th} = \left(\frac{\pi}{d}\right)\sqrt{\frac{K_{22}}{\varepsilon_o \Delta\varepsilon}},$$

but for the IPS display mode, the influence of the electrode's interdigital distance l must also be considered, which in terms of the threshold voltage is

$$V_{th} = \left(\frac{\pi \cdot l}{d}\right)\sqrt{\frac{K_{22}}{\varepsilon_o \Delta\varepsilon}}.$$

Compared with TN and VA, in the early IPS displays, in spite K_{22} being generally lower than the elastic coefficients for TN and VA, the threshold voltage was comparatively high, evidently because of the (l/d) factor and the low anisotropy of the IPS liquid crystals, which implies a higher rotational viscosity, requiring a greater driving force for in-plane turning. Modern IPS displays use specially formulated high-positive anisotropy liquid crystals to efficiently effect the in-plane turning and lower the threshold voltage; and as to be described below, the interdigital distance to cell gap ratio can be adjusted. As in TN displays, the IPS liquid crystal is driven by an alternating current square wave.

From the threshold voltage equation above, it can be seen that unlike the TN and MVA modes, the threshold voltage depends on the interdigital electrode gap (l), as well as the cell gap (d), and so IPS cells requires precise electrode gaps, making manufacture more difficult and time consuming. Also, to avoid light leakage from the sides, the interdigital electrodes in early IPS displays were opaque, and this reduced the aperture ratio to about 50% less than that for TN displays.

But the major drawback of the IPS display mode was the comparatively low transmittance and lack of image uniformity. This was ascribed to the region directly above the electrodes, where the fringing electric field is mainly vertically oriented, and so rather than twisting the molecules, tilts them toward the vertical. This vertical fringe field results in less phase retardation in those regions, and thus decreased light transmittance relative to other regions of the display, particularly in comparison with the space immediately between the electrodes where the horizontal component of the electric field is strongest. This discrepancy in horizontal molecule-turning effectiveness was seen as the cause of the poor transmittance and uniformity in the IPS display mode.

Figure 24.6 is a computer simulation of an IPS cell showing how the IPS liquid crystal molecules orientation are affected by the electric field produced by the interdigital electrodes situated below. At the top of Figure 24.6 is the curve of the calculated transmittance; the sectoral variation in the transmittance is clear, and this presents obvious image uniformity problems [5].

Fringing Field Switching

In order to alleviate the adverse vertical fringing effect on the transmittance, it is then necessary to increase the regions of horizontal electric field and

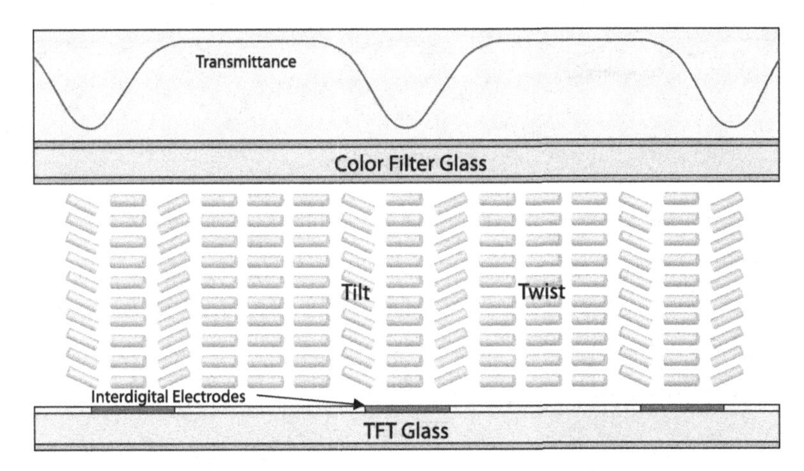

Figure 24.6 IPS liquid crystal molecules orientation fringing field from the interdigital electrodes.

decrease those of the vertical electric field. Since the regions in-between and at the edges of the electrodes produce the vertical fields, bringing the electrodes closer together and optimally adjusting the length of the electrodes should decrease the extent of the vertical electric field, and thus will help the horizontal component of the electric field overcome the vertical component of the electric field at the electrode edges [6].

Further, since there is strong anchoring at the glass substrate surfaces, the thickness of the liquid crystal layer (the cell gap d) will also affect the twisting efficacy of the molecules, that is, the thinner the cell gap, the more energy is required to twist the molecules, since the mid-layer molecules will be more greatly influenced by the anchoring.

So the width of the interdigitated electrodes (w), the distance between the electrodes (the electrode gap l), and the cell gap (d) are the dimensions that can be adjusted to optimally produce a more uniform transverse electric field to twist, rather than tilt, the molecules.

Typical *fringing field switching* (FFS) mode parameters are $w = 1–3\,\mu m$, $l = 0–5\,\mu m$, and $d = 2–5\,\mu m$. Adjusting these parameters, simulations have shown that when w is small, and l is near d, the fringing field is able to almost eliminate the vertical electric field component and effectively in-plane twist the molecules even directly above the electrodes, as shown in Figure 24.7. The simulation used a width (w) of $3\,\mu m$, length (l) of $4.5\,\mu m$, a cell gap (d) of $4\,\mu m$, and 4.5 V; the results is that that the molecules are almost completely in-plane, and the transmittance is much more uniform [7].

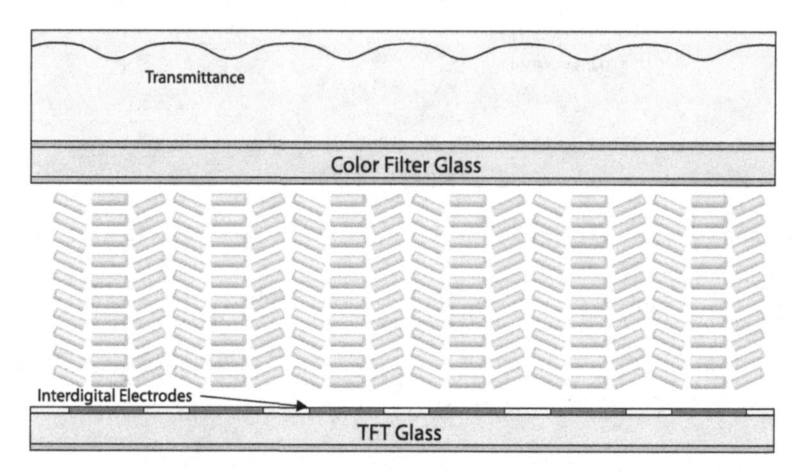

Figure 24.7 Fringing field switching (FFS) where molecules are almost completely in-plane with uniform transmittance.

Thus the FFS mode, while turning the molecules in-plane as in IPS mode, eliminates the molecules' tilt at the interdigital electrode fringes and thus achieves greater transmittance uniformity and intensity. In addition, the interdigitized common electrodes and grid electrodes used in modern IPS displays are transparent ITOs instead of the opaque electrodes used in early IPS displays, thereby considerably improving the aperture ratio.

Because the charge polarization of positive anisotropy ($\Delta\varepsilon > 0$) liquid crystals is along the long molecular axis, there is a stronger anisotropy (than that for negative anisotropy molecules) and a subsequent lower viscosity, and so using specially synthesized high-positive anisotropy liquid crystals in the FFS display can lower operating voltages and produce faster response times.

There is, however, a trade-off. Because positive anisotropy directors tend to tilt upwards to align with the vertical component of the electric field, in the field-on state, they contribute less to phase retardation, and consequently there is lower transmittance, as described above for the IPS display.

Negative anisotropy liquid crystals, on the other hand, turn toward the perpendicular to the electric field, which would push the directors down towards the horizontal in regions where there is a vertical component of the electric field, thus mitigating against the mid-layer tilt to improve wide-angle view and field-on transmittance. For this reason, in spite of the low viscosity attributes of the positive anisotropy liquid crystal just described, negative anisotropy ($\Delta\varepsilon < 0$) liquid crystals are preferably used in the FFS display mode, and can achieve transmittances as high as 98% of the TN LCD transmittance.

In the wake of the refined FFS techniques by Korea's LGD and China's BOE-Hydis, particularly to reduce the viewing cone asymmetry described in Chapter 23 that results in a rather bluish color shift, Japan's Hitachi developed a chevron-shaped interdigital electrode pattern called the *Super-IPS* that divides each pixel into two domains, and like the MVA, compensates the unwanted phase retardation by oppositely oriented molecules [8]. The fringing field and chevron pattern are among many lines of research for improving all aspects of the liquid crystal display image quality, including wide-viewing angle, contrast, colors, grayscale, and response time.

Oblique angle viewing in IPS display modes can suffer unwanted phase retardation from the different optical paths that as described in Chapter 23, can be compensated for by positive and negative a-plates; for tilt angle-produced oblique viewing angle problems, c-plate compensation schemes can be used to further improve the originally good wide-view angle.

In summary, the in-plane switching display mode conceptually has no mid-layer tilting molecules, so in principle, its wide-viewing angle quality is superior to the TN and MVA display modes, but the transmittance is less, and the aperture ratio is smaller. Further, to improve response time, IPS displays must use specially synthesized liquid crystals, and fabrication is more complex and requires higher precision. But with advances such as the FFS and chevron, together with progress along the learning curve and the economies of scale brought by mass production, the trade-off for wide-view angle quality is increasingly favorable. The IPS display mode for television has attained commercial success as the highest quality liquid crystal television, and Hitachi, Chimei (from its former Japanese subsidiary IDTech), and LGPhilips (now LG Display) all manufacture IPS displays.

Response Time

Particularly for television and computer game displays, with their dynamic, fast-moving images, the response time for the display has become a significant product differentiating feature. Early displays with 120 ms response times could not definitively capture rapid screen changes, but were fast enough for the early LCD computer monitors where there was not a great deal of motion in the word processors and spreadsheets (even in a volatile stock market). With more video applications, the monitors and notebook computers needed to be faster and quickened to 16 ms, but for television and modern computer screens for video and games, the dynamic motions of sporting events, computer games, and general video required response times of 4 ms or less to be competitive in the marketplace.

Generally, the response time of a liquid crystal display depends on the efficacy of the molecules turning in response to the field and its relaxation back to a field-off equilibrium state, and that efficacy depends on the nature of the liquid crystal itself, that is, its rotational viscosity, elastic coefficient, dielectric anisotropy, and external engineering factors such as the cell gap, energy density of the field, and the pre-tilt angle of the molecules.

The principal molecular turning modes in respect to the response of liquid crystal molecules to an applied vertical electric field are the *bend* in the VA (the negative anisotropy molecules tilt downward bending away from the field), the *splay* in the TN (the positive anisotropy molecules tilt upward towards the field), and for the horizontal field used in IPS, the *twist* in the IPS (either anisotropy molecules twist in the horizontal plane to align with the field). Of course each display mode also has some motion in the other turning modes as well (e.g., the twist and bend in the TN in addition to its splay). The relative magnitudes of the elastic coefficients are:

$$K_{33} > K_{11} > K_{22}$$

$$\text{Bend} > \text{Splay} > \text{Twist}$$

$$\text{VA Elasticity} > \text{TN Elasticity} > \text{IPS Elasticity}.$$

A greater elasticity implies a faster snap-back to a lower or field-off state, so with respect to elasticity, the larger K_{33} of the bend VA display mode gives the VA the fastest response to an applied electric field; the splay (K_{11}) of the TN display mode is the next fastest, and the IPS twist (K_{22}) is the slowest. Following are dynamic continuum expressions for the response times of the major display modes, plus the super-fast blue phase.

VA Response Time Is Good

From the Ericksen–Leslie dynamic continuum theory equations introduced in Chapter 12, the tilt angle (θ) in the homeotropic (VA) mode under the influence of an external electric field (E), including the effect of liquid crystal viscosity by means of a time derivative multiplied by the coefficient of rotational viscosity, is given by

$$[K_{11}\cos^2\theta + K_{33}\sin^2\theta]\cdot\left(\frac{d^2\theta}{dz^2}\right) + (K_{33} - K_{11})\cdot\left(\frac{d\theta}{dz}\right)^2 \sin\theta\cos\theta$$

$$+\,\varepsilon_o\Delta\varepsilon E^2\sin\theta\cos\theta = \gamma\frac{d\theta}{dt},$$

where the splay (K_{11}) and the bend (K_{33}) elastic coefficients come into play in the response of the VA molecules to an external field. As before, ε_o is the permittivity of free space, $\Delta\varepsilon$ is the electric anisotropy, $\varepsilon_o\Delta\varepsilon E^2$ is the electric field energy density, and γ is the coefficient of rotational viscosity. As noted in Chapter 12, the equation is nonlinear and can only be solved by numerical analysis, but if a composite single elastic coefficient and the small angle approximations are made, that is, $K_{33} \sim K_{11}$ and $\sin\theta \sim \theta$, then the nonlinear differential equation can be simplified to a tractable linear form as

$$K_{33}\frac{\partial^2\theta}{\partial z^2} + \varepsilon_o\Delta\varepsilon E^2\theta = \gamma\frac{\partial\theta}{\partial t}.$$

Now the differential equation is susceptible to a closed-form solution, and for the case of crossed polarizers where the rise and fall times are defined as the time for the transmittance to increase and decrease between 10% and 90%, the solutions to the simplified Ericksen–Leslie equation above for the rise time τ_{rise} and the fall time τ_{fall} are as follows [9]:

$$\tau_{rise} = \frac{\tau_o}{\left[(V/V_{th})^2 - 1\right]}\ln\left\{\frac{\left[\dfrac{\delta_o/2}{\sin^{-1}(\sqrt{0.1}\sin(\delta_o/2))} - 1\right]}{\left[\dfrac{\delta_o/2}{\sin^{-1}(\sqrt{0.9}\sin(\delta_o/2))} - 1\right]}\right\}$$

$$\tau_{\text{fall}} = \frac{\tau_o}{2}\ln\left\{\frac{\sin^{-1}(\sqrt{0.9}\sin(\delta_o/2))}{\sin^{-1}(\sqrt{0.1}\sin(\delta_o/2))}\right\}.$$

where δ_o is the net phase change for change in voltage, V is the applied voltage (where the effect of the cell gap d is felt) and V_{th} the threshold voltage, the latter two of which are derived in Chapter 12 for the VA Freedericksz transition as,

$$V = E_{th}d \quad \text{and} \quad V_{th} = \pi\sqrt{\frac{K_{33}}{\varepsilon_o\Delta\varepsilon}}.$$

A useful parameter is the director reorientation time (τ_o) providing the relationship of the liquid crystal orientational response time with the coefficient of rotational viscosity γ, cell gap d, and the liquid crystal coefficient of bend elasticity K_{33} as follows

$$\tau_o = \frac{\gamma \cdot d^2}{K_{33}\pi^2}.$$

From the equation immediately above, one can see that the greater the viscosity, the slower the reaction time, and the stronger the restoring bend elasticity of the liquid crystal, the faster the reaction time. Importantly, the reaction time is dependent on the square of the cell gap, so that decreasing the thickness of the liquid crystal layer can significantly reduce reaction time. So in the vertical alignment (VA) mode with normally black (NB) state, since the VA mode liquid crystal layer thickness can be smaller than that for TN and IPS, its response time is consequently faster. The VA alignment layer rubbing compound can induce a pre-tilt angle in the preferred field-on tilt direction to give the molecules a tilt head-start, but if the tilt angle is too great, or too many molecules are tilted, as described in Chapter 23, there will be oblique angle light leakage and subsequent loss of contrast, so the pre-tilt is usually only a few degrees. The VA mode is inherently the fastest of the three principal display modes with TN next fastest and IPS lagging behind.

IPS Response Time in Slow

For the in-plane switching (IPS) display mode, the general Ericksen–Leslie dynamic equation with the twist (φ) as the principal field-on turning mode is

$$K_{22}\frac{\partial^2\varphi}{\partial z^2} + \varepsilon_o \Delta\varepsilon E^2 \sin\varphi\cos\varphi = \gamma\frac{\partial\varphi}{\partial t},$$

where K_{22} is the twist elastic coefficient, and the other factors are the same as in VA case above. If it is assumed that the anchoring is very strong so that the bottom and top boundary conditions are fixed as $\varphi(0) = \varphi(d) = \Phi$, where Φ is the angle between the alignment layer rubbing angle (directors' orientation at the liquid crystal surface) and the interdigital electrode orientation, finding solutions to the nonlinear ($\sin\varphi\cos\varphi$ term) differential equation above for arbitrary Φ is fairly complicated. Taking only the $\Phi = 0$ case and making the small angle approximation (so that the electric field is only slightly above the threshold), the simplified rise time solution for the IPS display mode is

$$\tau_{\text{rise}} = \frac{\gamma}{\varepsilon_o |\Delta\varepsilon| E^2 - \dfrac{\pi^2 K_{22}}{d^2}},$$

where the absolute value of the anisotropy takes into consideration negative anisotropy IPS liquid crystals. The approximate fall time solution ($E = 0$) is

$$\tau_{fall} = \frac{\gamma \cdot d^2}{\pi^2 K_{22}}.$$

The early IPS displays had abysmal response times sometimes exceeding 50 ms. Significant improvement clearly was critical to the further development of the IPS display mode for television use. From the simplified rise and fall time equations above, decreasing the cell gap d in IPS displays will give the biggest improvement in response time, but recall that the IPS threshold voltage has a factor of (l/d), so decreasing d also undesirably increases the threshold voltage requirement. Simulation and trial and error studies have set an optimum IPS cell gap (d) at about 4 μm. As for the anisotropy, recall that to help avoid undesirable above-electrode molecular tilt, negative anisotropy liquid crystals are preferable for IPS and FFS displays, so strong negative anisotropy liquid crystals to decrease viscosity have been developed specifically for IPS displays, with the result that the IPS response time has been considerably improved. Further, IPS liquid crystals having the best K_{22} can be synthesized and chosen for use to improve the response time. These developments have made the IPS display mode response time acceptable for television use.

TN Is In-Between

Because of the wide-viewing angle problems arising from the mid-layer tilt described in Chapter 23, the TN display mode has not been used much for television. Recall from Chapter 11 that splay (K_{11}), twist (K_{22}), and bend (K_{33}) elasticities all come into play in a TN display mode Freedericksz transition, and that the turning of the TN molecules depends on the relative values of the three elastic coefficients. The TN rise and fall times can be expressed in analogy with the IPS case in simplified form as

$$\tau_{\text{rise}} = \frac{\gamma}{\varepsilon_o |\Delta\varepsilon| E^2 - \dfrac{\pi^2 K}{d^2}}.$$

and

$$\tau_{fall} = \frac{\gamma \cdot d^2}{\pi^2 K}.$$

with a composite elastic coefficient K.

With the response time expressions the same, the TN display mode's speed then depends on the value of the composite coefficient of elasticity and the typically higher positive anisotropy TN liquid crystals (so the viscosity is lower than that for the IPS mode), but the cell gap is more or less set because of the half-wavelength twist requirement for the white state. The TN response times have been of the order of 10–50 ms, faster than IPS, but not significantly so.

Blue Flash Is Fastest

A later developed liquid crystal display mode is based on Reinitzer's earliest observation of a *blue flash* in his cholesteryl benzoate sample that is the response speed champion of all the liquid crystal modes. The blue phase was observed in a tiny temperature range of just 2–3°C as a brief flash of blue light. The fast response time is due to the blue phase's position just between the chiral (cholesteric) nematic and the isotropic fluid phases, allowing the liquid crystal to rapidly switch from a very slightly ordered anisotropic phase to the isotropic phase, primarily because there is little difference between the two phases. This allows refresh rates to be driven at 240 Hz and even higher, which can provide extremely natural movement in fast-moving images. Such naturally fast transitioning obviates the overdrive systems (discussed below) commonly used in the VA mode, and the blue phase also does not require rubbing alignment layers on the glass substrate, thereby reducing process steps to streamline fabrication. However, the very narrow temperature range of 2–3°C for blue phase makes its stability a major challenge to be overcome for commercial display development.

Overdrive

In an effort to improve response times, the major LCD panel manufacturers have almost all developed some form of an *overdrive* system. As shown in the schematic Figure 24.8 of a typical drive system at left, a first voltage is sent, which causes the liquid crystal to respond in a parabolic curve that takes 3/5 of the frame to reach the gray level desired. In the overdrive system shown in at the right in Figure 24.8, the voltage is first boosted above the voltage required for the gray level (and then lowered to that voltage), thereby giving a boost to the liquid crystal response, allowing it to reach the desired gray level in only 1/5 of the frame; this overdrive effect can improve the image as shown in the sample images in Figure 24.9.

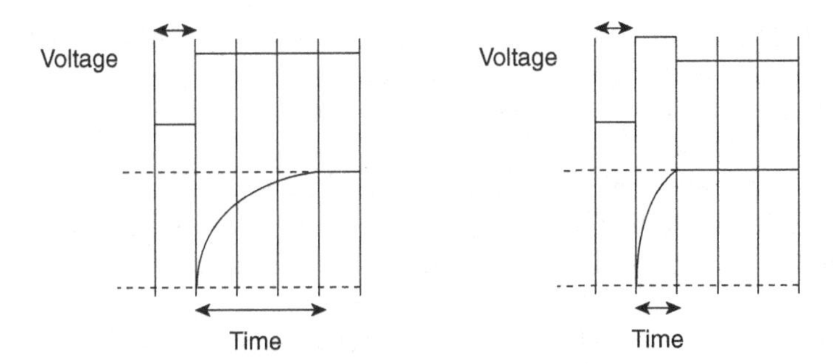

Figure 24.8 Drive system (left) taking 3/5 of the frame to reach the gray level desired and overdrive system with voltage boost to reach the gray level in 1/5 of the frame.

Figure 24.9 Improved image from overdrive.

The simplest way of implementing the overdrive system would be to use a first in-first out (FIFO) frame buffer to store a frame's signal voltage, then comparing the signal with the signal voltage of the next frame according to the relative values of the rising edge and falling edges, the proper signal boost voltage can be obtained from a data look-up table, and an overdrive voltage (or lower voltage undershoot) can be applied so that the desired grayscale voltage will be attained within the frame in question. The use of overdrive and undershoot can increase the response time by two to three times, and almost every major panel manufacturer has its own overdrive patents.

Flicker

The visual fatigue often felt when watching traditional cathode ray television sets for extended periods of time is caused by small, rapid changes in luminance of the images on the screen, a phenomenon called "flicker." If a CRT television uses a standard line voltage 60 Hz frequency signal to drive either interlaced or progressive scans of the screen, the positive/negative full period has a refresh rate of 30 Hz. Physiologists have found that the human eye can differentiate even small changes in luminosity at refresh rates of 40 Hz and below, so that luminance changes at the refresh rate of the 30 Hz CRT signal can be distinguished, resulting in a sensation that the image is flickering.

The early liquid crystal television sets operated at the 60 Hz line voltage, so in the transmittance (T) versus driving voltage (V) curve shown in Figure 24.10, and given the odd frame at positive (increasing) voltage and the even frame at negative (decreasing) voltage so shown, at every gray level, there are two values for the γ grayscale voltage, and the common voltage (V_{com})

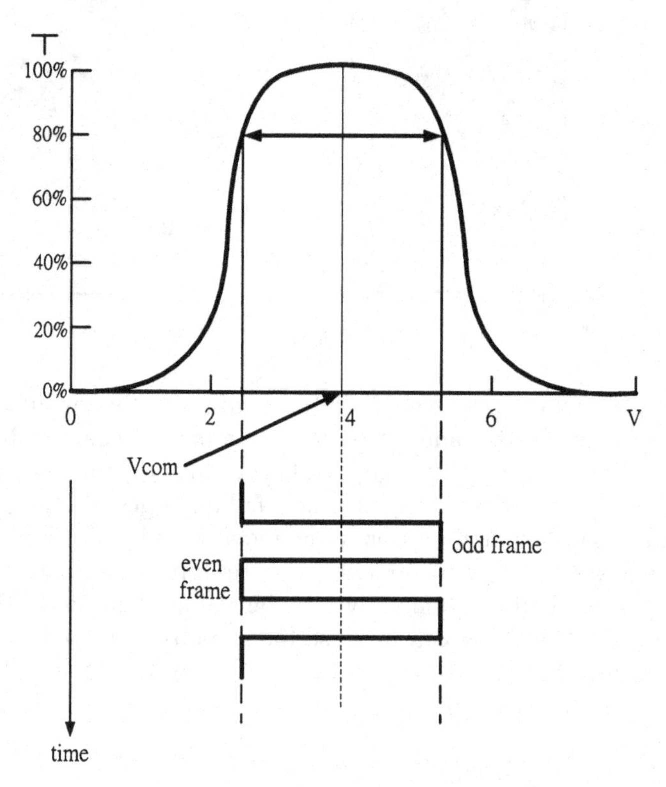

Figure 24.10 The odd frame at positive (increasing) voltage and the even frame at negative (decreasing) voltage and the common voltage (V_{com}) at the middle between the two γ-values.

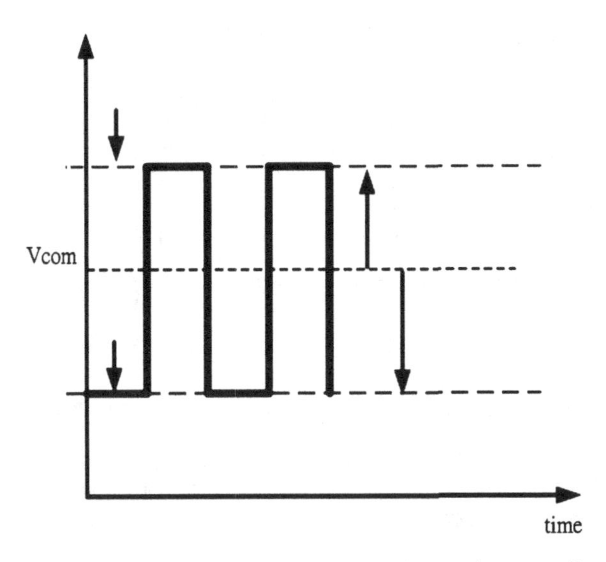

Figure 24.11 Different odd and even frame base voltages.

set at the color filter should be at the middle between the two γ values, as shown in Figure 24.10.

Because unwanted direct current voltages are very difficult to entirely avoid in any electronic system, the common voltage V_{com} may intermittently change, so the odd and even frame voltages having V_{com} for the base voltage will be different, as shown in Figure 24.11, and also change intermittently. This different voltage will produce a different polarization of the light passing through, and thereby change the transmittance, as shown in Figure 24.12, resulting in small intermittent changes in the luminance, which, at the low refresh rate of 30 Hz, can be discerned as the flicker causing visual fatigue.

The simplest solution for this problem apparently would be to just readjust the common voltage V_{com} back to be just in the middle between the two γ values, but as noted, it is difficult to precisely maintain a uniform common voltage throughout the system so the V_{com} would have to be continually readjusted, making such a remedial measure unfeasible.

The most commonly used method to resolve the flicker problem was to employ spatial averaging to alleviate the effects of the time-based flicker. If the voltage polarities on adjacent pixels are alternated from positive to negative, although the flicker in individual pixels could still be seen close-up, from a normal viewing distance, the eye would perceive a spatial average of the transmittance from adjacent pixels rather than the time-based

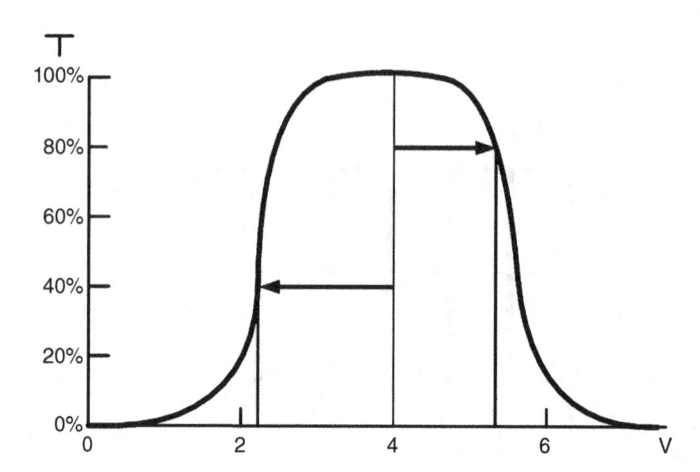

Figure 24.12 Changes in transmittance resulting from the different odd and even frame base voltages.

Row	Odd frame	Even frame	Odd frame	Even frame
1	+ + + +	- - - -	+ - + -	- + - +
2	+ + + +	- - - -	+ - + -	- + - +
3	+ + + +	- - - -	+ - + -	- + - +
4	+ + + +	- - - -	+ - + -	- + - +
Column	1 2 3 4			
	Frame inversion		Column inversion	
	+ + + +	- - - -	+ - + -	- + - +
	- - - -	+ + + +	- + - +	+ - + -
	+ + + +	- - - -	+ - + -	- + - +
	- - - -	+ + + +	- + - +	+ - + -
	Line (row) inversion		Dot inversion	

Figure 24.13 Inversion drive applied to individual frames, columns, rows (lines), or dots (pixels) to reduce flicker.

variations in an individual pixel, thus eliminating the perception of flicker. An added advantage of this *inversion* method is that since the voltage of the adjacent pixels are of opposite polarities, crosstalk is also reduced. This so-called *inversion drive* can be applied to an individual frame, column, row (line), or dot (pixel), as shown in Figure 24.13.

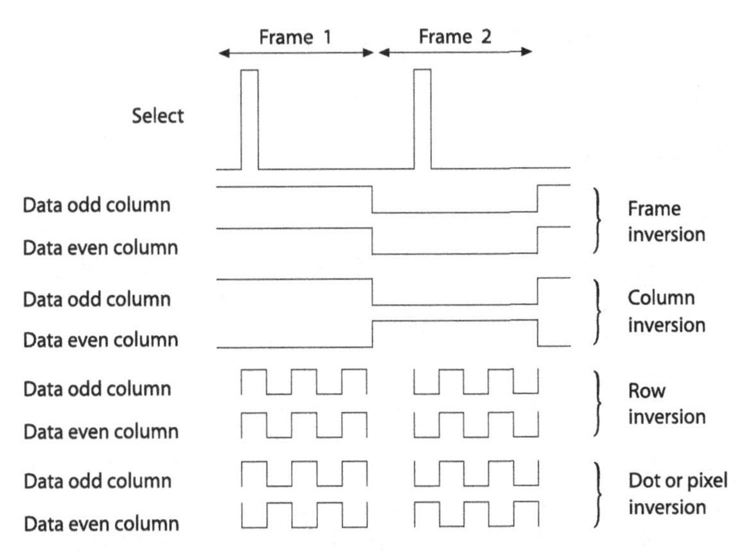

Figure 24.14 Timing diagram for the inversion for each element, frame, column, row, and dot.

To produce the inversion for each element, frame, column, row, and dot, the timing is performed in accord with a typical timing diagram as shown in Figure 24.14. From the pulse timing and widths, it can be seen that in frame inversion, all the data signals to the pixels within a frame are the same polarity, so the inversion effect is only manifested in the next frame having an opposite polarity, so there is not much alleviation of flicker in the frame inversion scheme. In column inversion, within one frame, the positive polarity signal will be sequentially alternated through the columns, and the polarity scheme will be alternated between frames as well. In row inversion, the polarity likewise will be sequentially alternated in every row as well as for every frame. Dot inversion combines column and row inversion, and because the sequential alternation of the voltage polarity will be at the finest resolution, the flicker alleviation using dot inversion is the best of the various inversion schemes.

The finest spatial resolution is from individual pixels, so unsurprisingly, dot inversion produces the best flicker reduction. For colors, the polarity of the individual primary color subpixels (subdots) can be alternated as well, thereby decreasing the flicker in the individual colors, and since the subpixels are even smaller than the pixel, the spatial averaging is finer and thus more effective to alleviate flicker. As shown in Figure 24.15, the vertical stripe

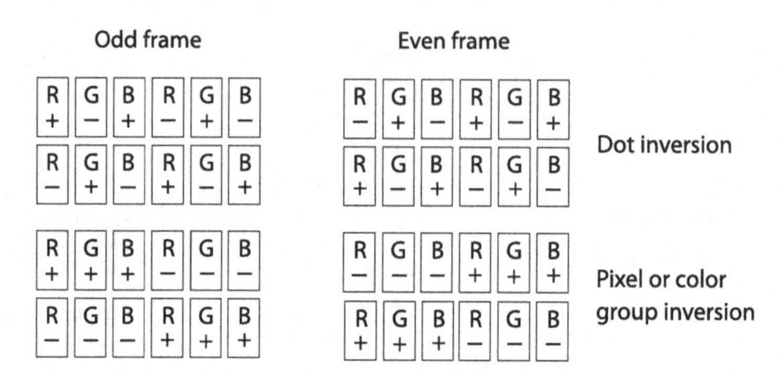

Figure 24.15 Vertical stripe subpixel color group inversion scheme.

subpixel pattern using dot inversion (also called *color group inversion*) alternates the voltage polarity of adjacent stripes [10].

Another solution to the flicker problem would be to increase the driving frequency to 90 Hz, which would give a 45 Hz full-period frequency, which would be beyond the 40 Hz threshold of human eye perception. Indeed, the later models of liquid crystal television can operate at 120 Hz and have gone up to the super-fast 240 Hz level.

In summary, liquid crystal displays having the best contrast, widest viewing angle, good color saturation and gamut, grayscale sensitivity, fast response time, and lowest flicker, as can be seen from the above discussions, require many technical innovations and the optimal adjustment of many parameters. The Maier–Saupe and Ericksen–Leslie theoretical models, together with the Freedericksz transition experiments, and the employment of physical optics analysis using the extended Jones matrix, altogether can produce rather comprehensive models of the phase retardation and polarization which, together with the compensation and timing schemes, are then used to calculate the transmittance of a liquid crystal display system.

The results of experiments and simulations can then be demonstrated in iso-transmittance contour figures showing the relationship between viewing angle and equal transmittance values, and iso-contrast contour figures showing the relationship between viewing angle and equal contrast values. These contours can be used to analyze the effect of the various techniques employed and thus allow optimal adjustment of the physical parameters and provide guidelines for additional techniques to improve the image. Further, using computer simulations, together with empirical results, the manufacturing production factors, such as the process, lifetime, costs, and so on, can be optimized to finally produce the highest quality and most cost-effective liquid crystal display.

References

[1] Ohmuro, K., et al. 1997. *SID Tech. Dig.*, **28**, 845.
[2] Kwag, J.O., et al. 2000. *SID Tech. Dig.*, **31**, 256.
[3] Soref, R.A. 1973. *Appl. Phys. Lett.*, **22**, 165; Soref, R.A. 1974. *J. Appl. Phys.*, **45**, 5466; Kiefer, R., et al. 1992. *Japan Displays*, **92**, 547; Oh-e, M., et al. 1995. *Asia Display*, **95**, 577.
[4] Oh-e, M., Yoneya, M., and Kondo, K. 1997. *J. App. Phys.*, **82**, 528.
[5] Yang, D.K., and Wu, S.T. 2006. *Fundamentals of Liquid Crystal Devices*. Wiley, Chicester, p. 205ff.
[6] Lee, S.H., Lee, S.L., and Kim, H.Y. 1998. *Appl. Phys. Lett.*, **73**, 2881; Jeon, Y.M. et al. 2005. *SID Symp. Dig.*, **36**, 328.
[7] Yang, D.K., and Wu, S.T. 2006. *Fundamentals of Liquid Crystal Devices*. Wiley, Chicester, p. 205ff.
[8] Asada, W.S., et al. 1997. *SID Tech. Dig.*, **28**, 929; Mishima, Y., et al. 2000. *SID Tech. Dig.*, **31**, 260.
[9] Wang, H., Wu, T.X., Zhu, X., and Wu, S.T. 2004. *J. Appl. Phys.*, **95**, 5502.
[10] den Boer, W. 2005. *Active Matrix Liquid Crystal Displays*. Newnes, Oxford, p. 87ff.

25

Glass, Panels, and Modules

All the science and technology are now in place for construction of the final product, the liquid crystal display. The manufacturing process begins with the glass substrate, which is provided by primarily two companies, America's Corning and Japan's Asahi.

Glass Generations

It has become an LCD industry convention to mark progress by glass substrate size in *generational* increments, and the plants that manufacture the LCD panels are so designated as *nth-generation* plants. The table below lists the generation, sizes, and representative optimum LCD panel "cuts" from the glass substrate. The earliest glass sizes were for digital watches, calculators, and small computers, and were manufactured in what is today called the first- to third-generation plants; the processing technology was relatively rudimentary and quite unlike the large-scale, automated manufacturing in later generation fabrication plants ("fabs"), so they are not included in the table. The table shows the tremendous increase in glass sizes [1].

Liquid Crystal Displays, First Edition. Robert H. Chen.

Generation	Glass size (mm)	Optimum cut (inches × panels)	
3.5a	550 × 670	12.1 × 6	15 × 4
3.5b	610 × 720	13.3 × 6	17 × 4
3.5c	620 × 750	14.1 × 6	18 × 4
4.0	680 × 880	15 × 6	20 × 4
4.5	730 × 920	17 × 6	22 × 3
5.0	1100 × 1300	27 × 6	22 × 8
5.5	1300 × 1500	32 × 6	56 × 2
6.0	1500 × 1850	37 × 6	32 × 8
7.0	1870 × 2200	46 × 6	37 × 8
7.5	1950 × 2250	47 × 6	42 × 8
8.0	2160 × 2450	50 × 6	46 × 8
8.5	2200 × 2500	55 × 6	46 × 8
9.0	2400 × 2800	57 × 6	52 × 8
10.0	2580 × 3050	65 × 6	57 × 8

Figure 25.1 8.5-generation glass panel. Courtesy of AU Optronics.

An example of an 8.5-generation AUO glass panel is shown in Figure 25.1; from the photograph the scale of the glass size is apparent, a more than two meters screen size can certainly begin to compare with movie theater screens.

The 3.5- and 4.0-generation fabs built by Japanese companies (including IBM-Japan) were designed to produce notebook computer screens and monitors. However, today, these "ancient" fabs can turn out glass than can be cut in what is called *small- and medium*-size panels for uses in mobile phones, MP3 players, automobile navigation systems, airline passenger display screens, small advertising displays, cash registers, handheld game players, photograph frames, and the like.

The precise size of the glass chosen by different manufacturers varies, and the panels are cut according to the most efficient allocation of the glass surface area and the companies' projections of the market for various screen sizes. For example, to provide television sets for the Beijing Olympics held in 2008, many manufacturers emphasized large-screen 56-inch panels, and Chimei's 5.5-generation plant could choose between the almost equally efficient cuts of 56 or 32 inches so that before the Olympics, more 56-inch panels were produced, and afterwards the cut could return to the lower cost, more mass-appealing 32-inch panels.

Accurately forecasting the market demand quickly became the key game to be played for profit or loss in television sales. For example, the early 30-inch liquid crystal televisions were quickly eclipsed by the 32-inch models, and AUO's briefly successful 38-inch panel, although winning out over the then popular 37-inch panels, when compared with the 42-inch and Chimei's later 47-inch models, quickly became excess inventory. It is just this inventory control that determines success or failure; betting on the wrong size glass horse would lead to immediate inventory problems, and a good bet would hand the excess inventory problem to competitors. An accurate forecast then would not only clear one's own inventory, but also provide a platform for long-term profitability and even the prospects for establishing a specification standard. For example, Chimei's 19/22-inch wide screen monitor not only could be produced most efficiently at Chimei's unique 5.5-generation fab, it also became the market standard for desktop computer monitors. Late entrants would have to conform their panels to the already "designed-in" Chimei monitor specifications, confirming Chimei leadership and giving first-to-market advantage to Chimei's sales force.

By 2010, Japan's Sharp, Korea's Samsung and LGD, Taiwan's AUO and Chimei already were planning 10th-generation and constructing 8th-generation plants, and although Taiwan and Korea often engaged in none too healthy plant-building races in large part only to prove their respective manhood, the macho adage in the electronics industry,

> get big or get out

just because of the sheer number of panels that can be cut from larger mother glass, unfortunately, has become all too true.

The TFT Array Plate

The thin film transistors used for addressing liquid crystal displays are formed directly on the glass substrate by the planar process described in Chapter 21. The successive layering of semiconductor material and metal forms the million-strong array of transistors called a *thin-film transistor array* (TFT array). The process steps for a typical TFT array glass plate are shown schematically in Figure 25.2.

Following Figure 25.2, first the TFT array is tested, cleaned, and then, depending on the type of panel (e.g., STN, MVA, or IPS) coated with a 50-nm thick polyimide layer, which is then rubbed using a velvet-covered rotating drum to produce the alignment layer, and baked at about 200°C.*

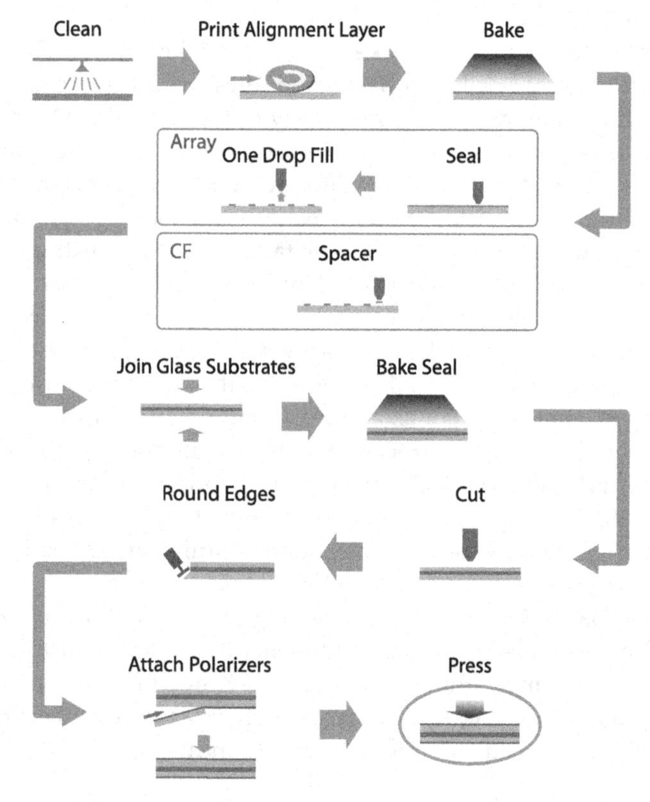

Figure 25.2 Process steps for a TFT array glass plate.

* A motion video of the panel assembly process can be viewed at the AUO website www.auo.com.

After baking, in preparation for the filling of the liquid crystal, a protective epoxy sealant is applied along the edges of the glass substrate. The liquid crystal can be filled in as individual droplets deposited directly on the TFT array glass plate from a dispenser in a technique called *one drop fill* (ODF), which is used for practically all larger-size liquid crystal cells. Thereafter, the color filter array glass is joined with the TFT array glass and sealed and baked. The completed liquid crystal sandwich is then cut according to the predetermined panel sizes, the edges are rounded, and the polarizers attached. Finally, the glass panels and polarizers are pressed together.

Another method of liquid crystal injection called *side injection* has been used in panel manufacture, and is performed after the TFT array plate and the color filter plate have been joined.

The Color Filter Plate

In the earlier days of display manufacture, color filter production was usually subcontracted to a specialty manufacturer and the completed color filter then shipped to the LCD maker for panel assembly; now most major panel producers manufacture their own color filters in-house to ensure supply, facilitate compatibility, and reduce transport risks and costs.

Figure 25.3 is a drawing of a typical color filter assembly. Figure 25.4 schematically shows a typical manufacturing process [2]. The color filter is made by first patterning and depositing a black matrix layer on the glass substrate for registration with the pixel layout. The black matrix blocks light in the non-light throughput pixel areas, and also shields the TFT from light that may cause undesirable photoelectric effects within the transistors. In the early days, the black matrix was made from chromium metal with a chromium oxide layer to reduce reflectance, but the toxicity of chromium compounds raised environmental concerns, so the chromium was replaced by a black polymer resin, which is generally used today. Unlike chromium, the polymer resin is photoimageable, and therefore easier to pattern using

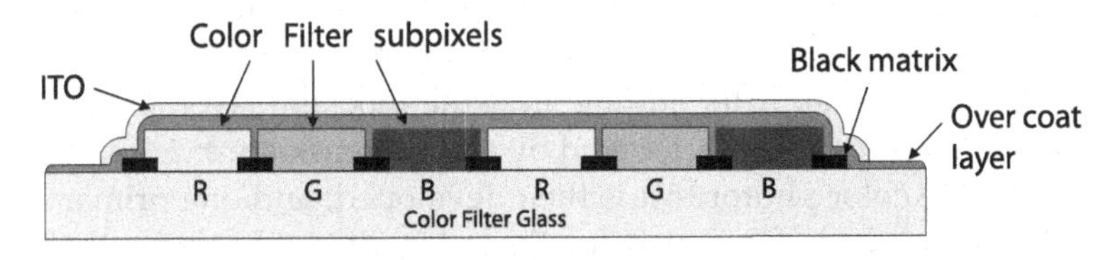

Figure 25.3 Color filter assembly.

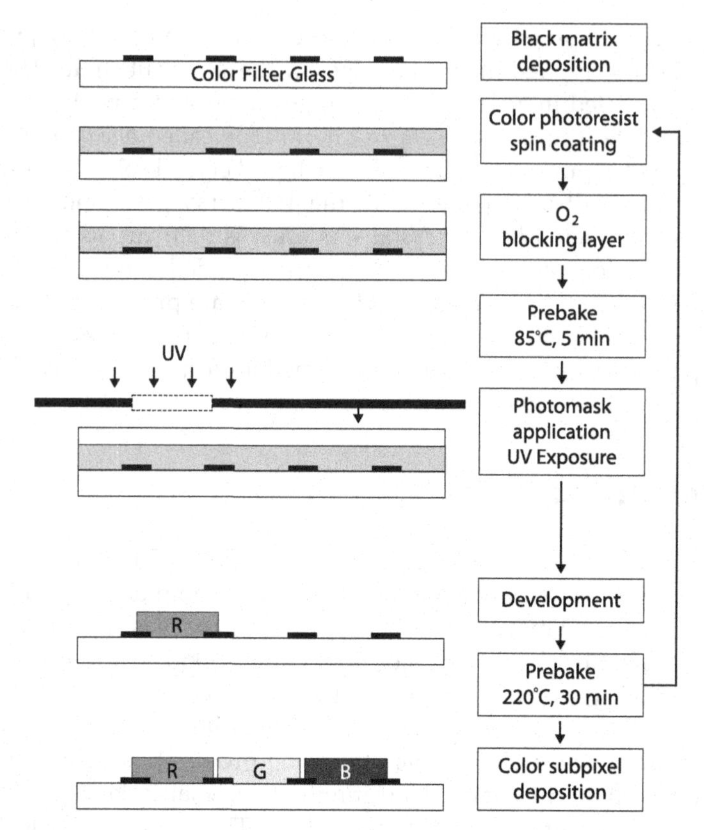

Figure 25.4 Color filter assembly manufacturing process.

deposition methods, and its index of refraction can be made to be almost the same as glass, thereby reducing undesirable optical effects *vis-à-vis* the glass substrate.

After application of the black matrix, the red, green, and blue color filter materials are deposited and patterned on the glass plate. The filter material comprises dispersed pigments with photo-initiators that are deposited by spin-coating on the glass plate with the black matrix already applied. The pigments also contain an acrylic ester binder, which hardens and stabilizes the color filter material. After adding an oxide blocking layer, the assembly is prebaked, and the areas not covered by a photomask are exposed to ultraviolet light. The color photoresist is then developed, and one primary color subpixel thus has been fabricated for the color filter. After a postbake, the next primary color filter subpixel is fabricated in the same way with a different color dispersed pigment, and so on for all the primary colors chosen for the color filter. The newer dispersed pigment materials have special photoimageable qualities under exposure by ultraviolet light, so that some photoresist and solvent stripping steps can be reduced or eliminated. A transparent protective layer then is deposited to protect and planarize the surface over the filters.

Incidentally, a newer process used particularly for larger displays is to pattern the color filters directly on the TFT glass plate in-between the transistors. This *color filter-on-array* (COA) can cover the light channels of the TFT glass plate, obviating the black matrix; COA further does not require the precise registration between the top glass plate (which now is not patterned) and the TFT glass plate, making the process easier. The trade-off is that patterning the color filter on the TFT plate requires two more patterning and deposition steps [3].

Returning to the separate color filter and TFT array plates, completing the process, the ITO electrode is deposited over the viewing area and skips over the nonviewing areas by employing a shadow mask through which the ITO material is sputtered.

Side Injection and One Drop Fill

In the early days, the fabrication of the TFT array and the color filter was performed in separate parallel production lines with the color filter glass plate and the TFT array glass plate then brought together and sealed to hold the liquid crystal between them to form LCD panels. More recently, the two glass substrate production lines have been combined where possible and where efficiency may be improved.*

There are two principal methods of filling in the liquid crystal. In the side injection method, the TFT array and color filter glass plates were first registered and sealed, and the liquid crystal injected from a hole in the side. The drop-fill method was developed later whereby the liquid crystal was deposited drop-by-drop onto the TFT array plate, and the two plates were sealed after deposition of the liquid crystal.

Side Injection

The side injection liquid crystal filling procedure requires aligning, pressing together, and sealing the two glass plates holding the TFT array and color filter, but leaving a small hole in the seal (the fill port), as shown in

* Even the combining of the color filter and TFT array manufacturing into common lines is also the subject of several patents.

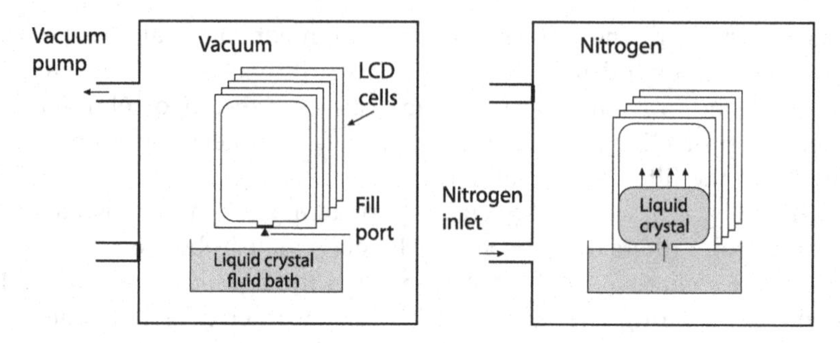

Figure 25.5 Liquid crystal side injection.

Figure 25.5. The plates were then placed in a vacuum chamber where the space between the glass plates would be evacuated, forming a vacuum. Lowering the plates including the fill port into a liquid crystal fluid bath then causes the liquid crystal to slowly enter the space in between the plates to fill the vacuum. By backfilling the vacuum chamber with nitrogen gas, the liquid crystal will more quickly enter the space between the plates due to the pressure of the nitrogen on the surface of the liquid crystal fluid bath, but the process still requires up to an entire day to complete. Finally, the fill port is sealed and the liquid crystal injection procedure is completed [4].

This kind of nonsymmetric injection from the side naturally leads to uniformity problems in the liquid crystal layer, and so the process is often stopped, the liquid crystal layer inspected, and any necessary remedial measures taken, and if necessary the whole procedure repeated. For a 40-in display, the process could often take several days to complete.

One Drop Fill

In this more recent process, as shown schematically in Figure 25.6, the liquid crystal is filled in drop-by-drop onto the TFT array plate before the glass plates are aligned and sealed, in the planar process-like fashion that is used in semiconductor fabrication. The ODF procedure produces a more uniform filling of the liquid crystal layer, and since performed before the alignment, pressing, and sealing steps, the inspection, adjustments, and remedial actions can be taken without re-doing the whole process, as required in side injection. Furthermore, the amounts of liquid crystal dropped into the pixels can be precisely controlled and delivered by computer-controlled automatic dispensing equipment, making the ODF procedure much faster and more

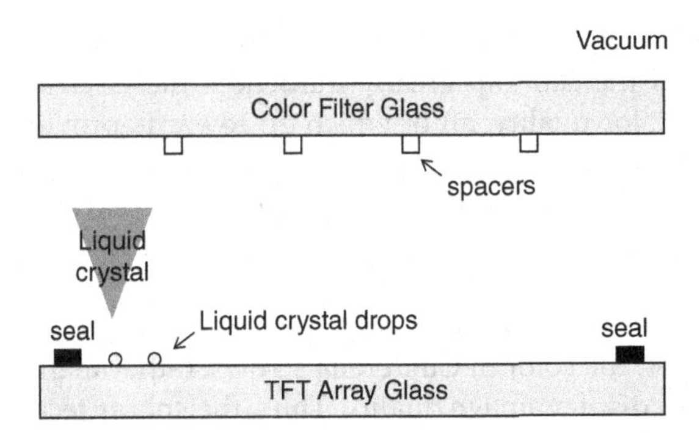

Figure 25.6 Liquid crystal one drop fill (ODF).

precise than side injection. The resulting liquid crystal layer also has greatly improved uniformity.

The ODF procedure is measured in minutes rather than the days required by side injection. ODF is particularly efficient for large-size display production of panels for televisions and large-screen monitors.

Spacers

After being cleaned and its alignment layer applied and rubbed, in order to maintain a uniform thickness between the glass plates, spacers made of balls coated with gold film are deposited like gold dust from a fairy godmother on to the color filter plate in nonviewing areas of the glass plates. Now the color filter glass substrate can be registered with the TFT array glass substrate to together hold the liquid crystal.

Conductive epoxy dots also are set in place in the epoxy sealant on the edges of the color filter plate to contact with the metal pads on the TFT array plate. This provides the contacts for the flow of a common voltage (V_{com}) between the two plates. Then by decreasing the pressure between the glass substrates, the surrounding atmospheric pressure will cause the glass plates to come together, and at the same time also compress the spacers in-between, thus ensuring a liquid crystal layer of uniform thickness. Finally another sealant is applied to the edges and the assembly is baked again, this time at about 120°C. The completed assembly is called a *liquid crystal cell.**

* The definition of "cell" sometimes includes the driver ICs and sometimes even the polarizers.

The gold dust helps to keep Cinderella's image uniformly bright, that is, the constancy of the cell gap greatly influences the electric field strength, grayscale, and color quality, all of which go towards providing a true and pleasing image. Apart from the mystical gold dust, cell gap uniformity also can be augmented by a mundane lookup table (LUT) that stores supplementary voltage information for adjustments of the electric field to account for variations in the cell gap thickness [5].

Because the human eye is much more sensitive to color variations than to intensity changes, the color of Cinderella's gown rather than its glitter is the best indicator of display image quality. Thus, the spacer technology of gold balls and LUTs is critical to display quality.

Sealing, Cutting, and Inspection

After the TFT array and color filter glass plates are registered and kept a precise distance apart by the spacers, they are sealed together, typically using a UV-cured inert polymeric cement to hold the liquid crystal inside and to prevent contamination. The completed panels then can be cut to size in accord with the glass generation and projections of market demand. The cutting process will cause some rough edges, which will have to be machine smoothed.† Thereafter, the polarizers are attached to the plates, pressed to remove any bubbles that may have formed, and the panel assembly is completed.

The bigger and bigger panel sizes brought with them greater production challenges. For example, the now-common 7th-generation glass is 4 m^2 in area, the load cassette can weigh up to 500 kg, and requires huge but precisely moving robot carriers. The scale of the robot can be seen in the photograph of Yaskawa's 10th-generation delivery robot in Figure 25.7. It can be seen that the scale of the LCD panel glass substrate, reaching 2.5-m length, is order of magnitude greater than the maximum 300 mm wafer size of semiconductor chip manufacturing, and the size of the LCD panel clean rooms are also order of magnitude greater.*

After the completion of the manufacture, the panel's silicon and metal layer mask lines and the ITO are inspected using *automatic optical inspection* (AOI) instruments, such as high-resolution charge-coupled devices that,

† Even the smoothing of LCD panel glass edges is the subject of several patents.

* A video of LCD manufacturing robots in action can be seen on tw.youtube.com/movie diginfor.tv.

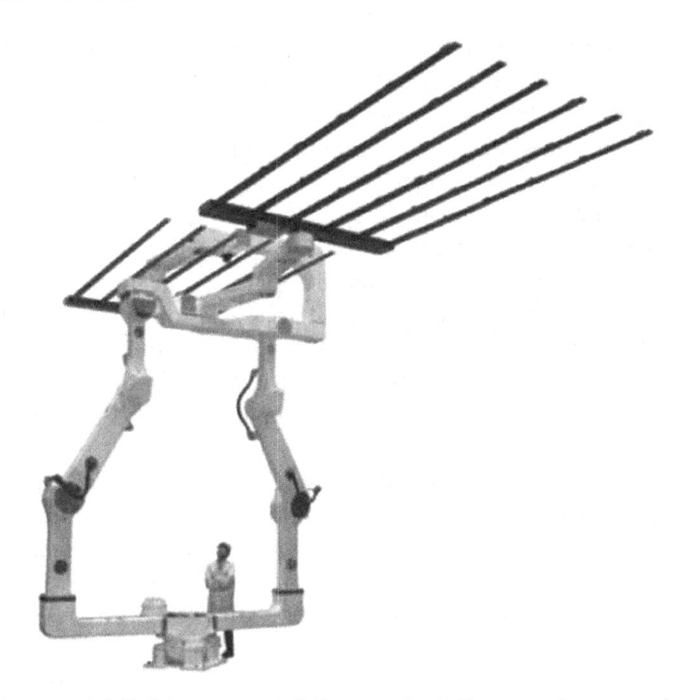

Figure 25.7 Glass panel delivery robot. From yaskawa.co.jp.

together with pattern recognition software, can perform high-speed optical scanning to look for and find defects. After inspection, the wires can be connected, the pixel storage capacitors charged, and the panel unit tested. Impurities in the alignment compound and contamination of the liquid crystal material will lead to discrepancies in luminosity, contrast, and color; any image defect or lack of uniformity is called generally *mura* (Japanese for "spots"). The occurrence of mura is common and it is a recurring migraine headache for volume manufacturers who must go through the laborious process of finding an often not-so-easy to find cause for the image floaters, and then find the fix for the problem.

Electrostatic Damage Protection

Static electricity causes many problems in microelectronic fabrication and system manufacturing. Static electricity charges are created mostly by the simple separation of matter from a surface of a material, thereby creating two surfaces; one of which will have more positive charge because it has lost electrons, and the second surface having more negative charge because it has gained electrons from the other surface. This actually is the same phenomenon of the "dangling bonds" created in the slicing of semiconductors, causing the "surface effects" encountered during the invention of the transistor (Chapter 19) and the reason for doping hydrogen atoms in amorphous silicon to increase the free electron mobility by capping the dangling bonds (Chapter 20). This charging by material separation is called *triboelectricity*, and commonly arises from simple contact, as well as material separation and materials rubbing together, and can build up a significant static electric charge. Of course, this has been demonstrated to generations of students all over the world by rubbing cat fur on a glass rod, and then showing the shocking effects of static electricity.

In the fabrication of electronic devices, the static charge will attract particles that can enter the photolithography process and contaminate electric circuits.* The smaller the scale of the circuit, as in submicron electronics, the smaller are the particles that can cause problems; this then requires ever more stringent contamination control procedures. Static charges and contamination particles can build up on clothing, equipment surfaces, and processing chemicals so the processes of microelectronic fabrication must be performed in a *clean room*, with the scale of the *design rules* determining the class requirement of the clean room. For example, the fabrication of liquid crystal panels must be performed in a fewer than 10 particles per cubic foot *class 10*-type clean room.

A schematic drawing of a typical fabrication clean room is shown in Figure 25.8. Four fans circulate fresh air from the outside through a service chase, and then by way of the plenum through *high-efficiency particulate attenuation* (HEPA) filters, the air flows to the clean room interior, and then out through the perforated floor to the service chases. A nonshedding, conductive fiber, antistatic protective "bunny suit" is worn by clean room personnel, as shown in Figure 25.8.

In electronic circuits, static charge build-up can cause serious damage to the component upon discharge, creating significant electrical surges. This *electrostatic discharge* (ESD) can be as high as 10 amperes in the gate electrode region of an integrated circuit, more than enough to destroy the whole circuit. There are many ways to protect a circuit from ESD, a typical one is shown in the circuit diagram of Figure 25.9, where four protection transistors

* Incidentally, this charge separation creating positive and negative ions is just the idea behind the popular ion-based air cleaners; the ions clean the air by attracting dust particles in the air.

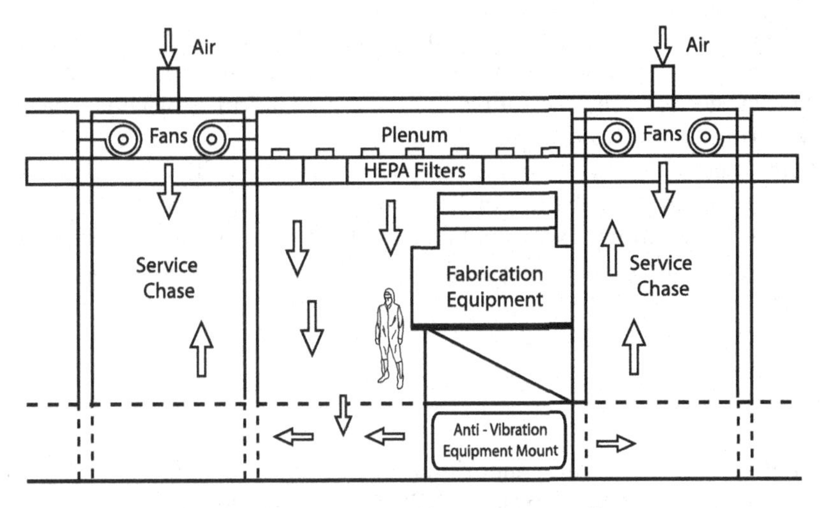

Figure 25.8 Clean room schematic.

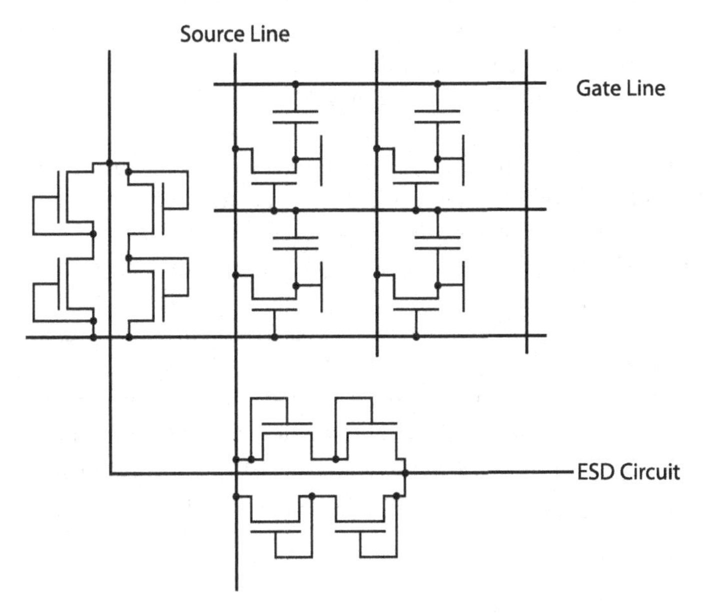

Figure 25.9 Circuit for electrostatic discharge (ESD) with four protection transistors.

are placed as shown at left and below at the edges of the circuit with their gates shorted to the source, which will conduct current when there are voltage surges between the source lines and gate (addressing) lines, particularly during the fabrication process [6].

Many manufacturers also use temporary ESD protection circuits, which will be removed after testing of the panel. As might be expected, because there are many different ways of handling ESD, there are many ESD protection patents, with almost every major panel manufacturer having several on hand.

Laser Repair

After the complicated and laborious fabrication and manufacturing processes have produced a completed LCD panel, upon testing if even a single bright spot appears on the dark screen, because it is so apparent, it is considered a fatal defect, rendering the entire panel useless. Particularly in the early days when processes were not as refined and automated, there were many such fatally defective panels, and if those panels were not salvageable, LCD panel manufacturing costs would remain unacceptably high. So the industry developed a rather negative remedy: a defect bright pixel would be zapped by a laser, effectively killing it. Now the dead pixel is a permanent dark pixel, and while still there as a defect, it will not be so apparent. Albeit far from an elegant solution, laser repair nonetheless has been widely used, but by advancing along the learning curve through ever greater volumes of mass production, and with improved process controls, the number of defective bright spots has decreased, and laser repair may eventually pass into desuetude.

Laser repair also can be used more positively to save panels that have open-line defects in the gate lines. As shown in the simple circuit diagram of Figure 25.10, there are repair lines placed around the periphery of the panel, so that if inspection and test show a defect in a source line, a laser can make a connection so that the signal can be rerouted through the repair line. In this way, an otherwise defect-free panel can be salvaged [7].

Yield

The yield, or percentage of good workpieces, can be said to be the critical factor determining profit and loss in the LCD panel industry. The level of

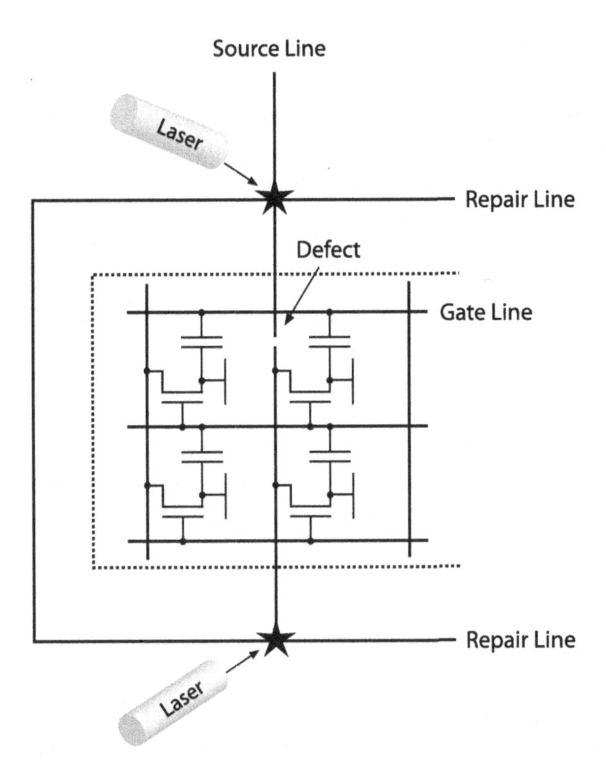

Figure 25.10 Laser repair lines placed around the non-viewing area periphery of the panel.

the manufacturing technology, its efficient use, and the manufacturer's know-how are all reflected in the production yield. Among the millions of pixels constituting a panel, only one or two defective pixels can cause the entire panel to be unfit for sale and simply discarded.

The difficulty of obtaining high-process yields caused the big Japanese electronics companies to doubt that the manufacturing capability in Korea and Taiwan was sufficiently advanced to produce high-quality LCD panels efficiently and cost-effectively. However, events proved that the panel manufacturers in Korea and Taiwan, through their quick mastery of the learning curve, not only eventually attained yields greater than 90% (as high or higher than in Japan), but at lower cost and greater efficiency.

The result, now known to all consumers of electronics products, is that prices of the LCDs dropped significantly, allowing the new electronics product to reach into far larger markets. The use of inexpensive LCDs in computers, mobile phones, digital cameras, personal navigation devices, advertising displays, and on and on, made those products accessible to almost everybody, and in particular, the hitherto exotic large flat-screen liquid crystal television almost unimaginably became a mass consumer product rather than just a plaything of the rich.

Of course, the Japanese contribution to the lower costs was critical in the early stages of LCD production, but later the big semiconductor equipment manufacturers contributed primarily through the production of improved manufacturing and automation equipment rather than panel manufacture itself. The improvements in process technology, and the standardization and quality of materials and parts brought by mass production, as well as the later production of less expensive production equipment, were the primary contributions of the manufacturers in Korea and Taiwan.

The yield (Y) formula used in the LCD industry, borrowed from the semiconductor industry, is this simple defect density equation,

$$Y = \exp\left(-A\sum_{n=1}^{N} D_n\right).$$

where A is the viewable area, D_n is the *nth* step's fatal defect density, and N is the number of processing steps in the procedure in question. From the formula, it can be seen that a large screen area directly causes the exponential decrease of yield to be even more acute.

The different factors in determining yield are usually categorized as to their frequency of occurrence, and then using a *Pareto chart*, the data is recorded, and the calculus of variations extrema methods described earlier in Chapter 7 may be used to perform optimization calculations. Finally, *statistical process control* (SPC) monitors the key process parameters, and from that is determined the remedial measures required.

An example of a key process parameter is the semiconductor fabrication *patterning critical dimensions* that include every layer's overlay, thickness and uniformity, conductive line width and spacing, process temperatures, and many equipment-operating parameters. Statistics from monitoring actual processing can be used to determine the parameter changes that resulted in improved or worsening yield, and all the process steps can be systematically adjusted for optimization of the entire process [8].

After the panel cuts and final testing, the finished panel is packaged for transport to the module assembly plant.

LCD Module Assembly

The liquid crystal cell has been completed and is now ready for the mounting of the addressing line and column drivers, a printed circuit board, light diffusers, the backlight unit, control board, and other small components depending on module type; the components are shown in the exploded view of Figure 25.11.

The connection of the driver ICs to the liquid crystal cell in the early days primarily used *tape automated bonding* (TAB), wherein a *tape carrier package* (TCP) holding the driver ICs used a "flex" tape to connect the driver ICs to the printed circuit board, and thence to connect with the liquid crystal cell addressing lines and columns. The so-called *chip on glass* (COG) and *chip on film* (COF) methods of driver IC and addressing system connection gradually have taken over the connection functions. As shown in Figure 25.12, COG or COF bonds the driver IC directly onto the edge regions of the glass (or film) to connect directly with the liquid crystal cell addressing lines, thereby eliminating the tape and also reducing costs.

Finally, adding the backlight unit (BLU), bezel, light diffuser, light guide plate, control board, and other components depending on the module type, and the labor-intensive assembly of the LCD module is finished. The module

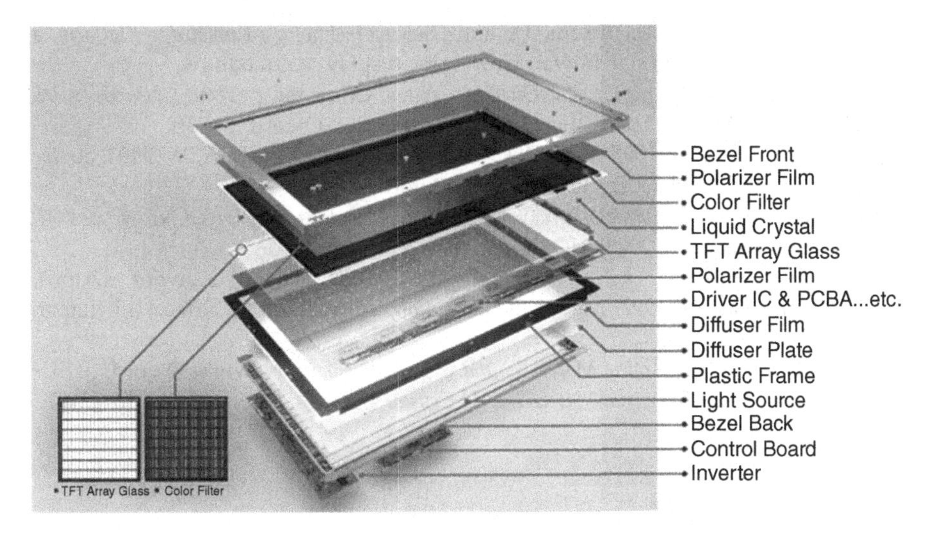

Figure 25.11 Liquid crystal display module in exploded view. From AUO website www.auo.com.

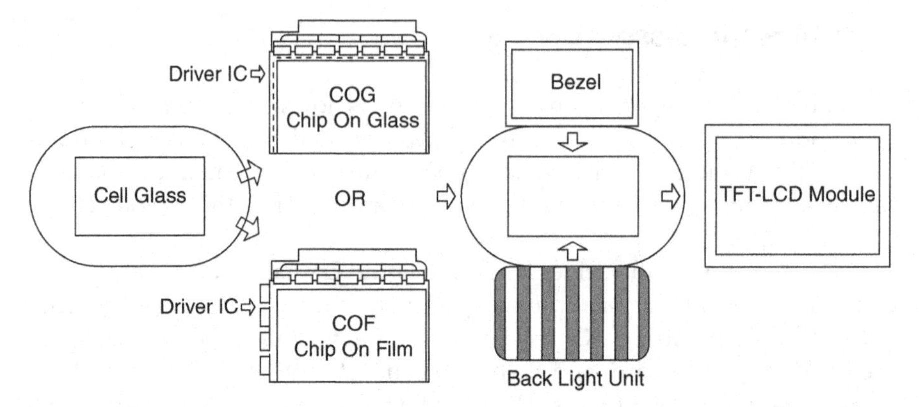

Figure 25.12 COG or COF bonds the driver IC directly onto the edge regions of the glass to connect directly with the liquid crystal cell addressing lines.

now can be shipped to the system integrator for assembly into the final consumer electronics product.

References

[1] Chimei website www.cmo.cmo.tw and *Display Technology Knowledge Platform*, a website run by National Taiwan University, display.ee.ntu.edu.tw.
[2] Lueder, E. 2001. *Liquid Crystal Displays*. Wiley, Chicester, p. 289ff.; den Boer, W. 2005. *Active Matrix Liquid Crystal Displays*, Newnes, Oxford, p. 73ff.
[3] Zhong, J.Z.Z., et al. 1998. *Proc. Asia Display*, **98**, 113ff.; den Boer, W. 2005. *Active Matrix Liquid Crystal Displays*, Newnes, Oxford, p. 144.
[4] den Boer, W. 2005. *Active Matrix Liquid Crystal Displays*. Newnes, Oxford, p. 76ff.
[5] Hashimoto, S., et al. 2005. *SID Digest.*, **36**, 1363ff.
[6] den Boer, W. 2005. *Active Matrix Liquid Crystal Displays*. Newnes, Oxford, p. 82; van Zant, P. 1997. *Microchip Fabrication*, 3rd edition. McGraw Hill, Taipei, p. 8ff.
[7] den Boer, W. 2005. *Active Matrix Liquid Crystal Displays*. Newnes, Oxford, p. 81.
[8] den Boer, W. 2005. *Active Matrix Liquid Crystal Displays*. Newnes, Oxford, p. 80.

26

The Global LCD Business

After the discovery of liquid crystals and the development of its science in Europe, the liquid crystal display prototype was invented in America, and with the synthesis of a stable liquid crystal compound in Britain and the advent of the hydrogenated amorphous silicon integrated circuit active matrix addressing system, America and Japan developed the super twisted nematic liquid crystal display for digital watches and calculators in the early 1970s.

A black-and-white twisted nematic television was demonstrated in New York in 1978 by Westinghouse's Luo Fang-Chen who had developed the first feasible liquid crystal active matrix driving system , but a month before the demo, in an all-too- familiar scenario, the Westinghouse liquid crystal group had been disbanded and all research terminated. Western corporations withdrawing and Eastern companies eagerly filling the void was effecting a paradigm for the liquid crystal display industry, exemplified by Luo himself eventually returning to Taiwan to become the chief technology officer at AU Optronics, and the once famous "Westinghouse" brand television devolving into an operation run by, and supplied with panels from, Taiwan's Chimei Optoelectronics.

Liquid Crystal Displays, First Edition. Robert H. Chen.

RCA's Legacy

Perhaps the most compelling example of the paradigm at work is RCA. At RCA, the brilliant and far-sighted researchers, although the originators of and the driving force behind the new liquid crystal display, and in spite of technical breakthrough after breakthrough, RCA's upper management terminated the program. Why would RCA give up on what appeared to others to be so promising? Apart from the obscuration wrought by the legacy of the shadow mask mentioned in Chapters 13 and 18, there was another factor: the temptation of a dramatic new technology beckoning.

While RCA was in the early stages of developing its liquid crystal displays, IBM was turning out its earth-shaking computer mainframes, and in the process fast becoming the *supremo* of high technology. RCA in 1972 could not resist having a go at the new computers game, and rashly jumped into the capital-intensive mainframe computer business. To make matters worse, heeding perhaps the business school call for diversification, the son of RCA Labs great founder David Sarnoff steered RCA into the car rental, food processing, and *chicken farming* businesses. Unsurprisingly, in venturing away from the now-vogue business school prescription of core competence, none of Bobby Sarnoff's ventures worked out, and RCA quickly faced suffocating operating losses. A severe cost-down program was initiated, and, as happens at many companies under severe cash flow stress, research and development was among the first to be cut. One of the victims was the pioneering liquid crystal display program.

Under the stultifying budget constraints, RCA's liquid crystal research group soldiered on bravely, but in 1985 gave up the ghost and the whole department was sold to General Electric. A big new employer should spark a revival, but any hopes of the liquid crystal researchers were soon dashed. GE not only did not continue with RCA's LCD research, it sold the entire television division, including the LCD group, to France's Thomson. A few years later, Thomson was to sell RCA's television division, including rights to the RCA brand name, to China's TCL, soon to become the world's largest television manufacturer. RCA's liquid crystal group has long since dissolved, and George Heilmeier's dream of an RCA brand wall-hanging television never came to pass.

Not only did RCA miss the liquid crystal display express that it launched itself, many related inventions, such as the complementary metal oxide semiconductor (CMOS) thin-film transistors (TFT) used in the active matrix addressing system, the charge-coupled device (CCD) for digital video cameras, the light-emitting diode (LED) later to be used in the LCD backlight unit, and other commercial successes, such as the video recording tape and compact discs, all were developed at RCA, and each invention went on to create huge consumer electronics markets, from none of which did RCA gain much sales, leaving the spoils as it were to companies in Japan, Korea, Taiwan, and China.

But RCA was not alone in its inertial malaise; the end of the twentieth century also saw the iconic high-technology giant IBM, in spite of its computer prowess, also leave the liquid crystal display industry. However, as related in Chapter 18, to try to capitalize on the rapidly developing notebook computer market, IBM re-entered the fray, but in order to defray costs and risks, moved all its LCD development and manufacturing to Japan, allying with Toshiba to share costs and risks.

But with the worldwide dotcom recession as backdrop, Big Blue retreated once again, and in a fateful decision, seeking royalty revenue, IBM in 2001 licensed its liquid crystal display technology to Taiwan's AU Optronics. Less than a year later, IBM quit LCD manufacturing altogether, selling its entire display division, including the state-of-the-art 3rd generation mass production facility at Yasu, to Taiwan's Chimei Optoelectronics.

In the period immediately following, further reengineering away from the manufacture of consumer electronics and focusing on information services, IBM sold its computer hard disk manufacturing business to Japan's Hitachi, and the personal computer manufacturing division to China's Lenovo, further crystallizing the West-to-East paradigm.

An epilogue to the RCA liquid crystal saga comes from a clause in the RCA shadow mask license. The license grants the right for all licensees to freely attend RCA's technology events. No doubt RCA's aim was to increase the opportunities for further licensing of its technology, but at the same time many companies from Japan, Europe, and the United States, including IBM, all received first-hand knowledge and guidance from RCA's pioneering liquid crystal display research group. The technical knowledge did indeed spread to others, and in the end, RCA's fundamental liquid crystal technology could be said to have been passed on to Taiwan by way of the licenses and technical assistance from IBM.

Optical Imaging Systems

The first mass-produced liquid crystal product was the digital watch. By 1981, the Japanese manufacturers, by dint of precision workmanship and high efficiency, had already taken 45% of the digital watch market, and 80% of the market for other liquid crystal products. American companies, among them Intel's predecessor Fairchild Semiconductor, understood that to be competitive with the Japanese, they had to cut costs, and that meant moving manufacturing to East Asia. So Fairchild's 200,000 units/month assembly line was moved lock, stock, and barrel to Hong Kong. But only three years later, Fairchild's Hong Kong operation was bought out by Hong Kong's Conic Semiconductor, and in 1982, China's state-owned China Resources Corporation, in what was then the first instance of a Chinese government corporation acquiring a Hong Kong company, bought out Conic. A plan was put in place to further increase efficiency and yield, so that by 1987, production had reached a staggering 200 million pieces per year [1]. Such volume production of course destroyed the market price, so the eastwards production move not only failed to solve the pricing problems, because of the technology and know-how transfer, it irreparably exacerbated them.

Earlier, from 1978 to 1982 still unable to compete, the American companies Intel, Fairchild, National Semiconductor, Motorola, Westinghouse, RCA, and Texas Instruments one by one quit the liquid crystal display business. By the middle of the 1980's, the main players in the LCD industry were Germany's Merck providing the liquid crystal, and Japan's Sharp, Seiko, and Hitachi, Holland's Philips, and America's soon-to-depart IBM in Japan making the displays. At this juncture, a sea change in the LCD manufacturing industry was in the offing, heralded by the appearance of a few very eager companies in Korea and Taiwan on the horizon.

The mass exodus of American companies from the LCD manufacturing business understandably raised concerns about the implications for national defense and security of the entire LCD industry moving to East Asia. So American Telegraph and Telephone (AT&T) and Xerox, while seeking government research funding support, took the opportunity to urge the government to establish an American company to manufacture LCDs in the United States. AT&T and Xerox believed that since the American companies could not compete with East Asian companies on cost, to prevent the supply of "strategic" liquid crystal displays being monopolized by Asian companies, the government should subsidize LCD manufacture.

The entreaties of the giant AT&T and Xerox were heeded, for the Pentagon's Defense Advanced Research Projects Agency (DARPA), quickly granted "emergency funding," earmarked for a new company called *Optical Imaging Systems* (OIS). OIS' mandate was to quickly develop and manufacture advanced LCDs for the American military.*

Notwithstanding the mandate, the first substantive act by OIS was to cooperate in a joint venture, through intermediation by Korea-born liquid

---

* DARPA is the research and development office of the U.S. Department of Defense; it was established to fund defense-related research in 1958 as a response to the launching of the first satellite Sputnik by the Soviet Union in that year.

crystal chemistry pioneer Oh Chan Soo, with Korea's Samsung to set up a thin film transistor liquid crystal plant on the outskirts of Seoul. Not only that, in 1990, OIS, in an ownership share payment in-kind arrangement, granted a patent license to Taiwan's United Microelectronics' liquid crystal subsidiary Unipac Optoelectronics, with OIS ultimately taking a 10% share of Unipac. These acts clearly were at odds with the mandate of the government-backed OIS charter to develop American capability in LCD manufacture in the United States; they even further strengthened East Asia's grip on the industry and exacerbated the production problems in the United States.

In mitigation, OIS perhaps acted in the belief that both South Korea and Taiwan, at the time stalwarts in the American East Asian military alliance, could be depended upon to supply LCDs. But the real reason was probably less *realpolitik* and more *realbusinik.* Military specification ("milspec" in the industry jargon)-compliant LCDs were difficult and costly to manufacture, and the market was limited to the military services. This made it extremely difficult for OIS to find any American company willing to take up the endeavor, so OIS had no choice but to turn to Korea and Taiwan for help. Although making realistic business choices, OIS still was roundly criticized by both government and industry.

The criticism stung, for OIS re-doubled its efforts, and finally persuaded an automobile glass company in economically depressed Detroit to try its hand at LCD manufacture. The terms of the venture were extremely favorable: DARPA would bear half of the cost of building the manufacturing plant, and Guardian Industries would only have to make a small-scale share investment, with the manufacturing target of only 50,000 units per year. However, as will be proved time and again, patriotism seldom trumps market reality, and even with such munificent terms, the expensive equipment, high labor costs, and an untrained and undependable labor force, together with the limited market, quickly brought the joint venture to the brink of bankruptcy. The conservative Guardian executives refused to inject further capital into the company, and in 1998 Guardian shut down the LCD plant. OIS was left with no more choices and would soon go out of business.

For South Korea and Taiwan, however, the OIS story did not end with the closure. In its re-organization proceedings in 2001, OIS was forced to liquidate its assets to pay its creditors, and among those assets were several patents. So even *post mortem,* OIS continued to abet the Asian LCD manufacturers by offering to sell its LCD patents to South Korea and Taiwan panel manufacturers. Samsung bought some patents, but newly formed and cash-poor AUO and Chimei politely declined the purchase as beyond their means.

That was a mistake, for in 2005, Guardian Industries, having acquired some of the OIS patents, sued AUO, Chimei, and practically all the large LCD panel manufacturers and brand-name sellers in Taiwan, Korea, and Japan for patent infringement. Guardian had reluctantly entered the LCD business, performed half-heartedly and quickly gave up the ghost, but later audaciously demanded patent royalties on LCD technology it received from OIS. The Guardian story perhaps epitomizes the failure of American companies in the consumer electronics business, many ultimately relegated as it were to licensing patents for income instead of manufacturing products.*

In continuing efforts to respond to the government's pleas for a liquid crystal knight-errant, the defense contractors Honeywell and Xerox also tried to develop and manufacture their own milspec liquid crystal displays, but although technologically very successful, their manufacturing businesses fared no better than OIS and Guardian, and LCD panel production was phased out at those companies in 1998 and 2001 respectively. Today Honeywell's military-use LCDs ironically are made almost exclusively by Taiwan's Chimei.

It can be said without exaggeration that America's national defense strategy for indigenous LCD production was a complete and utter failure. Not only were the tangible liquid crystal displays never produced in volume, even the intangible intellectual property behind the displays were sold. For instance, beginning in 2003, Chimei purchased patents from Honeywell, Rockwell-Collins, and General Electric, three of the biggest defense contractors in America. And based on its early LCD research, Honeywell in 2006 successfully licensed its remaining patent of value to all the large liquid crystal panel manufacturers, and OIS as late as 2009 was continuing to sell its remaining patents.

The moral of this story might be that the innovative electronics companies in America were betrayed by the lack of foresight and perseverance of their leaders, but more likely America and Europe's liquid crystal fate was preordained by the ineluctable course of social and industrial evolution.†

---

* But in other quarters, Guardian is a respected automotive glass company, well-known primarily because its boss William Davidson was also the owner of the NBA's Detroit Pistons. When Chimei took the first license from Guardian, as part of the news release, the author noted that the licensor's president was also the owner of an NBA team; however, that part of the news release was excised by the reporters (although extremely popular in Taiwan, the commercial affairs reporters evidently have little interest in basketball). William Davidson died in 2009, the ownership of the now-struggling Pistons taken over by his daughter, but the franchise was soon to be sold.

† In witness thereof, many of the brilliant liquid crystal display research pioneers have been relegated to the ranks of expert witnesses at patent infringement legal proceedings.

The Electronics Manufacturing Paradigm

The archetypical development process for today's electronics products starts with a scientific specimen hatched from European academia, the curiosity being taken in by American technology companies and a prototype fashioned. That device is then seized on by consumer electronics companies in Japan, and refined to become a viable commercial product. Then huge production lines in Korea and Taiwan efficiently mass produce the basic components of the new product, which are then assembled by system integrators in China to sell under different brand names all over the world.

After achieving significant research breakthroughs, the American and European pioneering liquid crystal display companies certainly had the chance to seize control of the new industry; still, one by one, they left the LCD business and relinquished the market to the Japanese. Why did first RCA, then Hoffman-LaRoche and Westinghouse quit the field, followed by Intel, Fairchild, National Semiconductor, Motorola, and Texas Instruments? Were the bosses of these companies so lacking in vision and fortitude as to not see the tremendous potential of the liquid crystal display? Of course, any post-mortem analysis is greatly aided by 20/20 hindsight, but it is a fact that the Japanese companies Sharp and Hitachi at the outset saw things much more clearly than their American counterparts. The genesis of the problem, however, lies not entirely in a vacuum of leadership, it is also likely caused by the natural economic evolution of an industry and a country.

The older Western companies were operating under the "burden of successful products," that is, as mentioned in Chapter 13 in regard to RCA's shadow mask, the commercial success of earlier products brings an inertia that impedes the development of competing new products. Companies new to the market on the other hand have no such product legacy, and are thus freer to take risks on completely new products, with a compelling goal to capture new markets and gain a place among the industrial elite.

As for internal decision making, in contrast to the popular perception of America as the bastion of free-wheeling enterprise,* it was Japan's Yamasaki Yoshio of Seiko who said [2],

> Our company's culture is that, unless there is a good reason not to, top management usually approves what the lower levels decide is the best way to go. It's not our style for top management to say, "Do this." Rather [the direction] comes from the bottom up, like, that way looks good, why don't we do that?

* The enterprising spirit in America is of course more evident in the youth culture Internet-related companies, but some such as Google sought out the seasoned old-hand Eric Schmidt to lead the company.

In other words, big Japanese electronics companies are nothing like the caricature of the rigid hierarchy of old men making decisions for deferential young employees to obey in lock-step; big business Japan was rather much more open to young ideas than their ostensibly youth-oriented competitors in America.

In fact, it seems it is in big American companies where the bosses act like rockstars, with their words sanctified in management books, but they too often make terrible decisions that not only deprive the company of profit, but alienate capable and sometimes brilliant young employees. In the LCD business, the names Heilmeier and Helfrich come to mind. Contrast this with the Japanese electronics companies who profited from new products, and where Japanese engineers like Sasaki and Wada were encouraged by management in their pursuit of risky new technology products.

As for business strategy, in particular, RCA, generated income from licensing its technology, and in pursuit of the revenue gained, carefully avoided competing with its licensees, particularly those overseas and specifically companies in Japan. So RCA apparently attached greater importance to intangible intellectual property than to tangible manufacturing capability. The result was that in spite of being a world-beating high-technology company, RCA was more a domestic operation than a multinational conglomerate.

Another factor is of course the natural maturation of a society. Decades of industrialization and social benefits inevitably make workers less eager, and the powerful unions demands for higher pay and fewer hours resulted in high costs and low productivity in the labor-intensive mass-manufacturing industries. This together with the burden of the aforementioned legacy products, together with lack of leadership, resulted with few exceptions in little competition for the efficient, disciplined, and diligent labor force of East Asia and their hard-driving business leaders.

Although making a comeback, many of the scientific elite in Europe have emigrated to America, while the young educated elite of America, tempted by get-rich fast careers in medicine, finance, and law, shun science and engineering.* And while the Japanese appear to have maintained a national bent for new technology and an admirable diligence in the pursuit thereof, their

* For the last 30 years, more than half of the science and engineering PhDs in America were foreign born, and many of the technology entrepreneurs similarly born and raised outside the United States. With economic downturns now prevalent in the West, many of these immigrants are now returning to their countries, armed with American technological education and know-how; this means the West-to-East electronics paradigm will only prove to be more true.

country has been mired in a capital-starved 20-year recession. Korea and Taiwan seem to have filled the manufacturing void, but those societies now, and China later, cannot expect to avoid social evolution, so the natural question to ask then is whether emerging Korea, Taiwan, and China will also succumb to creeping manufacturing indolence, with the manufacturing industries continuing their journey through the East onwards to other countries and regions.

Will those at the top move up (and perhaps out) of the electronics food chain and those in the middle and bottom move up? And will new players emerge at the bottom? That is, will the high-technology electronics manufacturing paradigm apply to newer emerging economies in Eastern Europe, Southeast Asia, South Asia, Central Asia, the Middle East, or even South American and Africa?

A preliminary analysis of this momentous question might be found in examining whether a country or region can meet the following basic requirements for a manufacturing industry: (1) an educational system that values science and engineering so that it can produce competent engineers; (2) a culture that gives social status to engineers;* (3) some manufacturing tradition; (4) a diligent populace; (5) capital resources; (6) modern and efficient basic infrastructure such as water, electricity, transportation, and communications; and (7) a society organized to allow the development of technology and avoid political and social unrest.

But even if all these initial conditions are satisfied, a country or region may miss a technological wave through bad timing. For example, the peoples of Russia and India certainly possess the capability and educational system to partake in electronics manufacturing, and the oil-rich Middle East has hatched plans to develop an electronics industry, but politics, religion, the worldwide economy, and certain societal factors might cause them all to miss the high-speed technology express.

Based on the conditions listed above, it appears that the following regions and countries should have the best chance at developing a domestic electronics manufacturing industry: Russia, Eastern Europe, India, Iran, Egypt, Turkey, and Pakistan. But again, unpredictable factors of religion, tradition,

* For example, for many years, the first preference for university study for male students in Taiwan is electrical engineering, and before that, because of the influence of the Nobel Prizes awarded to C.N. Yang and T.D. Lee, it was physics. The social ascendency of science and engineering in East Asian countries is not likely to be matched in any other region of the world. Perhaps it can be ascribed to the reverence for education in the Confucian societies of China, Japan, Korea, and Taiwan.

and the residues of communism make predictions tenuous at best and likely imprudent.

It might be best to instead take a real example for more substantive analysis; following is a short history of the development of the liquid crystal display industries in East Asia, focusing on Korea, Taiwan, Japan, and China, and with particular emphasis on what it took to be successful in electronics manufacturing.

Korea, the Emerging Economy Model

South Korea is the epitome of the late twentieth-century newly-industrialized emerging economy; its fast-paced development was led by its flagship conglomerates, Samsung, Hyundai, LG, and Daewoo. The first of the foursome to try its hand at the new liquid crystal display technology was Samsung Electronics; the objective was modest: to produce an LCD computer monitor to challenge the entrenched CRT.

However, while Samsung could perform the basic manufacturing, the cost-effective, efficient volume production required at least an acceptable yield and presented problems. At the time, only the Japanese companies could mass produce LCDs with profitable yields. Samsung first turned to its CRT monitor manufacturing partner NEC for help, but NEC itself had not yet commenced LCD mass production, so had little to offer in volume production assistance. Samsung next turned to Sharp and Citizen each of whom had considerable experience in LCD mass production for calculators and watches, but both summarily refused to help. Of the other experienced Japanese LCD companies, Hitachi was in partnership with arch-rival LG, so that was not an option for Samsung.

With nowhere to turn, once again the liquid crystal chemistry pioneer Oh Chan Soo fortuitously turned up. Originally at RCA and then at Beckmann Instruments, Oh was facing imminent relocation due to Beckmann's recent decision to exit the LCD business. After being introduced, Samsung saw Oh as a savior and hired him on the spot to be a senior consultant. Returning after many years to his native land, Oh's first contribution was the aforementioned joint venture with the ignominious OIS; his second bit of advice was for Samsung to try to go it alone and overcome the LCD monitor yield problems completely in-house [3].

But Oh's expertise was in research and not in production, so his self-help sentiments and words, noble as they were, in deed could not solve the yield problems, and little progress was made. Daewoo, Hyundai, and LG in the meantime had also all set up their own liquid crystal display companies, but with Japan refusing to help, now all four conglomerates after nearly 10 years of effort, saw no light at the end of a very long tunnel; they began preparations to abandon the effort. But just then the farsighted Korean government came to the rescue, and in 1995, under the well-established *modus operandi* of Korea's insidious government–banking–conglomerate cooperation, the big companies were offered almost risk-free terms and conditions to continue the effort, at least until fate could intervene.

The Crystal Cycle and Korea

The principal obstacle to a good yield was the time-consuming and fault-ridden process of the side injection of the liquid crystal into the space between the glass substrates described briefly in Chapter 25. Samsung, LG, and Hyundai were trying mightily to overcome the technical problems, but could not break through. Ironically, the transformation of the *crystal cycle* from a liquid crystal death star into a star of light for Korea was the impetus for a yield breakthrough.

In the LCD business, the so-called "crystal cycle" refers to the periodic large amplitude swings between profit and loss in panel manufacture. It comes about because the manufacturers simultaneously though not in concert overreact to the vagaries of the market, that is, the manufacturers, responding to a panel shortage sellers' market, will naturally increase production and open up new production plants. The strong demand also entices new manufacturers to enter the profitable new business. A new manufacturing plant typically takes two years to complete, and when all the new plants begin almost simultaneous mass production, with the increased capacity plus the new players all contributing to new supply, the resulting overproduction leads to severe oversupply, producing a market glut of LCD panels.

The buyers' market now produces a precipitous drop in panel prices, but in the capital equipment-intensive LCD industry, the manufacturers cannot simply stop or even decrease production without incurring idle capacity losses from depreciation and fixed costs, so they soldier on hoping to weather the storm by decreasing their prices to gain market share, but of course this only exacerbates the problem, accelerating the vicious cycle's race to the bottom. When recognition of their dire straits finally sets in, the manufacturers begin cutting production and putting off plans for new plants; some companies go so far as to leave the industry altogether.

At the time, the only volume manufacturers of LCD panels were all in Japan, but either unable or unwilling to endure the deep trough of the crystal cycle, and in the face of a worldwide recession, they not only decreased production, but began leaving the industry. Although a few companies did indeed see the potential of the LCD market and wanted to continue, because of the bursting of its enormous real estate bubble, Japan was in the depths of a long-term domestic recession, and those companies could not secure capital for building new plants to supply the growing LCD market, and eventually they too were forced to cut production.

In the lifetime employment tradition of big Japanese companies, there thus appeared a coterie of skilled and experienced engineers who were suddenly left with little or nothing to do. South Korea, seeing an opportunity, surreptitiously invited the idle engineers to Korea to assist the Korean LCD panel manufacturers with the liquid crystal injection problem. And so there came to be a samurai "weekend warrior," crossing the Sea of Japan to Korea on his days off to provide technical assistance to the Korean conglomerates (and on the side earn a tidy moonlighting fee).

Just as in all high-technology endeavors, experience and skill counts in the LCD industry, and soon the injection problems were overcome and the yield began to rise. Thus it was that the dreaded crystal cycle and the debilitating real estate bubble, while devastating Japan, ironically provided a timely and salubrious balm for the ailing liquid crystal display industry in Korea.

Crisis and Fortune in the LCD Industry

Just when the LCD monitor business in Korea was beginning to look good, the 1988 East Asia financial crisis hit, and it hit Korea particularly hard, leveraged as its conglomerates were by huge loans from their pliant national banks. But once again, not only was the crystal cycle turned on its head in Korea, even the financial crisis miraculously propelled Korea's LCD industry to greater heights. The effect of the crisis on ordinary people was that suddenly their purchasing power, demarcated in the won, had precipitously depreciated; the banks were in dire straits, and the entire country fell into a deep recession, with the government humiliated by essentially being in the receivership of the International Monetary Fund.*

But in the electronics industry, Samsung and LG in particular, were perversely reveling in the crisis. With the help of the Japanese weekend warriors, the LCD panels were being produced with ever higher yield, and with

* Television news watchers perhaps remember the scene of the young man holding a sign "IMF = I am fired," and young women patriotically lining up to sell their gold jewelry for foreign exchange.

the depreciation of the won, the panels were now very, very inexpensive for foreign buyers. Korean LCD monitors were soon flooding the market just in time for the rapidly growing worldwide personal computer market. The sales payment for the monitors was in strong US dollars from foreign buyers, and the production costs to Samsung and LG were in depreciated Korean won. It was cash flow Heaven, Samsung and LG were enjoying a "high yield/low exchange rate" paradise in the midst of the misery of their country's financial crisis.

Flush with profits, Samsung began to expand, and in 1999 completed the world's first large-scale 3.5 generation liquid crystal panel production line; soon after, South Korea surpassed Japan's ever dwindling output to become the world's biggest supplier of LCD panels.

Samsung Is the Lucky Goldstar

As children, we would often be admonished by elders that only hard work can bring success, but upon growing up, many encounters have also told us that sometimes it's better to be lucky than good. Here, the dreaded crystal cycle and the calamitous real estate bubble bursting in Japan, combined with the economic crisis in East Asia, produced the perfect storm for invigorating Korea's moribund LCD industry. Of course, there were other factors: the vision of the conglomerates' leaders to pursue the capital-intensive LCD manufacture, the dogged determination in the face of difficulty characteristic of the Korean people themselves, and the government and banks' benign complicity with the conglomerates. But still the recession in Japan and the East Asian economic crisis ironically were like manna from Heaven to a beleaguered industry, and so "Samsung" ("Three Stars") and "LG" ("Lucky Goldstar") were clearly more than just quaint names for the companies, the stars in their names also determined their fate as the new global commercial stars.*

And their lucky stars continued to bring forth good fortune, for just at the descent into the first deep trough of the crystal cycle with customers fleeing and monitor production slowing, the notebook computer raised its lovely liquid crystal head above the horizon. The beautiful image presented, combined with an indeed very useful new portable computer device, gave impetus for an exceptionally strong new demand for liquid crystal screens, and LCD panel prices began to rise once more.

---

* It is perhaps fitting that after centuries of actual and perceived humiliation by its larger neighbours China and Japan, and even subjugation by far-away America during the Korean War, South Korea has indeed stood up and quickly joined the ranks of the advanced countries.

Clearly appreciating the potential and eager to seize the opportunity, LCD manufacturers once again ramped up production lines and built new plants to meet the new demand, and new players energetically entered the fray, eager to catch the wave of a huge new market. And once again the crystal cycle set forth on a precipitous upturn into the next period of the dangerous crystal cycle.

Once the successes of the South Korean LCD manufacturers were manifest, companies in other countries could not help but notice. Holland's Philips in 1999 formed a joint venture with LG appropriately but prosaically named LG.Philips LCD (LPL), in which Philips invested more than US$1.5 billion dollars. After just one year of construction, the first LCD panel came off the production line, and in only six years, LPL gained the mantle of the world's largest producer of liquid crystal display panels.

However, just at that height of production, the crystal cycle reared its frightful head (or rather lowered its head into a rapidly propelling downwards curve), and the market rendered unto LPL the dubious distinction of suffering a record US$800 million dollar business loss. This loss drove Philips, in a surprisingly abrupt decision, to pull the plug and sell its entire share in LPL to LG; the company now to be renamed LGDisplay (LGD), leaving only one parent, LG, and severing the Philips connection altogether.

But there were other companies in Korea and Taiwan that, although similarly bleeding, were able to endure. Philips perhaps was just following in the steps of other companies like RCA and IBM, enchanted by the glitz and glamour of global brand names, but not willing to do the nitty-gritty manufacturing. The question then is, can a company do both? Korea's Samsung, LG, and Hyundai, although gigantic conglomerates, somehow were able to do successful mass-manufacturing and at the same time vigorously promote a global brand.

Their perseverance was rewarded, for the downturn of the crystal cycle that caused Philips to so quickly turn tail was this time fortuitously countervailed by the coming of age of the beautiful new flat-panel liquid crystal television set.

The inflection point where the advantages of screen size, sleekness, and fashion overcame the price differential with CRT sets was at last reached, and with the decline of the production of competing flat-screen plasma television, the beautifully large yet slender liquid crystal television provided the impetus for yet another journey up the sharp slope of the crystal cycle. Again, the big manufacturers engaged in a race to build capacity, this time new 8th, 9th, and 10th generation plants. The building flurry ominously portended yet another down cycle, but for the moment, record production, revenues, and profits were there for all to take, generated as they were by the televised crest of the virtuous crystal cycle, now in the bloom of full color.

Particularly enjoying the ride was Samsung, which quickly became the world's largest electronics manufacturer and at the same time overtook Sony as the world's biggest electronics brand (measured by sales). Just how Samsung could manage such success in the wake of the failures of others offers a very interesting case study.

The almost completely vertically integrated super conglomerate Samsung is somehow able to supply LCD panels to other brand companies such as Sony and Toshiba, and at the same time push its own global brand. Even more curiously, Samsung would purchase panels from competing manufacturers such as Chimei and AUO to install in their Samsung brand products. Just how does Samsung manage to successfully integrate all of these disparate functions and at the same time avoid conflict, both within the company and with its customers?

Samsung's great depth and width requires either unsurpassed intelligence to work it all out, or else supreme internal discipline to keep all the inherent conflicting forces at bay. Different interests within the company surely will ultimately collide, generating business contradictions. That is, from the point of view of competing brands, the brand name customer company (such as Sony) will complain that the panels purchased from Samsung, being the same as those used by Samsung's own brand, will be in the same sales channel and thereby directly compete with Sony's brand, and Sony at the same time will always suspect that internal panel sales at the internal transfer price (ITP) to the Samsung brand division will surely be on better terms than sales to outsiders such as Sony. From the point of view of Samsung's brand name sales force, the overlapping Sony/Samsung sales channels will produce channel conflict, where a channel has both Samsung panels in Sony products and Samsung panels in Samsung products. Samsung's manufacturing and brand divisions should be at constant loggerheads.

Further, from the point of view of Samsung's manufacturing divisions, the Samsung brand division's purchase of panels from other sources (such as Chimei) should leave a sour taste, and when the manufacturing divisions sells panels to Samsung's brand competitors, this would leave the brand division fuming. Of course, things can be worked out where only those panels that are in short supply are externally sourced and/or only those panels which are in oversupply are sold externally, but what happens in other supply/demand times? To make it all work, preternatural intelligence is required to assuage offended sensibilities and encourage individual sacrifice for the common good; either that or an iron discipline wrought from a militaristic leadership ethos ("sit down and shut up").*

Taiwan is the consumer electronics *original equipment manufacturer* (OEM) capital of the world. In the computer business, practically all of the well-known brands, such as Apple, Dell, HP, Sony, Toshiba, and so on, are actually manufactured by Taiwan companies, to the extent that some 80% of the world's notebook computers are made by Taiwan OEMs either in Taiwan or China.† But in Taiwan, the brand name companies (such as Acer and Asus), to assuage the misgivings described above of their brand name customers but lacking the Korean-style paternalistic familial power, tend to break up into two, separating brand name sales from manufacturing (for example, the Acer Group's Acer/Wistron and BenQ/Acer Peripherals in a cascade of breakups). One of the new companies concentrates on product design and brand sales and does no manufacturing (that is, present-day Acer), while the other does only manufacturing for brand companies (Wistron).

The success of the division rests on a tenuous peaceful coexistence between the new companies, upheld by the mutual exclusion from each other's business that is exacted from a "balance of terror." The terror is that if an OEM manufacturer threatens to break the coda and create its own brand, the brand company can retaliate by doing its own manufacturing. And vice-versa, if the brand company wants to manufacture its own parts, the manufacturing company may then create its own competing brand.‡

In Taiwan, however, the norm is that the newly-minted boss of a designated manufacturing company finds it hard to resist the cachet of an own brand and chafes under the restriction, furtively promoting and ultimately pushing his own new brand (for example, BenQ). As for the designated

* The collective good over the individual interest is part of East Asia's Confucian culture, and is particularly strong in Korea where uniting and sacrificing in the face of foreign aggression is almost a national ethos. In businesses, the authoritarian coercion can work, helped by the fact that all the big Korean conglomerates are family concerns, that is, only a leader with inherited absolute familial power would be able to impose iron discipline on the heads of the competing divisions.

† The term "OEM" is used differently in Europe and sometimes in America as referring to the brand name company; in the computer business, they do little or no manufacturing of the product, so "original equipment manufacturer" for computer brand companies is largely a misnomer; its genesis likely from the brand companies desire not to publicize the fact that they do not in fact manufacture their own products.

‡ This did indeed happen when Acer Peripherals began selling its own BenQ brand notebook computers soon after separation from Acer Brand, much to the consternation of the latter and its eventual use of rival notebook manufacturer Compal as OEM supplier.

brand company, while happy with the glamour of an own brand, the designers are often unhappy with the manufacturing counterpart's parts and service, and the allure of vertical integration impels the boss to similarly furtive and ultimately blatant manufacture of its own parts. This devolution typical in Taiwan likely stems from the new bosses having tasted both the heady wine of branding and the compelling efficiency of vertical integration while working in the mother conglomerate; one then can hardly blame them for trying to recapture that old spirit when they become bosses of their own companies. After all is said and done, the result is that the problem has not been solved but has only shifted to the newly integrated company, which now has evolved into a conglomerate.

If the breakup on the other hand is only to ostensibly placate customers, it is unlikely that the brand customer will not notice, and in most cases, the outcome rests upon the powerful verity of supply and demand. That is, in times of short supply, the customer will not dare to broach the subject, but during oversupply, the internal conflicts of the manufacturer will surely be an issue in the business negotiations for lower prices and higher performance.

Returning to Samsung, its success clearly demonstrates that it is indeed possible for a vertically integrated whole, and horizontally spread semi-independent divisions, to solve the conflicts and prosper. However, perhaps it is just the sheer size of the conglomerate and the economic power it brings to bear that makes the whole thing work, or maybe it is the Korean people's historical unity of purpose and highly competitive spirit that makes it possible. If that is the case, such overarching conglomerates are only possible in Japan and Korea, and perhaps in the historically family-run businesses in Europe (particularly Italy); it seems unlikely to happen again in America.

Another aspect, compellingly contradictory, is despite the internal harmony wrought by a strong nationalism, there is nonetheless an intense internecine rivalry between conglomerates such as Samsung and LG. In analogy with sibling rivalry, the competition may have provided the impetus for their phenomenal growth, but brutal competition also can be debilitating. Interestingly enough, the government of Korea has taken note of the competition and after besting Japan, in the face of growing competition from Taiwan, in 2007 the Korean Ministry of Economics called together the leaders of Samsung and LG Electronics and informed them in no uncertain terms to cease the internecine warfare and work together, especially share each other's technology, cross-license patents, and above all, stop buying panels from Taiwan and concentrate on supplying one another [4].

Whether the blatant nationalism of *Buy Korean* and a strategic alliance of historically bitter foes can overcome the commercial ethos of *price/performance* and strategic international cooperation remains to be seen. But recent years have seen not only no decrease in Korean panel purchases from Taiwan, but rather increased outsourcing to Taiwan. South Korea's inward turn likely would meet the same fate as Japan's attempts at technology lock-out (to be discussed below).

Taiwan's Twin Stars

The earliest liquid crystal panel manufacture in Taiwan was by Chung Hua Picture Tube (CPT). From its origins as a cathode ray tube (CRT) producer for monitors and televisions, CPT opened Taiwan's first super twisted nematic (STN) plant in 1985 [5]. Shortly thereafter, the governmental-backed Industrial Research Technology Institute (ITRI) in 1986 announced an information technology (IT) initiative, to promote industries in displays, compact discs, image processing, and image transmission. Many of Taiwan's optoelectronics executives and technology officers were in fact spawned at ITRI.

In the 1980s, Hitachi and Sharp also opened up small passive matrix addressing liquid crystal panel facilities in Taiwan's southern city of Kaohsiung. Later in 1991, Robert Tsao's United Microelectronics Corporation (UMC) group started Unipac, which in cooperation with Prime View International began small- and medium-size TFT LCD manufacture.

However, at the outset production yields were unacceptably low, and just as in Korea, assistance sought from Japan's LCD manufacturers was summarily denied. The TFT LCD manufacturing industry's fits and starts nonetheless did not deter Acer founder Stan Shih from starting Acer Display Technology in 1996 with a mission to manufacture liquid crystal screens for Acer's notebook computers. In those days, the LCD screen accounted for one-third the total cost of the notebooks, and an Acer-made screen could provide notable cost-savings.

Practically all notebook computer screens at that time were made in Japan by Sharp, NEC, and Toshiba, and although (or perhaps because) Acer was a major customer of those companies, they adamantly refused to assist or transfer technology.

Since the Japanese companies refused to help with technology transfer, three of Taiwan's big notebook computer manufacturers, Quanta, Arima, and Mitac, turned to South Korea to try to obtain LCD know-how by acquiring Hyundai Display (Hydis). Hyundai, unlike sister conglomerates Samsung and LG, was unable to make the LCD manufacturer Hydis viable (probably because of Hyundai's emphasis on heavy industries that were hard-hit by the global recession), and the mother company decided to alleviate the pain of the 1988 East Asia economic crisis by disinvesting, and among those let loose was Hydis, who struggled on but soon found itself in dissolution. However, when the parent Hyundai kept raising the acquisition share price, the takeover failed to materialize, and the supply of liquid crystal notebook screens was still controlled by the big Japanese companies.

Taiwan's liquid crystal display companies, just like those of South Korea, were of course offended by the refusal of the self-interested Japanese LCD manufacturers to transfer technology and know-how. The reluctance of the Japanese companies, however, was well founded, for once Taiwan's companies entered a mass electronics consumer industry, their low-cost production efficiency would inevitably drive out the competition. Japan's intransigence put the development of Taiwan's LCD industry at risk; however, once again, one's misfortune may be another's fortune. The bursting of the real estate bubble in Japan was to spread an improbable beneficence over nearby Taiwan's nascent notebook computer screen industry, giving it sustenance in its despair, just as it had however inadvertently nurtured South Korea's LCD computer monitor industry.

Just what is a "real estate bubble"? Put very simply, because many bank loans use real estate as collateral for loans, when the assessed value of the real estate precipitously rises (the "bubble"), many home-owners in their irrational exuberance can obtain unreasonably large loans to spend as they wish, which is all well and good in prosperous times. But when the bubble bursts (for whatever reason, for example a recession), the real estate values drop also precipitously, and when the overcollateralized borrowers' loans are called in, they encounter difficulty in repaying beyond-their-means debts, and the banks suddenly encounter a surfeit of "bad debts," that quickly permeate the whole banking industry. Defaults on debt payment resulting in foreclosure and repossession, however, see little return on the banks' auction sales in a bad economy, leaving the banks with huge write-offs and subsequent shortages in capital reserves. Making matters worse, in trying to pay off their debts, the borrowers have no money left to spend, resulting in an ever-worsening recession. With no money in circulation, domestic consumer demand drops like a rock, and adding on a global recession, foreign demand falters as well.

The LCD industry in Japan in particular experienced a long and deep trough in the liquid crystal cycle during the recession, with the result that as revenues contracted, there was little desire or money to invest in new production. At the time, a new generation facility costs about a billion dollars, so even if the LCD manufacturers wanted to expand to supply the burgeoning monitor and notebook market, the banks had little or no capital to lend. These *zombie banks*, even in syndication, were just a file of the walking-dead, of no use to anyone, but striking terror in all. Large capital loans for expansion were almost impossible to obtain in Japan.

With demand collapsing, the Japanese manufacturers began to tighten their belts and cut back production. In a stark example, in 2000 the hundred million dollar prepaid order for notebook computer screens from Taiwan's Arima Corporation was abruptly and unilaterally canceled by NEC, who shortly thereafter exited the notebook computer screen business altogether.

Bleeding losses and with nowhere to turn, it was just at this time that Taiwan's newly-started LCD manufacturing companies were literally begging Japan's LCD manufacturers to grant them costly patent licenses and provide expensive technical assistance. The Japanese companies, thinking that Taiwan would never be able to master the high technology know-how of LCD manufacture, with little to lose and much-needed income to gain, decided to accept the patent royalties and provide the technical assistance.

And so CPT, Unipac, and Acer Display, beginning in 1999 were granted patent licenses and technology assistance by respectively, Mitsubishi, IBM-Japan, and Matsushita. IBM-Japan, not being a pure Japanese company was perhaps a little more willing to provide technical assistance with fewer reservations. The technical boycott was breached, and Taiwan's LCD manufacturers could embark on a journey that would take them to become the biggest liquid crystal display producers in the world.

Chimei Jumps into the Liquid Crystal Sea

The transition from traditional industry to high technology manufacture in emerging economies is exemplified by Taiwan's PVC giant Chimei Industries' (CMC) entrée into the burgeoning LCD business.* Based on the expertise of its parent as the world's largest acrylics manufacturer, the start-up Chimei Optoelectronics was conceived as a producer of the plastic color filters made from PVC and acrylics. Starting off by purchasing a second-hand dry thin-film filter production line from Japan, and re-constructing it at CMC's plant in Tainan, Chimei thus, albeit utilizing its traditional plastics expertise, began a new business.†

* PVC is short for polyvinyl chloride, a product from the petroleum plastics industry. The major "traditional industries" in Taiwan are textiles, steel, and plastics.

† Recall that as described in Chapter 25, the manufacture of the color filter used color pigments with an acrylic ester binder which hardens and stabilizes the color filter material.

It was not to be a propitious beginning, for although burgeoning, the LCD panel industry had not yet blossomed. At the time Taiwan's LCD industry had not begun volume production, Japan was cutting back on LCD production (hence the sale of the color filter line), and the Koreans had their own vertically integrated color filter suppliers. With the global recession of the late 1990s, even the relatively low cost high-quality Chimei color filters could find no buyers. CMC's Chairman, the plastics magnate Xu Wenlong,* instead of shrinking back, decided instead to build the whole liquid crystal display, and so become its own color filter customer.

And so Chimei Optoelectronics (CMO) jumped headfirst into the roiling waters of upstream (color filters, polarizers, and IC drivers), middle stream (TFT fabrication and panel manufacture), and downstream (module assembly and Chimei™ brand).

Chimei's initial technology assistance was from Japan's Fujitsu, but Chimei soon acquired IBM-Japan's pioneering Display division in a divestiture that reorganized the latter as Display Technologies International (DTI, later to be renamed IDTech), including its 3rd generation state-of-the-art production facility at Yasu. IBM's corporate re-engineering mentioned in Chapter 18, thus helped Chimei to quickly acquire display manufacturing know-how and a major IPS LCD production facility.

The key technologies licensed from Fujitsu were the multiple-domain vertical alignment (MVA) and the one-drop fill (ODF) liquid crystal injection process. As described in Chapters 24 and 25, MVA and ODF were designed and well suited for LCD televisions, thus giving Chimei a distinct competitive advantage in the LCD television-induced upward swing of the crystal cycle.

One-drop fill technology was developed by Fujitsu in 1986 as a more precise and efficient liquid crystal filling method, and the MVA mode for the liquid crystal was implemented for mass production some 10 years later. Fujitsu transferred this technology to Chimei because of the 2001 dot.com global recession and the continuing domestic recession in Japan.

At that time, with little cash to invest and no hope of reducing production costs, Fujitsu had to find a partner, and through its parent CMC's close business ties with Japanese companies, it was Chimei who providentially came calling. Fujitsu's plan was to sub-contract fabrication of TFT arrays and MVA

* The Romanized spelling system formerly used in Taiwan is different from the now universally used *pinyin;* Shu Wen-lung is the old spellings. The *pinyin* system will be used in this book unless the otherwise tradition spelling has been culturally or commercially established (for example, Chiang Kai-Shek and Kaohsiung).

cells to the lower-costs Chimei production lines, with most of the arrays and cells to be sold back to Fujitsu. In this way, Fujitsu did not have to invest in new production facilities and could gain capacity and reduce costs of fabrication. At least that was the thinking, but Fujitsu in so doing unwittingly unleashed a tiger into the liquid crystal jungle.

Chimei's ODF and MVA successes provoked a reordering of the LCD strategic alliances as well. In 2004, AUO, acknowledging the technical superiority of the ODF MVA cells, took a 20% share in Fujitsu's LCD Fujitsu Display Technology Corporation (FDTC), which at the time was suffering severe business losses. In less than a year, AUO's new big-screen MVA LCD panels were in mass production, and AUO quickly rose to challenge Chimei for LCD television panel supremacy in Taiwan.

Two Tigers, Three Cats, and a Monkey

Soon after Chimei's jump into the liquid crystal sea, the world's biggest manufacturer of notebook computers, the original equipment maker of almost all the well-known brand name notebook computers, Quanta Computers in 2000, with assistance from Sharp, established the new company Quanta Display with the goal of making Quanta self-sufficient in notebook computer liquid crystal screens. The capital investment came from Quanta, while Sharp provided technical assistance in the expectation that Quanta Display would be a second source for notebook computer screens at favorable (almost ITP) prices, and at the same time Sharp could spread the risk of a new plant to Quanta. Soon thereafter, Taiwan's communications cable conglomerate Walsin Lihwa, with a license and technical assistance from Toshiba, also joined the fray forming a smaller LCD panel maker called Hannstar Display.

Taiwan's LCD industry began to grow in earnest, but the global recession and cutthroat pricing competition also brought considerable losses. So with the aim of decreasing the number of competitors, and in hopes of a business-school type synergy, the computer maker Acer Group's Acer Display and the semiconductor maker United Microelectronic's Unipac merged in 2001 to form "Acer United Optronics" (AUO). The company immediately became Taiwan's largest LCD panel manufacturer. Soon thereafter in 2002, Chimei acquired IBM Japan's IDTech, and the two new companies were now the world's numbers three and four biggest LCD panel manufacturers, after only Korea's Samsung and LPL.

The fast-growing new LCD industry now began to draw the attention of Taiwan's lively news media, which coined "Five Tigers" for the five LCD companies. But in an interview, AUO's chairman K.Y. Lee dismissively pointed out that it was really two tigers and three cats; one of the tigers obviously being AUO, the other somewhat reluctantly acknowledged as Chimei. Lee's remark was based on the fact that the output capacities of the three little cats (CPT, Quanta Display, and Hannstar) were insufficient to influence the market.

As Taiwan's LCD prospects grew, the giant parts manufacturer Honhai (listed in Hong Kong as "Foxconn") formed Innolux as both a manufacturer and a distributor of LCDs, and banking on a wealthy parentage, the chairman Terry Kuo anointed Innolux as Taiwan's "Golden Monkey" in an allusion no doubt to the nimble warrior "Monkey King" of Chinese mythology.*

Taiwan's liquid crystal display jungle now had two tigers, three cats, and a golden monkey roaming about. But the law of the jungle soon claimed its first victim when Quanta Display was acquired and merged by AUO in 2006; Taiwan's LCD jungle now was less one cat, and one of the tigers had gotten considerably larger. Then in late 2009, in response to the global financial tsunami, Chimei and Innolux merged, creating another large tiger with a monkey on its back. The roster in 2010 is two tigers and two at-risk cats, where will it all end?

Taiwan's semiconductor and liquid crystal display industries had been anointed by the government as national treasures each worth a trillion Taiwan dollars; but by 2006 the capitalized value of the LCD industry had already surpassed its trillion dollar goal.† Taiwan's LCD panel players were almost always in a rather unhealthy race to build capacity, and often engaged in fierce competition to gain market share. They were no doubt motivated by an observation by Chimei's Xu [6]:

> I have never seen an industry where only Japan, Korea, and Taiwan are investing and the demand is enormous.

* Westerners may find complimentary references to monkeys (and dragons) puzzling, but in Chinese culture and literature, the monkey is revered as quick-witted and intelligent, easily outmaneuvering his foes, as chronicled in the Chinese epic novel, *Journey to the West* (the "West" meaning India, just as for Columbus). The same can be said of the mythical dragon, who represents the emperor and is a propitious symbol as opposed to the malevolent and fearful dragons of Western mythology.

† The "two trillion" originally had biotechnology as one of the trillion, but because of the long time for return on investment, biotech has not taken off in Taiwan. The LCD industry attainment of its trillion was chronicled in 2008 by the *Economic Daily News*, 2008.01.14. The Taiwan dollar exchange rate is usually about 30:1 US dollar.

Japan's Closed Shop

The rapid growth of Taiwan's LCD panel production, however welcome in Taiwan, was viewed in Japan with alarm. Inexpensive and high-quality liquid crystal displays from Taiwan began to appear in the Japanese market. Japan's LCD makers had hitherto enjoyed a near-monopolistic 90% market share in Japan, and they now felt threatened by an invasion of foreign makers. The Japanese Government placed the blame squarely on the "leakage" of Japan's LCD technology to Korea and Taiwan. Indeed the weekend warriors had surreptitiously helped Korea, and Mitsubishi, IBM-Japan, Matsushita, Fujitsu, Toshiba, and Sharp had all, for reasons of their own, provided technology transfer to Taiwan.

So in the summer of 2004, the Ministry of International Trade and Industry (MITI) called in the leaders of the big LCD manufacturing companies, and in the manner of an imperial edict, directed them to coalesce their power and resist the intrusion of particularly the liquid crystal display panels from Taiwan; and MITI suggested that patents were the best weapons to defend the technology border. In late 2004, in an apparent response to the new policy, Sharp wholly acquired FDTC, in the process also buying out AUO's 20% share. The acquisition was an attempt to keep technology inside Japan, but the cat had already been let out of the bag by Fujitsu, and Chimei and AUO could already produce MVA panels using ODF as well or even more efficiently than Fujitsu and Sharp. With the license agreements already in place, the border-sealing policy was doomed to fail, and in the ensuing years, not only Fujitsu, but also Sony and Sharp subsequently increased their procurement of LCD panels from Taiwan.

MITI's policy also met with resistance in the industry. In the beginning, the plan was to have Sharp organize and lead the LCD companies in the MITI plan, but Sharp balked, apparently fearful of having to share its leading-edge technology with its competitors. In counterpoint, the old-line conglomerates, for their part were not keen to be led by Sharp, whom they regarded as an upstart.* Stumbling at the gate, the would-be consortium never materialized. There were some attempts at cooperation however, for example, Seiko-Epson and Sanyo Electric formed the joint venture Sanyo-Epson Imaging, and in 2004, Hitachi, Toshiba, and Matsushita (Panasonic) hurriedly formed the joint venture IPS-Alpha to produce IPS panels, although with an initial planned annual production of only two and half million panels.

* Indeed, Sharp began life as a pencil manufacturer while the old-line companies were mostly heavy industry and/or electronics pioneers in Japan.

Unwilling to lead a consolidated LCD industry to protect Japanese technology, but deeply concerned with the leakage of its own technology, Sharp cynically instituted the industry's most stringent and onerous security procedures at its Kameyama 8th Generation plant in the hills of central Japan. The top-level security clearance and all-body scans and searches upon entering and leaving the facility were legend among those wishing to visit, but some of Sharp's measures have in fact been implemented in other LCD fabs, such as the prohibition and confiscation of camera-function mobile phones in the plant, and car trunk searches on exit from the plant even for employees. In addition to entry and exit security measures, Sharp also attempted to force manufacturing equipment vendors to agree to prohibitions in its equipment procurement contracts, including such restrictive clauses as "similar equipment may not be sold to other companies." Of course such clauses would have the effect of hindering the advance of other Japanese LCD manufacturers, as well as those of Korea and Taiwan.

As for the use of patents to protect their technology, because Japanese companies had severe cash flow problems, foregoing the royalty income from patent licenses was not really an option (although some companies tried to exempt certain key technologies, such as Sharp's transflective method patent, from the licenses). There were lawsuits from time to time, but the legal fees were steep and the legal remedy for infringement was usually a license on reasonable terms anyway. So it appeared that patents could not be used to control technology for proprietary use and could only be used to collect royalties.

On the plasma display front, the big Japanese companies, perhaps because of their precarious position in the competition with liquid crystal television, were more receptive to MITI's entreaties. NEC and Pioneer joined to form NEC Plasma Display, and there were instances of mergers, restructuring, exiting, and so on. Further, Japanese litigants called off the lawsuits against each other, and settlements were reached, while at the same time, litigation against foreign companies, particularly Korean, were more aggressively pursued, for example, the Fujitsu patent infringement suit against Samsung. But the effect of the lawsuits was to hinder plasma television expansion, making plasma TVs even more vulnerable to the competition from the fast-growing liquid crystal TVs, and ever more companies left the plasma display manufacturing business.

Although there were later instances of consolidation, for example, Sony and Sharp's proposed cooperation for 11th-generation panel manufacturing, they were largely in response to the financial tsunami (discussed below) rather than MITI's policy of closing the barn door after the cows had left.

Indeed, soon after promulgation of the policy, Japan's brand name makers increased panel procurement from Korea and Taiwan, and Sony joined with Samsung in a joint venture, named S-LCD, to produce the latest generation LCD panels in South Korea primarily for Sony. Before that however, the panels for Sony's price-busting Bravia® liquid crystal television were largely procured from Korea and Taiwan, and although Sony in 2008 announced the joint venture with Sharp for 11th generation panel manufacturing, at the same time, the procurement policy was changed to become more diversified, meaning a more multisupplier strategy was in the offing. Japan's closed-shop policy, intermittently tried on occasion, was always plagued by the reality of the globalization of the liquid crystal display industry and thus had little chance of success.

The transactional maze of the liquid crystal display panel manufacturing business as it appeared in 2008 is shown in Figure 26.1. The many overlapping and parallel investments (shown as single lines), panel supply relations (dotted black lines), and strategic alliances (thick gray lines) betray the complexity of the industry, and the relationships among manufacturers are more-

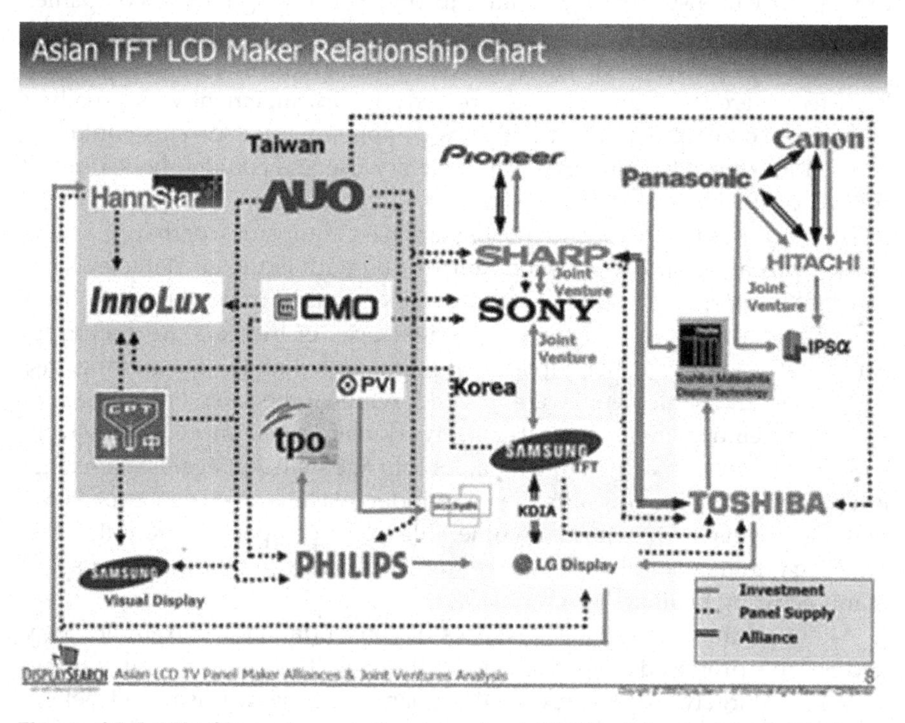

Figure 26.1 TFT-LCD maker relationship chart (2008). Adapted from Display search.

over dynamic and ever-changing. Although the Japanese companies did engage is some mutual investments, the Japanese and Korean procurement practices not only did not appear to respond to their governments calls for closing the border to foreign panels, but the supply dotted lines show that the procurement from other countries was varied and rampant. The many interconnecting lines clearly show that any government policy of self-consolidation and exclusion of other's panels would be extremely difficult to implement, persuasively demonstrating once again that *price/performance* ultimately will always triumph over *nationalism.* This is something that perhaps the world's leaders ought pay more attention.

The Worldwide Financial Tsunami

The liquid crystal display panel manufacturers of course operate in the context of the world economy, and its fate once again particularly impacted Korea. The subprime defaults and subsequent collapse of Lehman Brothers in the third quarter of 2008 brought down on the world a calamitous global financial tsunami that with all else, also almost completely leveled the LCD panel industry in Japan and Taiwan.

Although all companies suffered record operating losses, in Korea, once again overlending by national banks led to a 40% depreciation of the Korean won, and "wonce again," Korea's electronics companies found themselves in the "cost in won, price in dollars" cash flow Heaven. The now extremely cheap Korean LCD panels allowed Samsung and LG to succeed in mutual procurement and essentially drive out the competitors in Japan and Taiwan. Now the Korean conglomerates found that they now could indeed heed their government's patriotic admonitions to cooperate to "buy Korean"; in other words, they could have their cake and eat it too. This nationalistic sibling cooperation, however, would have an unexpected adverse impact in the days to come in neighboring China.

Facing a worldwide financial crisis of unprecedented scale and the concomitant collapse of Western consumer spending, together with an ascendant Korea's aggressive price-busting, the prospects for Japan's LCD industry looked grim. Hitachi forthwith announced that it was quitting the LCD manufacturing business and transferring all of its shares in the joint venture IPS-Alpha to its partner Matsushita (now Panasonic) for large screens, and to Canon for small and medium screens.*

---

* At the time of Hitachi's announcement, the plans of the other IPS-Alpha partner Toshiba were not clear.

While announcing severe manufacturing cutbacks, Sharp's new executive director Katayama Mikio went so far as to predict that because of the high costs of manufacturing, volume production in Japan would no longer be feasible and announced strategic changes, among them moving production to China, concentrating solely on high-end differentiated products, and selling technology rather than products from Japan.

These strategies however well reasoned, carried inherent risks: First, China's manufacturing abilities are well documented and the production follower may well quickly become the production leader; second, marketing high-end expensive LCD televisions in a down economy quickly allowed the hitherto little-known upstart Vizio brand, armed with management and panels from Taiwan and Korea, to become the LCD television sales leader in America, relegating the heavyweights Sharp and Sony to single-digit market shares; and third, selling technology will naturally blur any product differentiation advantage Japanese products may have.

However at the same time, Sharp and Sony announced their joint venture for the 11th-generation LCD plant to be built near Osaka *in Japan*. From these contradictions, it appears that Japan's response to the financial tsunami is still evolving, and it will be interesting indeed to see just what course Japan's LCD industry will take to meet the challenges brought by the financial crisis and the competition from Korea, Taiwan, and China. And of course, the 9,0 earthquake and real tsunami of 2011 in northeastern Japan may have serious long-range effects on Japan's recovery, particularly in diverting short resources to the re-building effort.

As for Taiwan, the financial tsunami and the depreciation of the *won* again virtually inundated Chimei and AUO, each of whom recorded record operating losses. And once again mergers were raised to gain market power and leverage over suppliers, and to "get big" to compete with Samsung and LG. For instance, a merger between AUO and Chimei would produce the world's largest LCD manufacturer, and this allows considerable control over the specification standards and model dissemination in the market, and further brings to bear pressure on the manufacturing equipment and component suppliers to accept terms and conditions for fear of losing a customer that cannot be lost. However, the solitary-hero entrepreneurial ethos in Taiwan is not conducive to mergers on that scale which would require some sublimation of personal ambition, and instances of successful big mergers are precious few.

However, the early LCD days did produce AUO and Chimei, each of whom acquired other LCD companies, portending mergers forced by the increasingly dire economic circumstances wrought by the financial crisis.

And indeed, there was a merger, but instead of the industry leaders AUO and Chimei merging to form a world's No. 1, it was Chimei and Innolux who announced their merger in late 2009 to form a No. 3.*

Another strategy for Taiwan to survive the financial crisis is to find ways to use the cultural and linguistic relationship with China to exploit a vast and growing Chinese consumer market that was not particularly affected by the financial tsunami; in particular using the dearth of a global television brand from Taiwan and China's ambitions to establish one, a downstream–upstream cooperation would appear to be quite attractive.

Is China a Rising Liquid Crystal Star?

As in almost all current affairs, the role that China will play is an issue of global significance, and the liquid crystal display industry is no exception. In an earlier section, it was mentioned that a consortium of companies from Taiwan had unsuccessfully tried to acquire Korea's Hyundai Display, which was experiencing financial difficulties. After the talks broke down, China's Beijing Eastern Electronics (BOE), the Beijing city government, and Taiwan's Jiandu Color Filter company combined to make a successful bid, and a new joint venture company called BOE-Hydis was formed.

China now had its first liquid crystal display panel manufacturer of scale; the new company consisted of three 3.5-generation production lines in Korea, and a new 5th generation plant in Beijing was completed in 2006. Following closely, China's Shanghai Electric also completed another 5th generation plant in Shanghai in collaboration with Japan's NEC called SVA-NEC, and the former President of the Chimei subsidiary IDTech, Hashimoto Takahisa, established Infovision Optoelectronics (IVO) in Kunshan. These three LCD panel manufacturers are the biggest in China, and although all three plants had plans for television panel manufacturing, they have been stuck at the 5th generation glass size, with the main product still being computer monitors.

Unlike the other major manufacturers in Korea and Taiwan, the Chinese fabs were not vertically integrated and had no corporate group supply chain or local components manufacturing cluster to speak of, making their costs so high that the accumulating losses were threatening their existence. Under government pressure, an effort was made to consolidate BOE-Hydis and

* An AUO-Chimei merger did not come to pass in spite of the efforts of AUO's Chairman K.Y. Lee, and the title of the world's biggest LCD manufacturer remains with Samsung.

SVA by merger to achieve economies of scale, but since each company was not itself competitive, there was little reason to believe that a merged company would do any better, and the merger/cooperation talks lapsed.

While the domestic LCD companies were having their problems, any visitor to Shanghai could not help but see signs of the cultivation of the market by the Japanese and Korean conglomerates. From Nanjing East Road's huge "SAMSUNG" sign, to the Bund's view of "HITACHI" across the Huangpu River in Pudong, then to LG's department store LCD advertising displays in elevators and the subways, the symbols of the incursion of East Asia's electronics manufacturing giants into the China market were all too evident.

As for the ethnically identical and culturally similar Taiwan, things were not so simple. Up until 2010, Taiwan's government policies have limited Taiwan investment in front-end (TFT array and cell) panel manufacture to four-inch screens and below, and to 40% of total investment capital, constituting an almost complete barrier to Taiwan's LCD manufacture investment in China. Since Japan's NEC and Korea's Hyundai Display have already entered into joint ventures, and Sharp, Samsung, and LG all have plans for front-end LCD manufacturing in China, Taiwan's Mainland policy unfortunately only harmed Taiwan's LCD manufacturers, that is, a vertically integrated large-scale LCD plant close to the gigantic and growing Mainland market should be able to attract supply chain companies to form the desired "cluster" that can achieve cost-efficient production, but this may accrue to the Japanese and Koreans instead.

The saying in Chinese is "when there are policies from above, there are countermeasures from below," meaning that the resourceful businessmen in Taiwan, to get their share of the Mainland market, will think of something, government policy notwithstanding. For example, virtually all of the Taiwan LCD manufacturer's labor-intensive LCD modules assembly is already performed in gigantic plants on the Mainland, and panel manufacturers have formed alliances or invested in Chinese system integrators and brand-name television manufacturers. With the loosening of rules beginning in 2008 by the more Mainland-friendly Kuomingtang in Taiwan, AUO and Chimei have already increased their investment in China with the hope that the China world factory can upgrade to high-technology products. And in the end, who will benefit most from China's LCD consumer market? Will it be Japan, Korea, Taiwan, or China itself?

The effects of the severe worldwide economic recession of 2008 on the LCD industry of course cannot be ignored, but the market appeal of big-screen flat panel televisions is undeniable, and liquid crystal screen computers, be they personal, notebook, netbook, or tablet, are proliferating. In addition, there are the smart phones, pads, e-books, personal and car navigation systems, portable music and video players, computer game players, and many other new products, as well as industrial use meters and displays, all using newer and newer liquid crystal displays. And the LCDs themselves are increasingly feature full, with touch screens, 3D, and mini-projectors all coming into common use.

But in the interim between recession and recovery, the LCD industry has encountered realigning tidal forces. For example, after the biggest losses in its history due to the financial tsunami, Sharp announced that because of the double curses of Japanese labor costs and exchange rates, it was no longer feasible to volume manufacture in Japan, and that Sharp would be moving production outside of Japan with the headquarters devolving into pure research and development and the sale of technology as the principal businesses. Sharp thereupon approached the Nanjing City Government with an offer to sell its formerly top-secret, lock-down Kameyama 6th generation facility, and move it lock, stock, and barrel to Nanjing. Sharp also agreed to a joint venture to build an 8.5-generation plant also in Nanjing; the sea-change in the offing was the relocation of mass LCD manufacturing to China.

Indeed, the China market and potential for high-tech products manufacture already has attracted considerable foreign investment, but Sharp's transfer of liquid crystal display manufacturing technology to China is a harbinger, induced no doubt by China's increasing requirement of technology transfer as a condition for entry into its domestic market.

To remain competitive, the "China card" no doubt must be played by all the LCD manufacturers, but in South Korea, it was badly parlayed. Notwithstanding its aggressive incursion into the brand name consumer electronics market in China, in response to the Korean Government's nationalistic admonitions, Samsung and LG blundered by unilaterally canceling panel supply to its Chinese systems-integrator customers, reserving the panels for Korean companies to capture market share with low prices in the financial tsunami-induced down economy. This understandably angered the Chinese companies, but more importantly, made them realize that the Korean manufacturers could not be depended upon, subject as they were at times to government *diktat*.

The fallout from Korea's commercial provincialism was that the Chinese government, in its huge economic stimulus programs, "Electric Appliances to the Villages" and the later "Electric Appliances to the [second tier] Cities," designated Taiwan's LCD panel manufacturers as the preferred source for color television LCD panels. Taiwan for its part, responded by considerably loosening the restrictions on investment and LCD front-end manufacturing in China. China's huge systems integrators and electronics brand-name companies could now not only be supplied by Taiwan's huge LCD panel manufacturers, the production technology and know-how will eventually allow them to efficiently produce panels themselves. The huge Mainland market, in accord with Government policy, then can be exploited by local companies, and hopefully from that springboard launch the Chinese brand name products into Europe and America.

Seeing the China market possibly slipping away, the Japanese and Korean LCD panel manufacturers re-doubled their efforts in China with plans for 8.5 generation fabs. The race for China's LCD market has officially begun, can China be a manufacturing player, brand name seller, as well as a consumer? If other products are taken as a guide, though admittedly far less technology driven, it is predictable that eventually ever lower-priced liquid crystal televisions will be produced in China and marketed all over the world.

Not only in China, as the global economy recovers, the allure of liquid crystal displays and particularly televisions, should bring the LCD panel-makers back to profitability. But not all of them; the winners will be those who can emerge from a cruel contest of survival of the fittest, and the financial tsunami of 2008 clearly demonstrated that whomever the victors turn out to be, China, as the only major economy continuing a high rate of growth during the financial tsunami-induced downturn, will be the key.

The Solar Cell

At the beginning of the twenty-first century, the price of a barrel of oil reached an astronomical US$146, threatening to disrupt economies all over world and compelling leaders of the industrialized countries once again to beg its citizens and businesses to conserve energy and to call on the technology companies to redouble their efforts to develop renewable energy sources. The most alluring new energy source once again was clear as daylight—solar power.

As mentioned in Chapter 21, Sharp developed thin film transistors originally for use in solar cells in response to the first oil shock of 1974, but as oil prices stabilized and demand for solar cells fell, the TFTs were changed over for use in the addressing systems of liquid crystal displays. The pattern has been that at every steep oil price rise, the oil cartel OPEC will rake in cash, there will follow dire warnings of energy shortages and calls for energy independence, and everyone will think *solar energy*. Whereupon cooler heads in OPEC, even in the intemperate revel of oil-begotten gains, will realize that solar energy has the potential to cook their golden goose. The cartel colludes then to increase supply to drive down oil prices, and although lowering their profits, they once again nip solar power in the development bud.

Solar energy has yet to become a viable substitute for fossil fuels primarily because its energy conversion-to-cost efficiency ratio is too low to compete with reasonably priced coal and oil. Universities, government labs and technology companies in America, Germany, and Japan in particular are striving to increase solar cell efficiency, and when the per kilowatt prices can decrease to compete with normal oil prices, the solar energy age may at last dawn. But after so many instances of calls for solar energy only to see oil prices drop and solar energy once again relegated to the back burner, one has to wonder, is it just one more instance of an empty promise?

The early twenty-first century oil shock however has struck in a different commercial environment from the past. In addition to advances in the energy conversion efficiency, recalling that the hydrogenated amorphous silicon thin film transistor was originally developed for solar cell use and only later converted for use in liquid crystal displays, the giant LCD plants in Japan, Korea and Taiwan could be converted to solar cell production and their huge capacity and cost-efficient production could dramatically lower the costs of solar cells. This capability has been aptly demonstrated by the transformation of the once prohibitively expensive liquid crystal display into a mass consumer commodity affordable all over the world, even in the poorest countries. Indeed, the liquid crystal display, DRAM, smart phones, and personal computer, among many others, are all products of what might be called the "Taiwan effect" of turning high technology electronics into affordable everyday consumer products.

The gigantic size of the latest-generation LCD panel has brought a concomitant increase in the number of thin film transistors produced, such that if LCD production is converted into highly efficient solar cell thin film transistor production, and the high-capacity plants in East Asia brought to bear, costs could decrease to the point of meaningful competition with fossil fuels energy production. Solar cells then would be as ubiquitous as the other technology products that have come into every household through the Taiwan effect.

In the big picture, solar energy will be able to not only help solve the epochal energy shortage problem, it can also help to protect the environment and reduce carbon emissions to mitigate global warming. China, Japan, and Germany are leading solar cell module production, and with advances in energy conversion efficiency, these producers can be supplied by the billion-unit thin film transistor annual capacity of the LCD panel manufacturers in Korea and Taiwan.

In a more parochial view, solar cells production could alleviate the effects of the crystal cycle on the liquid crystal display industry, that is, in a downturn, more production can be allocated to solar cell manufacture and less to LCD panels. Maybe the business schools are right after all about the benefits of diversification.

References

[1] Castellano, J.A. 2005. *Liquid Gold*. World Scientific, Singapore.

[2] Johnstone, B. 1999. *We Were Burning*. Basic Books, New York, p. 126.

[3] Castellano, J.A. 2005. *Liquid Gold*. World Scientific, Singapore, p. 154.

[4] Chen Yongcheng. May 15, 2007. South Korea's TFT strategic initiative. *Daily Industrial News (Taiwan)*; Yang Mingwen. Jan. 14, 2008. Mutual panel sourcing in south Korea delayed. *Economic Daily News (Taiwan)*.

[5] Chen, Y.C. 2004. *Taiwan's Exclamation* (Translated title). (CPT Chairman Lin, C.H., ed.). The Times Publishing, Taipei, p. 15.

[6] Author's personal communication. Internal management meeting at Chimei Optoelectronics.

27

New Technologies and Products

In addition to the mainstream twisted nematic, vertical alignment, and in-plane switching liquid crystal display modes, there are other molecular-level orientation structures, such as the liquid crystal polymer composite, the cholesteric nematic bistable, the ferroelectric chiral smectic-C, and the blue phase, as well as a cousin to the liquid crystals, the electrophoretic liquid suspension display. Another major display mode is the organic light-emitting diode (OLED), which is an emission display rather than a transmissive or reflective display, and thus fundamentally different from liquid crystal displays. These display modes are suitable for small portable devices, such as mobile smart phones, GPS navigation systems, tablets, e-readers, mid-size flexible displays, and large-size optical shutters for advertising displays and privacy windows. Also, other than the direct-view transmissive display, particularly for handheld and near-eye displays, there are the reflective and transflective displays, and for large-screen applications, the projection display. The functionality and attractiveness of the different display modes are further enhanced by brightness enhancers, touchscreen user interfaces, and the thrilling 3D display.

Liquid Crystal Displays, First Edition. Robert H. Chen.

Following are very brief descriptions of the technologies and products listed above, meant only for an introduction to the subjects; the reader is encouraged to delve further by accessing the references noted.

Light Scattering

Many of the display types to be described below use the scattering of light for illumination. As described for polarization, the electric vector of light incident on a medium drives vibrations of the electrons in the atoms of the medium, generating dipole oscillations at the frequency of the light, and re-radiating light because of the acceleration of the electrons.

For rigidly ordered solid crystals and high orientational order liquid crystals, the re-radiated light will be for the most part in the same direction as the incident beam and because of an anisotropic structural order, the phase of the reradiated light will be different, and this produces birefringence (confer Chapters 3).

If on the other hand there is little structural order, and the atoms are randomly distributed in the medium, the re-radiating by the electrons in the molecules will be in many directions other than that of the incident light beam; this is called *scattering* of light, and the total intensity in any given direction of the re-radiated light is the sum of the intensities that are re-radiated by any one atom [1].

That scattering occurs depends on the size and disposition of the things doing the re-radiating. Since atomic dimensions are of the order of 10^{-10} m (1 Å) and visible light is 5000 Å Angstroms (or 500 nm), the atoms are very close together compared with the wavelength of visible light, and so there are many electrons in those atoms that will move together with the electric field vector of the light. Thus, in response to the light electric vector, the electrons will not only oscillate, they will oscillate *in phase* and at substantially the same amplitude, and this in-phase and equal amplitude oscillation produces re-radiation that is coherent (in-phase), generating constructive interference of the re-radiated waves that squares the intensity of the light (much like the onset of lasing action).* Further, the free oscillation of the

* Astronomical water masers (microwave amplification by stimulated emission of radiation) and CO_2 lasers have been detected in space; Townes, C.H. *How the Laser Happened* (Oxford, 1999). Townes won the 1964 Physics Nobel Prize for discovery and invention of masers and lasers, and endured a patent dispute with one of his graduate student regarding that invention.

electrons in response to the light electric vector causes the re-radiated light to be plane (linearly) polarized.*

From the constructive interference of coherent light by the individual atoms, the intensity of the scattered light is proportional to the square of the number of atoms in the molecule or particle, so the larger the molecule or particle, the greater the intensity of scattered light squared. Continuing in this way seems to imply that giant particles will scatter prodigious amounts of light, but there is a limit to the scattering phenomenon. When the particle size is larger than the wavelength of incident light, the electrons in the particle will no longer vibrate in phase because they are too far apart compared with the wavelength of the incident visible light. Thus the scattered light will have maximum intensity when the size of the particles is about equal to the wavelength of the incident light, and above that size, the scattered light will be incoherent and subsequently of lower intensity. Incidentally, this squared intensity of scattered light explains why clouds are so bright, the water droplets in the cloud scatter light very efficiently, but when they grow too large (and heavy) just before the rain falls, their size exceeds the wavelength of visible light, and the clouds grow dark, and the rain falls.

The scattering phenomenon also explains how the first liquid crystal displays using the Williams domain and dynamic scattering mode (DSM) described in Chapter 13 worked. The turbulence produced by hydrodynamic instability in the liquid crystal disturbed the orientational order, producing off-axis scattering from random atomic motions resulting in a bright image; by returning the liquid crystal to a semblance of order by the application of an external electric field, there was little or no off-axis scattering and the display was dark (with bright streaks where the PAA liquid crystal was focusing light polarized parallel to the director).

---

* The classical theory of light scattering by small particles is called "Rayleigh scattering," in which theory the intensity at angle θ of scattered light is proportional to $(1 + \cos^2\theta)$ and inversely proportional to the fourth power of the wavelength of the incident light. Energy is proportional to the square of the electric field, and the scattering cross section for air molecules is defined as energy scattered per second/incident energy per square meter per second; for light in air, the natural oscillation frequency of air molecules is higher than visible light, so the scattering cross section is almost directly proportional to the fourth power of the frequency (and inversely proportional to the fourth power of wavelength); which explains why the sky is blue since light of higher frequency is preferentially scattered (blue frequency is about twice that of red). This also explains why sunsets are red since the lower the sun is on the horizon, the more it travels through the larger size particles close to the earth (atmospheric pollutants), and those particle sizes are closer to the red wavelengths, which will be preferentially scattered with the blue color washed out.

Liquid Crystal Polymer Composites

Adding high molecular weight polymers to a low molecular weight liquid crystals can promote light scattering by the polymers and/or produce multi-domain random orientation liquid crystal structures that scatter light. There are two main kinds of so-called *liquid crystal polymer composites* (LCPC) for displays; the *polymer-dispersed liquid crystal* (PDLC) where micron-sized droplets of liquid crystal are dispersed in a polymer binder and the concentrations of each are about equal, and the *polymer-stabilized liquid crystal* (PSLC) where the concentration of the polymer is less than 10% of the liquid crystal, which is continuously distributed rather than dispersed as droplets in the composite mixture.

The liquid crystals and the polymers in the mixture are phase separated, meaning that instead of being a homogeneous composite at the atomic or molecular level, the constituents are distinct in what is called a heterogeneous mixture (like beef curry). Whichever state has the lower free energy is the preferred state, and the free energy not surprisingly depends on the temperature of the mixture. Therefore, LCPCs can be made by *thermally-induced phase separation* (TIPS), which is sort of a misnomer, since the phase separation actually is induced by taking away heat, and the rate of cooling determines droplet size.* Phase separation can also be induced by polymerization, which is promoted by using photoinitiators and ultraviolet light to photopolymerize the mixture in a method called *polymerization-induced phase separation* (PIPS); the rate of polymerization determines the size of the liquid crystal droplets, with smaller droplets formed from faster polymerization, and the polymerization rate depends on the photo-initiator concentration and UV intensity. At room temperatures, polymers are typically solid while display-use liquid crystals are liquid, so a solvent is required to promote the initial phase separation in a process called *solvent-induced phase separation* (SIPS); but because it is difficult to control evaporation of the solvent in the SIPS process, it is not commonly used except to prepare mixtures for TIPS phase separation.

The addition of polymers to a liquid crystal to form a phase-separated PDLC mixture scatters light because of the random spatial distribution of different indices of refraction of the polymer and the liquid crystal droplets, and the large sizes of the polymer molecules and droplets increases the intensity of the scattering, in accord with the theory of scattering outlined

* TIPS should perhaps be renamed "CIPS" for "cooling-induced phase separation," but since this can sound like "SIPS," a Germanic "kooling" might be substituted to become "KIPS."

above. In PSLCs, the polymer network creates differently oriented domains of the liquid crystal, and light is scattered from those domains, with the size of the domains determining the intensity of scattering.

To analyze scattering functionalities in an LCPC, the director configurations can be modeled, and in general LCPC director configuration is a function of the droplet shape and size, surface anchoring conditions, elasticity of the mixture, and the effects of the applied electric field. For rigorous calculations, a dielectric tensor must be used, and because of the different types of phase separation, shapes and sizes of constituents, anchoring effects, and the different reactions to an applied electric field depending the charge separation character of the droplets, and the fact of mutual interdependence and interaction, the analysis can become quite complicated [2].

Conceptually, in a switching PDLC in the field-off state, the directors of the liquid crystal droplets and the directors of the liquid crystals inside the droplets are randomly oriented, and light is scattered with intensity depending on the size of the polymers and the droplets. In the field-on state, the droplet directors line up with the field according to the polarity of the droplet, and since the index of refraction is uniform within the polymer and the liquid crystal, there is no scattering and the light passes through in the transmissive mode, thereby creating a light shutter controlled by the application of an electric field.

In another polymer-dispersed switching device, the PDLC is doped with a dichroic dye which resides inside the liquid crystal droplets; the absorption and transmission can be controlled by an applied field depending on the orientation of the dye molecules (which is just the definition of "dichroism"), and the device can operate as a light shutter.

In a PSLC, the liquid crystals near the polymers tend to align with the polymer network in a stabilized configuration. In an example of a PSLC light shutter, a polymer-stabilized homogeneously aligned nematic liquid crystal in the field-off state allows light to pass through unscattered because of the homogeneous orientation of both polymer and liquid crystal. An applied field, however, exerts a force on the liquid crystal molecules to align with the vertical field, but the polymer network tries to hold the liquid crystals in the horizontal homogeneous direction; the result of the contest is the formation of a multi-domain structure wherein liquid crystals within a domain are oriented uniformly, but the domains are oriented randomly. Therefore, incident light encounters the different indices of refraction of the domains and is scattered, but because of the polymer network orientation, scattering is dependent on the direction of linear polarization, which makes intensity control more difficult. To avoid this dependence, a polymer-stabilized homeotropic negative anisotropy liquid crystal in the field-on state will create multi-domains that tilt in different directions about the field, so regardless of the direction of polarization of the linearly polarized incident light, it will encounter different indices of refraction and be scattered. These two PSLC modes thus can act as light shutters, and are termed "reverse mode" because they transmit in field-off and scatter in field-on in contradistinction to the two PDLC shutter examples given immediately above.

For display use, the optical characteristics of the LCPCs are not as good as other type displays, but the polymers have other salutary effects, as shall be seen. Also, the addition of polymers can make the liquid crystal display flexible for more varied display applications, and the composites can be manufactured in rolls.

Cholesteric Bistable Reflective Displays

Cholesteric (chiral) nematic liquid crystal molecules, as described in Chapter 5, form a naturally stable spiral about a helical axis. If the helical axis is perpendicular to the glass substrate, the cholesteric liquid crystal is in a so-called *planar state* which naturally reflects light; when the helical axis is not necessarily perpendicular but almost randomly oriented, it is in the so-called *focal conic state*, which is orientationally multi-domain and therefore scatters light with no reflection. Both the planar and focal conic states are stable in the absence of an applied electric field. If an intermediate electric field is applied, the helical axis turns to be parallel with the glass substrate, and the cholesteric liquid crystal is in the *fingerprint state* (refer to Chapter 5 for a picture), and when a sufficiently high electric field is applied, the molecules are driven into a vertical orientation called the *homeotropic state*, and the chiral liquid crystal is transparent.

A bistable cholesteric display utilizes a color absorption layer on the bottom glass substrate. The planar state exhibits Bragg reflection and reflects circularly polarized light with the right- or left-handedness dependent on the spiral structure (just as Reinitzer reported in his seminal paper). In the focal conic state, there is no reflection, so only the absorption layer reflects light; if that layer is black, the display will reflect the color depending on the Bragg reflection of the planar area, and otherwise be black.

Thus switching between the planar and focal conic states can function as a black-and-white bistable display. The switching function is a contest between the elastic energy minimization versus the electrical energy supplied by the applied field (again calculated by the calculus of variations and the Euler–Lagrange equation). The planar state has zero free elastic energy because it is in the lowest energy natural state. The focal conic state has positive free elastic energy because of the molecular bending in the random orientation. In an applied electric field, the planar state has zero free electric energy because the directors are perpendicular to the field everywhere, while the focal conic state has negative free electric energy because the directors can be parallel to the field. In other words, the elastic forces are against the planar-to-focal conic transition, while the electric field promotes the transition. To switch from the planar to the focal conic stable state, a sufficiently high voltage is applied to push the molecules to the focal conic state where they remain, owing to surface anchoring or polymer stabilization, even when the voltage is turned off. To switch back, a higher voltage is applied to push the molecules to the homeotropic state where when the field is turned off, they relax back to the planar state.

With this state of affairs, original images can be maintained with no power, making the cholesteric nematic bistable a low-energy consumption switching display, but with a rather slow switching speed and thus rather poor response time. Varying electric field voltages can produce different reflectances to produce a grayscale that in conjunction with a color filter can produce color, or cholesteric liquid crystal layers of different pitch (that will Bragg reflect different wavelengths of light) are used to produce color (as done by a Fujitsu prototype). As might be expected, oblique angle viewing of the planar state will cause color changes due to the wavelength-dependent Bragg reflection. This can be compensated for by dispersing a polymer to create a multi-domain structure to cancel out the color variations (similar in principle to the multi-domain MVA), and the view-angle quality can be improved [3].

With these characteristics, it is clear that the bistable chiral nematic display inherently is best suited for monochromatic, static advertising displays, switching opaque/transmissive windows, and black-and-white electronic books, but new schemes and technical improvements may make it more competitive.

Ferroelectric Chiral Smectic-C Bistable Displays

If the natural-state ferroelectric chiral smectic-C liquid crystal introduced in Chapter 5 is surface-stabilized using parallel anchoring to the substrate to unwind the helical twist, there will be two stable states. An oppositely directed applied electric field then can switch between the two stable liquid crystal states in an operation similar to the twisted nematic display, but with neither stable state requiring an electric field to maintain. This makes the *surface stabilized ferroelectric liquid crystal* (SSFLC) display a bistable display that can operate much like the cholesteric bistable display just described above. FLCs have a fast response in a relatively thinner cell so the field sequential color wheel (FSC described in Chapter 22) can be used to provide color and pulse width modulation (PWM) can provide graycale. The manufacturing however is difficult to control over large areas, so the FLC has been used primarily in near-eye displays with a switching LED backlight [4].

However, because light is being generated from a backlight, and the reflection of ambient light off of a glass substrate, neither the bistable cholesteric display nor the ferroelectric smectic-C can provide the look and feel of a completely reflective piece of paper that is the traditional book experience; for that, it appears that electronic paper may be the best choice.

Electrophoretic Paper

The electronic emulation of the look of words and images on regular paper has long been an aspiration of display research. The goal of so-called *electronic paper* is to as closely as possible reproduce the white page and words in a book to produce an *e-book* using *e-ink* for the words in an ambient light-reflecting display mode.

Most e-books currently employ *electrophoresis*, wherein titanium oxide particles are suspended in a hydrocarbon polymer and black dye matrix, and after being charge polarized, the particles will be influenced by an external field (about ±15 V) to "swim" either up or down (or in other display modes right and left); those that go up to the surface of the display will be white and scatter light with a 30–50% reflectance, thereby producing a reflected white, and those that swim downwards will be dyed by the black dye and absorb light to exhibit black; those in the middle will exhibit different shades of gray. This then creates the black words delineated on a reflected white background that is the traditional book experience. Also, because of the intermediate grayscale, in principle, the addition of a color filter should make colored images also possible, just as in the liquid crystal displays.

Active matrix thin-film transistors can be used to address and control the voltage on the pixels, and because the various positions of the titanium oxide particles do not require an electric current to maintain, e-paper's power consumption is very low, and when the power is switched off, the last image state is retained. The electrophoretic display can be fabricated on glass or on flexible plastic substrates. The response times (less than 200 ms), while slow, and viewing angle, while limited, are more than adequate for use in electronic books [5]. Although Sony was the first to mass produce an e-book, the major commercial breakthrough came when the Amazon Kindle® provided the convenience of anytime, anywhere mobile phone network-downloading capability and voluminous content.

Another type of electronic paper uses the SiPix™ *microcup* technology. Inside each microcup domain are assemblies of white particles and colored liquid layers. When an electric field is applied, the white particles move upwards vertically to display their whiteness, or in an oppositely directed field, they move downwards to allow the colored layers to display their color. Microcup technology is thus suitable for direct color image displays that do not require color filters or wheels.

The Bridgestone-developed *electronic liquid powder* system utilizes black and white electrically charged particles in air that can be controlled to move vertically by an external electric field to display white, black, or grey shades, and the mounting of a color filter in principle can provide color.

The Organic Light-Emitting Diode Display

Particularly because of their very fast switching times and flexible plastic substrates, the *organic light-emitting diode* (OLED) display has become quite common for small and medium applications, and larger screen sizes have been in intense development.

The basic light emission phenomenon has been outlined in Chapters 19 and 22 in regard to LEDs. The OLED display exploits the electron-hole recombination in the conduction and valence bands in delocalized molecular orbitals of small-molecule (SMOLED) and large polymer molecule (P-OLED) amorphous organic solids. Different emissive materials, or fluorescent dyes doped in the recombination region that absorb specific wavelengths, can be used to produce different colors, and passive (PMOLED) and active (AMOLED) matrix-addressing schemes are used to drive the display [6].

The all solid-state, near-Lambertian emission (as opposed to the inherent limited viewing angle of LCDs), fast-switching, flexibility, and low-power consumption are attractive features of the OLED, but processing, particularly for the P-OLED AMOLED, is quite involved, and large-screen OLEDs, while feasible and attractive, will have difficulty displacing the economically entrenched large-screen LCD televisions, primarily because of costs of manufacture.

The Blue Phase Display

The brief flash of blue light seen by Reinitzer in his carrot sample was the stable double-helix counterpart of the chiral nematic, as described in Chapters 5 and 24. Because different colors could be produced by selective Bragg reflection off of the double helix, a color display not requiring a color filter based on the chiral nematic blue phase is theoretically possible.

Because the blue phase is in the very narrow temperature range (2–3°C) between the chiral nematic and an isotropic liquid, its temperature range must be widened before it can be used in consumer displays. Research has been conducted using a chiral dopant and polymer-stabilizing the blue phase to create a pseudo-isotropic state that has a wider temperature range [7]. Because the blue phase can be triggered by a quadrapole (four-charge separation, as opposed to dipole separation) electro-optical Kerr effect (electric field effect on an isotropic liquid), wherein a spherical isotropic liquid becomes an anisotropic birefringent ellipsoid, the switching can be done very rapidly, with the result that high-speed switching of 240 Hz is possible, ideal for fast motion movies and video games.

Reflective Displays

Liquid crystal transmission displays require a backlight for constant illumination and thereby always consume power; they also do not display images well outdoors because the reflected sunlight may be more intense than the backlight. Purely reflective LCDs are used both in direct-view and projection systems. In a drawing of a typical reflective LCD shown in Figure 27.1, a linear polarizer in conjunction with a quarter-wave plate, in reflection form effective crossed polarizers; a bumpy reflector reflects and diffuses incident ambient light back through the liquid crystal (thus serving as the backlight), and the TFT array is disposed beneath the reflector.

The vertical alignment and homogeneous display modes use phase retardation to modulate intensity just as in the transmissive display, but since the light is reflected, the cell gap can be just half, allowing the reflective LCD liquid crystal layer to be thinner. The optics of the reflective homeotropic cell and reflective homogeneous cell are identical to the respective counterpart transmissive display mode cells (described in Chapter 24) with allowance for different light paths due to reflection. The so-called *mixed mode twisted nematic* (MTN) utilizes both rotation of the plane of polarization as in the twisted nematic and the liquid crystal's inherent birefringence.

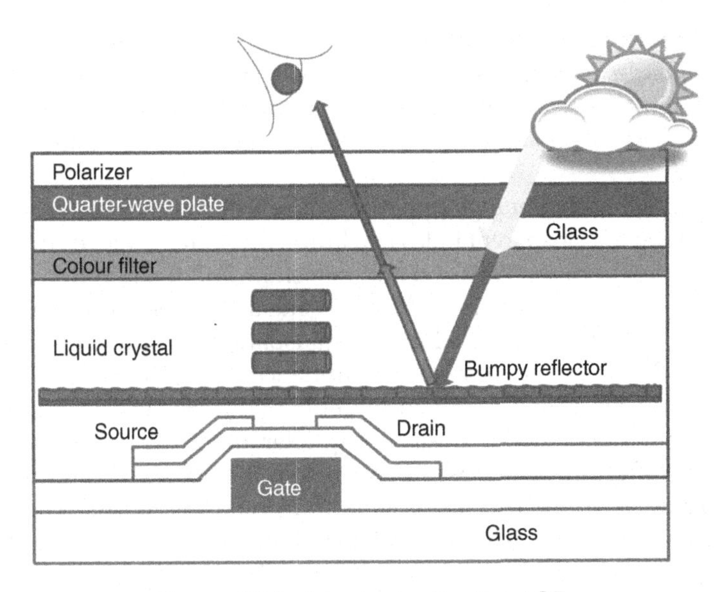

Figure 27.1 A typical reflective LCD.

A Jones matrix calculation, similar to the calculation for transmissivity of a TN cell given in Chapter 17, gives an expression for the normal reflectance $R_{\perp}$ of the MTN reflective display,

$$R_{\perp} = \left(\Gamma \frac{\sin X}{X}\right)^2 \left(\sin 2\beta \cos X - \frac{\phi}{X} \cos 2\beta \sin X\right)^2.$$

where ϕ is the twist angle, β the angle between the axis of polarization and the directors at the surface of the liquid crystal, $\Gamma = 2\pi d\Delta n/\lambda$ is just the phase retardation parameter, and $X = \sqrt{\phi^2 + (\Gamma/2)^2}$, all as given in Chapter 16. This reflectance calculation can be used together with the calculations for transmissive displays to analyze the display characteristics, just as in the TN mode. Twist angles varying from 45° to 90° have been used in the MTN direct-view and projection displays, and 75° and 90° twist angle MTN cells have been used in transflective displays.

Transflective Displays

The *transflective* display, logically named, both transmits and reflects by utilizing ambient light where possible and activating a backlight when the reflected ambient light is insufficient for viewing. This dual mode thus necessitates a *transflector* that both reflects incident ambient light and transmits backlight light. There are four major types of transflector, called respectively, *openings-on-metal, half-mirror metal, multilayer dielectric film,* and *orthogonal polarization.*

The aptly-named openings-on-metal transflector is self-explanatory; the simple etching of transparent holes in a bumpy reflector achieves the desired functionality of both transmission and reflection, and the low-cost manufacture allows common use. The half-mirror metal transflector is constructed by depositing a thin metallic film on a transparent substrate and adjusting the film thickness for the desired transmittance and mirror-like reflectance. However, its utilization `is limited by an extreme sensitivity to film thickness, which requires tight manufacturing tolerances. The multilayer dielectric film transflector is composed of successive layers of different refractive index half-mirror dielectrics to provide transmittance and reflectance, and its manufacture unsurprisingly also has tight tolerances. Orthogonal polarization transflectors such as the *wire grid polarizer* (WGP) transmit horizontally plane polarized light and reflect vertically plane polarized light using a grid mesh of wires at right angles to each other. A variation is the cholesteric orthogonal polarization transflector that transmits right-hand circularly polarized light and reflects left-hand circularly polarized light. As it is simple in concept and manufacture, the WGP eventually will be widely used.

The liquid crystal in transflective displays also can be classified as to function, that is: absorptive, scattering, reflection, and phase retardation. To effect absorption, a dichroic dye is added to the liquid crystal host, and as described in Chapter 4 for polarizers, the dye absorbs based on the light electric vector being parallel (absorbed) or perpendicular (transmits) to the molecular axis of the dye.

Polymer dispersed liquid crystals, polymer stabilized cholesteric texture, and other liquid crystal gels in the scattering type transflective displays all scatter light as described in concept above. For example, in a homogeneous nematic liquid crystal and monomer gel, the display is naturally transparent in the field-off state. In the field-off reflection mode, unpolarized ambient light goes through unpolarized, producing a good white state; in the transmission mode, the unpolarized backlight is linearly polarized by a polarizer, and after passing through a first quarter wave plate into the transflector and a second crossed quarter wave plate (in combination with a half-wave plate), the light remains linearly polarized, resulting in a bright white state.

In the field-on state, multi-domains are formed along the polymer chains by the applied electric field so that linearly polarized light in the alignment layer rubbing direction (the extraordinary wave) is scattered, as long as the sizes of the scattering domains are comparable with the wavelengths of the light; the ordinary wave passes through unscattered. In the field-on reflection mode, the unpolarized ambient light is linearly polarized after passing through the activated liquid crystal, then, passing through the quarter-wave plate twice after reflection by the transflector, the light is polarized as an extraordinary wave and thus will be scattered by the multi-domains. This slightly scattering of the light produces a translucent appearance. In the transmission mode, unpolarized backlight light passes through the polarizer, the quarter-wave plate, the transflector, and another quarter-wave plate, to be a linearly polarized extraordinary wave that will be scattered by the activated liquid crystal gel, again resulting in a translucent appearance. This transflective liquid crystal gel display requires only one polarizer, so transmission is good, but since the transmission states are translucent, the contrast ratio is poor.

As an example of both absorption and scattering, printing on white paper depends on the absorption of the ink and the scattering off of the blank parts of the paper, so a combination of absorption and scattering in a dichroic dye-doped, photo-polymerized liquid crystal gel can mimic paper with the dye absorbing light and the polymer network strongly scattering it [8].

In an example of a reflection type transflective display, a cholesteric liquid crystal with a focal conic texture in the field-on state exhibits strong Bragg reflection and utilizes optical elements to produce right- and left-handed circular polarization states for desired transmission, absorption, and reflection.

Phase retardation type transflective displays can utilize VA, homogeneous, and TN display modes, and since these are all well-studied modes, the design tasks of phase retardation transflective displays are primarily for improving brightness and contrast and avoiding parallax. Since the manufacturing processes for the mainstream modes are well established, the phase retardation transflective displays are the most commonly used in mobile phones and digital cameras.

As an example of a phase retardation transflective display, a 90° twisted nematic liquid crystal satisfying the Gooch–Tarry minimum condition operates the same way as a transmissive display except for an additional transflector placed between the bottom polarizer and the backlight. In transmission mode, the optical activity waveguiding results in a bright white state as described in Chapter 16 for the transmissive TN. In reflection mode, as the linearly polarized light from the top polarizer goes through the liquid crystal, its plane of polarization is rotated 90°, it passes through the bottom polarizer, and is reflected by the reflector back through the bottom polarizer again, and the plane of the linearly polarized light is rotated 90° once again, and the light passes through the top polarizer for the bright state. In the field-on state, the directors are vertically perpendicular to the glass substrate, and the polarization state of light passing through the liquid crystal does not change. In transmission mode, the backlight light is blocked by the crossed polarizers to produce the black state. In reflection mode, ambient light is linearly polarized by the top polarizer, passes through the liquid crystal with no change in polarization, and is blocked by the bottom polarizer, resulting in the black state. From the progression of the light in the field-off reflection mode just described, since the field-off state reflected light must travel through the polarizers four times, there is significant loss of luminosity. Different placement of the bottom polarizer, varying the cell gap and anisotropy ($d\Delta n$) in MTN mode, and/or placing the transflector right next to the liquid crystal (to reduce parallax) can alleviate the reflection luminance problem and provide a more uniform dark state [9].

Another approach utilizing the LCD homogeneous cell mass production techniques is the *dual cell gap* transflective display where each pixel is divided into a transmission region with one cell gap (d_T), and a reflection region having a different cell gap (d_R), which is half of d_T since the reflected light goes through the liquid crystal twice. At the boundaries of the different pixels for transmission and reflection, there will be some distortion of liquid crystal alignment, and this must be covered by the black matrix to improve contrast ratio. The response times through the different regions will be different, and the mid-layer tilt common to electrically controlled birefringence devices will leak light, thus requiring compensation films to improve wide-angle viewing. Other problems are unbalanced color owing to the reflected beam passing through the color filter twice, which can be corrected for by varying the color filter thickness for the reflected beam, punching pinholes in the color filter so that some of the reflected light is not affected the filter, or filling the reflection regions of the color filter with scattering materials. Since mobile devices do not require great contrast, fast response times, or wide viewing angles, all of these problems can be overcome satisfactorily using various different techniques [10].

Projection Displays

Liquid crystal projection displays fulfill the dream of ultra large-screen television and slide display for the home and business, and are already in common use. The projection mechanism has also been shrunk to allow incorporation into handheld devices, such as pocket displays and mobile phones, for large display multi-viewer uses.

A projection display can use one, two, or three very small (inch size and smaller) LCD panels to modulate light from a light source, and through polarizers, mirrors, beam splitters, and lenses project an image near-eye or onto a screen. The LCD panel can be VA, TN, PDLC, FLCD, or LCOS.*

Particularly for large-screen projection, the source and screen are two important new factors to consider in LCD projector design. A parameter useful for evaluating performance is the *luminous flux*, which is the rate of flow of radiant energy measured in *lumens*. From the point of view of the source, one lumen is the luminous flux emitted within a unit solid angle (one steradian) from a point source having a uniform intensity of one candela;† from the point of view of the screen, one lumen is the luminous flux received on a unit surface, all points of which are at a unit distance from such a source. Describing a projector using lumens allows comparisons without regard to image size or screen details. Raising the lumens in the source, however, are subject to collimation limits from *étendue conservation*; étendue (E) is given by‡

$$E = n^2 \iint \cos\theta dA d\Omega,$$

where étendue depends on the refractive index of the medium (n) and the integral of the cosine of the propagation angle θ over the area extending through the solid angle Ω (the projection). As can be gleaned from the equation above, étendue increases in an entropy-like way as light propagates through the medium, that is, if the output luminous flux of a lamp is L, the lamp luminance is L/E, where E is calculated for the particular lamp; so a high luminance requires a lamp with low étendue.

Projection light sources for cinema projectors use xenon arc lamps that operate at 6 kW with small (~1 mm) arcs to provide high luminous flux and low source étendue. For commercial and home theater projectors, high pressure metal halide and ultra-high pressure mercury discharge lamps are

* LCOS is an abbreviation for liquid crystal on silicon (to be discussed below); FLCD is an abbreviation for ferroelectric liquid crystal displays.

† A candela is defined as 1/60 of the luminous intensity per square centimeter of a black-body radiator operating at the temperature of freezing platinum, obviously an historical experimentally derived definition of a unit of luminous flux.

‡ The French *étendue* means "expanse" or "extent".

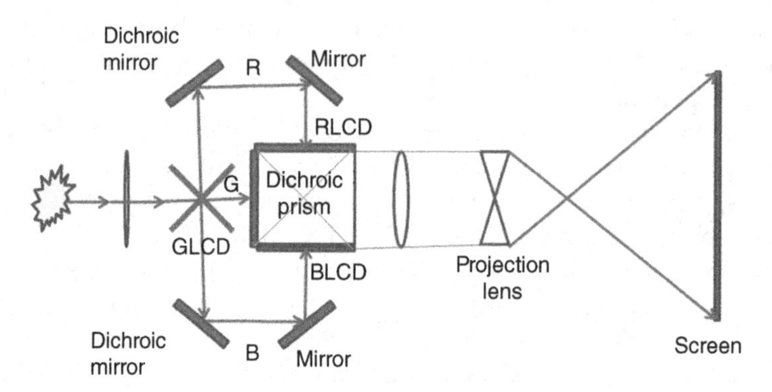

Figure 27.2 A three-LCD panel projection display.

used.* Other light sources are the LED and laser. The small size and high light-producing efficiency of LED sources make LEDs very suitable for mini-projectors, but the laser diode can produce compact, even higher efficiency luminosity, high color saturation and gamut, low étendue, and long life. The laser diode will eventually provide the ideal light source for liquid crystal displays of all types.

An example of the above-mentioned "screen details" is where a screen is designed for more luminance in a given direction ("screen gain") by sacrificing luminance in other directions in a departure from pure Lambertian refection (reflected luminous intensity follows cosine law angle scattering).

A schematic of a three-LCD panel projection display is shown in Figure 27.2. The three LCD panels for each primary color provide greater luminance than a single LCD panel system. Light from the source is focused through a condenser lens; dichroic mirrors divide the light into the three primary colors red, green, and blue; the liquid crystal panels modulate the primary color components, and the dichroic prism recombines them to send an image through the projection lens system. The aperture ratio can be increased to greater than 90% by utilizing reflective TN and VA three-panel systems that allow the placement of the TFTs behind liquid crystal panels [11].

Another LCD projection scheme uses a single normally black TN panel, two mirrors, and a polarizing beam splitter (PBS) as shown in Figures 27.3 (field-off state) and 27.4 (field-on state). The PBS divides unpolarized light from the source into two orthogonal linearly polarized *p-waves* (polarized in

* The smaller the arc, the greater the luminous flux; the halides are the fluorides, chlorides, bromide, iodides, all derivatives of the halogens Group 17 (except astatine).

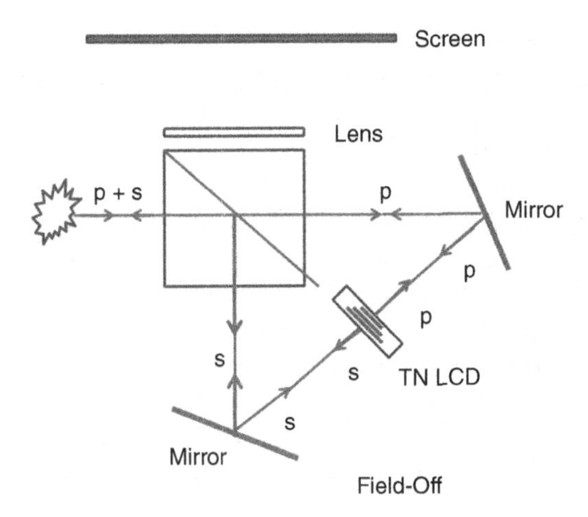

Figure 27.3 A single normally black TN panel with a polarizing beam splitter in the field-off state.

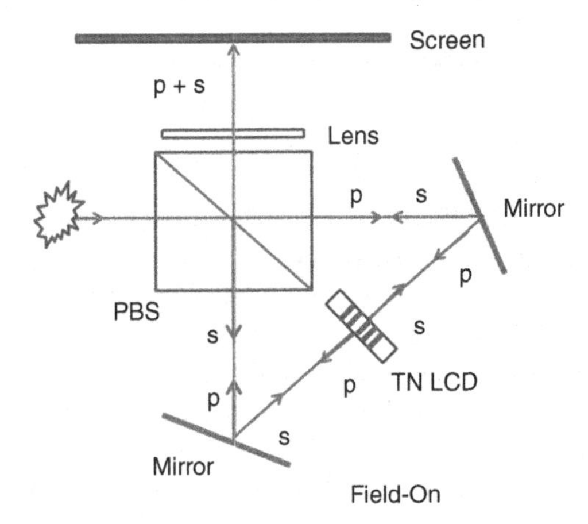

Figure 27.4 A single normally black TN panel with a polarizing beam splitter in the field-on state.

the plane of the incident plane-polarized beam) and *s-waves* (polarized orthogonal to that plane) that denote the plane of polarization of the reflected and transmitted beams. The s-wave and p-wave notation is particularly useful in reflective and transflective display analysis because of the changing planes of polarization between optical elements. The p- and s-waves impinge the TN panel from opposite directions, and in the voltage-off state, the TN of course rotates the planes of polarization oppositely to mutually transform the p- and s-waves (p ⇔ s), and the light is blocked to form the black state. In the voltage-on state, the NB TN liquid crystal does not change the polarization of the light, and it goes through to be reflected by the mirror to the PBS, and thence to the screen [12].

Polarizing beam splitters use dielectric layers, for example, SiO_2 and TiO_2 that use *Brewster angle* polarization requiring careful p- and s-wave analysis.* Alternatively, the PBS can utilize the birefringence of polymer films, where Cartesian coordinates can describe the polarization components, and thus no s- and p-wave notation is necessary; these are called *Cartesian PBS*.

Other projection liquid crystal modes, such as the *Bragg reflection cholesteric* and the subtractive mode *TripleSTN* three-panel system, have been constructed, as well as two LCD panel systems, which are cheaper than the three-panel systems but have lower resolution.

The polarizers used in projection displays can be the typical absorption type, or the newer orthogonal polarization wire grid polarizers (WGP) made of an aluminum parallel grid mounted on glass, wherein the reflected polarization is parallel to the grid lines, and the transmitted polarization is orthogonal to the grid. The WGP has been used as a polarizing beam splitter (PBS), as well as substitutes for traditional polarizers. To increase luminance, the rejected polarization component can be recycled by rotating the plane of polarization after the polarizer effect.

For color, in addition to the three-panel individual RGB system described above, as in the transmissive displays, typical subpixel color filters, field sequential, or direct LED lamps all can be used. In addition, particularly for projection displays, by rotating three prisms, synchronized to the addressing signals, a scrolling RGB band pattern with color moving off the bottom of the scroll and reappearing at the top, so light from each band is always projected, is shown schematically in Figure 27.5 [13].

The scrolling has high étendue because only part of the panel transmits light on one color at any time and so requires high luminosity lamps. Colors arrayed on a color wheel in the form of an Archimedes spiral (radius of curvature proportional to the angle of rotation; i.e., the more the rotations, the tighter the spiral) assures constant radial speed of the color bands, and

* Brewster's angle is the angle of incidence of light reflected from a partially reflecting surface at which the reflected light is fully plane polarized, that is, the light electrical vector induces electron oscillations in the surface that radiates a wave (the reflected light) that has only a component parallel to the surface since electrons cannot substantially oscillate perpendicular to the surface at or greater than that Brewster's angle of incidence (they would have to leave the surface). Brewster's angle is also called the "polarizing angle," and such reflection was used to produce plane polarized light, but polarizers as described in Chapter 4 are easier to use.

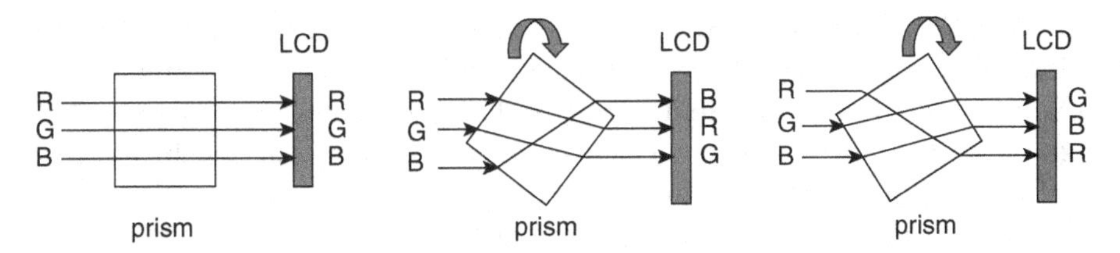

Figure 27.5 A scrolling RGB band color projection system.

reflected colors are recycled through an integrator to illuminate the output filters again in a procedure called *sequential color recapture* (SCR). These rotating type color schemes generally suffer from color breakup that is typically suppressed by high frame rates (450 frames/s), but these rates require very thin cell gaps (e.g., a TN cell gap of 1 μ or less), but large pixel sizes can reduce fringe field effects (see Chapter 24 on fringing field switching displays) [14].

The first application of the CMOS TFT in an active matrix addressing system was for the *liquid crystal on silicon* (LCOS) microdisplay. LCOS is an` LCD assembly formed directly on the silicon substrate that contains the addressing circuits for each pixel. Generally, amorphous-silicon TFT displays integrate the pixel array electronics only, polysilicon-addressing displays integrate the driving circuits,and LCOS integrates virtually all the IC electronics onto its backplane. Because of the small size of the LCD panels in projection displays, LCOS displays can use pure crystalline silicon or polysilicon transistors and thus derive the benefits of their high electron mobility. In fabrication, LCOS may use the well-established mass production DRAM-type analog addressing (pulse amplitude modulation), or SRAM-type digital addressing, similarly well-established, and having lower voltage and power requirements, but needing to reduce chip count and real estate in comparison with DRAM-type schemes [15].*

Brightness Enhancement Film

For portable devices, the brightness of the image might be the most important attribute, but since notebook computers, mobile phones, personal navigation devices, and music players operate on batteries, power consumption

* DRAM is an abbreviation for dynamic random access memory; SRAM is an abbreviation for static random access memory. They are used in computers and many other electronic device for memory storage, and are fabricated in the trillions of units.

must be minimized. Brightness enhancement films (BEF™) are plastic sheets with prism arrays formed on one side that refract light at small angles and reflect light with large incident angles back to the backlight to enhance that light. Multiple BEF sheets can also be employed to almost double brightness levels, all without consuming any power.

DBEF™ film transmits light of one polarization to the viewer and reflects light of the orthogonal polarization back into the backlight, where it is depolarized by the diffuser, and so is recovered to enhance brightness. The polarization recovery can operate in multiple cycles so brightness can be enhanced by up to four times, and better performance can be obtained by adding an enhanced specular reflector (ESR) behind the backlight. DBEFs, however, are more expensive because of their rather high manufacturing costs.

Touch Screens

The development of the human-display interface has advanced from drawing and writing with a pen to typing on a typewriter to the computer mouse, and now to the more intuitive touch screen and motion-capture sensors to facilitate the interaction between man and machine. The now-ubiquitous touchscreen user interface actually has long been developed for the traditional CRT screens, and much of that technology has been used on LCD screens, but with many new advances. Touchscreens are based on dielectric or conducting layers providing resistivity or capacitance, mechanical strain gauges, inductors, or transducers for transmitting, detecting, and reflecting waves, and a controller loaded with algorithms to process the signals to locate position. There are many different types of touchscreen systems; the main technology types are classified and briefly described following [16].

Resistive

Two conducting layers held a small distance apart by spacers are connected to a voltage; when pressed at a point, they contact at that point, forming a voltage divider and the point of touch location is calculated by the ratio of the measured voltage to the applied voltage for each coordinate in the horizontal *xy*-plane. Because of the mechanical deformation required, resistive touchscreens are subject to wear, but they are relatively easy and inexpensive to manufacture, and thus are used widely in consumer electronics that have shorter lifetimes than those required for industrial or retail display devices.

Strain Gauge

Similar to the resistive touchscreen, a detecting screen is spring mounted, with strain gauges to detect the point of touch; in this way, the force of the touch (z-axis displacement) as well as the position (xy-position) can be detected.

Capacitive

A transparent insulator and a transparent conducting sheet such as an ITO are disposed on the screen; when a finger touches the ITO, some of the charge on the conducting sheet is transferred to the finger, generating a current that is proportional to the distance of the contact point from the four corners of the conducting sheet, the difference being calculated by a controller to locate the position of contact. Since there is no mechanical deformation, capacitive touchscreens are more durable than resistive touchscreens, and can be used in repetitive-use applications such as retail cash registers and information kiosks.

In *surface capacitance* systems, one side of the insulator is coated with a conductive layer; when a finger touches the insulated side, a capacitance is formed between the conductive layer and the body belonging to the finger. The point of touch is determined from the change in capacitance measured from the corners of the screen. Since there are no moving parts, the surface capacitance touch screen is durable, but not terribly accurate and prone to parasitic capacitance; it is often used in outside applications, such as information kiosks.

In *projected capacitance* systems, the insulating layer is etched with a conductive material in either parallel rows or in a row/column electrode grid, and the grid is protected by a glass sheet. The projected capacitance system can offer high resolution using a fine grid that is suitable, for example, for signature capture in point-of-sale cash register screens. There are two kinds of capacitance in projected capacitance touch screens: self and mutual.

In the *self-capacitance* scheme, the capacitance between the finger + body and the electrode grid point is recorded as a current by current meters for each row and column.

In the *mutual capacitance* scheme, there is a capacitor at every point in the grid; when a voltage is applied, touching the screen changes the electrostatic field at that point, thus changing the capacitance between the capacitor at that point and the capacitance of the finger + body, that is, the mutual capacitance of two capacitors. Multiple touches can be simultaneously recorded as changes in voltage at the different points, making possible, for example, the control of expanding and contracting the images in smart phones using the thumb and fingers.*

Inductive

A sensor board in a grid pattern is placed behind the LCD; a stylus with an oscillating resonant circuit (coil and capacitor) marks the position, and the sensor board senses and marks the xy-position on the grid; the z-axis position can also be sensed for contact pressure, and hovering and variable line width can also be inductively sensed in a form of motion capture.

Surface Acoustic Wave

A controller sends a MHz burst to a transmitter, which converts the signal to ultrasonic mechanical waves that propagate over the surface of the screen as acoustic surface waves. Piezoelectric receivers are placed oppositely the transmitters, and reflectors are positioned between the receivers. When the screen is touched, the SAW is slowed by absorption that is measured by the reflectors and receivers; a location algorithm in the controller then processes time delay and amplitude changes to mark the spot. Since there are no metallic layers as in the resistive and capacitance touchscreen systems, the light throughput is good, and since there is no mechanical deformation as in the resistive system, the SAW system is durable.

Infrared

An array of LED beams and detectors around the periphery of the screen pointing inwards form an xy-plane grid that detects disruptions in the infrared beams at points on the grid. The infrared system can detect any type of pointing device, be it a finger, gloved finger, or stylus.

Optical Imaging

Infrared sensors are placed on the periphery of the screen, and an infrared backlight shines light through the screen from below it; a touch will create a shadow that is sensed by sensors, and an algorithm triangulates to find the position of the touch. The size of the touch can also be detected from the

* A particularly compelling demonstration of touchscreen technology is by Jeff Han of the Courant Institute of Mathematics showing the multipoint interaction-multiple control point system using a mathematical mesh (refer to computer.howstuffworks.com).

size of the shadow. Optical imaging touchscreen systems are being increasingly used in large displays.

Dispersive Signal

With contact and upon distortion, piezoelectric materials transform mechanical energy into electric signals, thus a grid of piezoelectric vibration sensors can detect mechanical vibrations caused by a touch and transmit the electric signals generated thereby to a controller to compute the position of a touch. Dispersive signal technology touchscreens are suitable for larger displays, but as a type of motion-capture, it cannot detect a motionless pointer.

3D

It was a certainty that the large-screen television made possible by the liquid crystal display would be enhanced from two to three dimensions to provide a more realistic (albeit exaggerated) image for viewing pleasure. Three-dimensional (3D) vision is possible because human beings have two eyes in the same plane but separated by a small distance. When viewing objects as shown in Figure 27.6, the slightly different angles of view from each eye engender a different perspective (called *parallax*), which when combined in the brain, creates a stereoscopic image.

Flat-screen 3D image perception technology has been in development since the 1940s. The first flat 3D image were side-by-side pictures placed to give a perception of depth by simply allowing the brain to simultaneously process the two images. To improve the effect, special lenses were used to simultaneously view photos taken from different angles at more or less those angles; these stereographic viewers are still in use, but now primarily for children's cartoons.

Stereoscopic images requiring glasses are divided into active and passive types. One type of active 3D display uses shutter glasses that are made of liquid crystals that alter the polarization state to block or transmit light in active synchronization with the images on the display, thereby providing dynamic dual images necessary for three dimensional motion perception. A passive display uses linear polarizing glasses made of two orthogonal polarizers (just like the polarizer and analyzer used in LCDs) so that if an image is separated into two linearly polarized images having mutually perpendicular planes of polarization, one of the lens in the linear polarizing glasses lets the light of that image through, and the other lens blocks that image (the

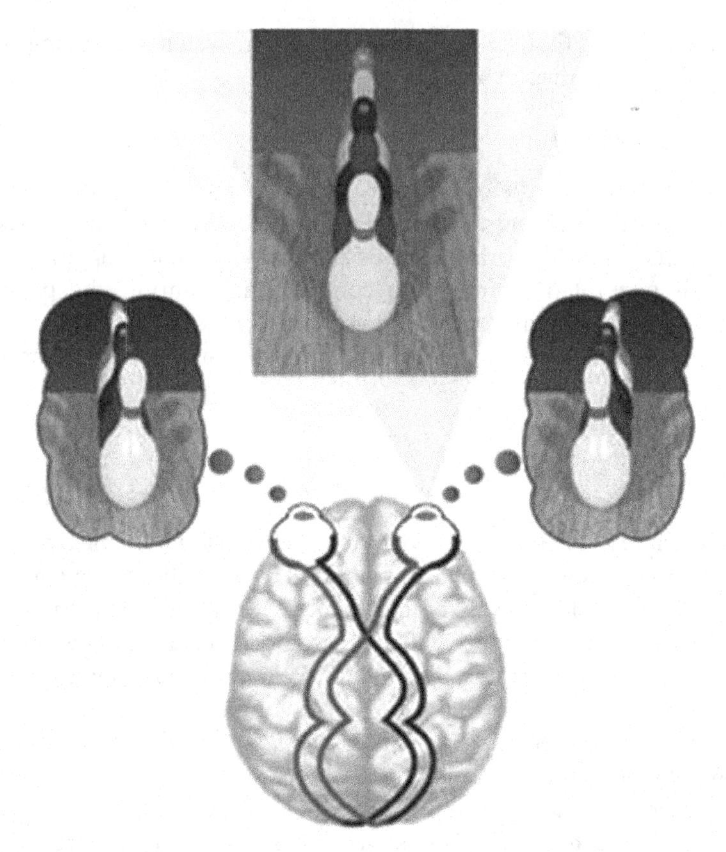

Figure 27.6 The different angles of view from each eye engender parallax. From vision3d.com.

dichroic absorption described in Chapter 4), and vice-versa for the other glass lens. So the composite 3D image is composed of two images each sent to a different eye, thereby creating the 3D effect just as in the parallax in natural viewing. As might be expected, the same can be done with right and left-hand circularly polarizing lens in the glasses, where one eye receives a right-hand circularly polarized image and the other eye a left-handed one. A more sophisticated scheme utilizes interference filters to divide the primary red, blue, green colors into two sets, one for each eye, producing high-quality color 3D images.

Autostereoscopy refers to techniques to impart a perception of depth to the naked eye without the use of shutter glasses. In a return to the two-picture formation of 3D images, the simplest (and cheapest) scheme is to align the pictures on a corduroy-like texture plate. Disposing lenses on the plate to direct the image at slightly different angles to the viewer imparts a perception of depth, although the effect is limited to a rather small viewing angle. This so-called *lenticular* technique is commonly used on postcards and credit cards, and unless it is enhanced by multiple layers, its 3D image effect is minimal.

In the spirit of the lenticular, an LCD screen can be simply covered by two transparent layers with black stripes to block certain pixels in the display and pass others, allowing the eyes to see different perspectives to simulate 3D by a *parallax barrier*. A more sophisticated scheme is to use arrays of prisms on the LCD screen to appropriately divert the light to the eyes to create the 3D image effect.

Refinements to the lenticular and parallax barrier schemes include designing arrays of parallax barriers to allow multiple 3D viewing angles, and employing eye-tracking from the display to follow the viewer's eye position to adjust for viewing angle. Other sophisticated techniques are to create a three-dimensional volume light field image (*volumetric display*) instead of just a planar field, and a fly's eye multiple lens array (*integral imaging*).

The volumetric display has been in development for years for the CRT display. In the swept-volume display, a time-sequenced series of sliced images projected from the display must be reconstructed by the "accommodation response" of the eyes and brain to form the 3D image. The parallax barrier arrays, if sufficiently deep to invoke an accommodation response, can then be considered a volumetric display [17].

Integral imaging utilizes an array of spherical convex microlenses that projects different images depending on the viewing angle, creating parallax as the viewer moves to create a dynamic 3D view.

LCD Products

New liquid crystal display products are being vigorously developed, but the mass production LCDs of the early twenty-first century already have been determined to be the twisted nematic (TN) and the super twisted nematic (STN) for small- and medium-size screen, and the multiple-domain vertical alignment (MVA) and in-plane switching (IPS) for large screen sizes. Efficient production, accelerated depreciation of expensive fabrication equipment, and full capacity high volume day-and-night production have created powerful economies of scale that make possible the annual production of hundreds of millions of low-cost but very high-quality LCD panels.

Some other display modes, such as the bistable cholesteric liquid crystals and electrophoresis electronic paper for electronic books are already in popular use, and the blue phase for super-fast response time televisions is in promising development.

The prime mover of liquid crystal display development was the notebook computer and the form factor has not changed appreciably, but it has become astonishingly light and thin primarily because of the use of edge LED backlights for illumination. The "diffusion rate" (in industry jargon) of LCD monitors replacing CRTs has almost reached 100%.

The many new products such as smart phones, video recorders, digital cameras, portable DVD players, photograph frames, GPS devices, vehicle rear-view and side mirrors, all use the small and medium size LCDs to display pretty and informative images.

Because the image-producing principles are essentially different from cathode ray tubes, liquid crystal displays are not subject to the CRT constraints of high voltage, weight, and thickness, and the screen size of the LCD television sets increased rapidly. But the sheer size and fragility of very large glass plates, the extremely expensive large-scale production equipment, increased transportation costs, and risks of breakage caused some to say that the 7th would be the last generation. And early "market intelligence" on liquid crystals reported that because of the small size of Japanese living rooms, large screen televisions would have little success in the Japan market. But it turned out that the 40-inch and above size LCD televisions were extremely popular in Japan, and the LCD panels produced from the 9th and 10th generation fabs used in the super-size television screen and the very large advertising displays are increasingly common in everyday life. At the other end of the size spectrum are the tiny near-eye microdisplay projector headsets for immediate information display and games playing.

The conventional rectangular LCD screen has also been shaped into curved, round, and elliptical screens for use in picture frames, mobile phones, watches, dashboard panels, and game displays. The early LCD projectors were relatively heavy and bulky, but have become indispensable for all kinds of presentations in schools, businesses, and almost any undertaking. Particularly because of the development of the LED backlight, the LCD projector has become smaller to the point of handheld microprojectors that will be the latest "killer application" for incorporation into smart mobile phones and other hand-held devices, such as digital cameras, e-mail communication devices, and video game players, solving the problem of too-small displays on those pocket devices.

For Internet and web access, the success of the advanced touchscreen panels using small and medium size LCDs are exemplified by Apple's iPhone™ and iPad™ and their progeny.

Early e-books used bistable cholesteric displays and privacy windows used switchable bistable cholesteric liquid crystals. Later, the e-ink Kindle DX™ e-book developed by Amazon and manufactured by Taiwan's Prime View International (PVI) used electrophoresis, and a similar technology is AUO's SiPix™ Microcup® electronic paper. Sharp has also developed an LCD with memory (so turning off the power will not change the display) designed to compete with e-paper.

The "blue phase flash" in carrots first observed by Reinitzer himself has become a liquid crystal display mode itself, as developed by Samsung in a 15-inch blue phase television which features an incredibly fast 240 Hz image refresh rate driver that can display very natural image motion.

As for the thrilling 3D images, for example, the German VisuMotion and Sharp cooperative effort on the 65-inch *parallax barrier autostereoscopic* display. The three-dimensional effect is produced by five or six multiple layers oriented at different angles, so that the viewer will perceive two images and the brain will automatically resolve them into a 3D image, without the need for 3D glasses.

The parallax from three transparent liquid crystal panels arranged on the faces of a six-sided box can also provide a "real body" visual for use in education, industrial design, and computer games.

The evident reality of the liquid crystal display, from the precisely beautiful colored image to 3D and real body images, not only reflect the beauty of the world around us, but also displays information for mankind to use to improve that world.

References

[1] Feynman, R.P., Leighton, R.B., and Sands, M. 1963. *The Feynman Lectures on Physics*, Vol. I. Addison-Wesley, Reading, pp. 32–36.

[2] Yang, D.K., and Wu, S.T. 2006. *Fundamentals of Liquid Crystal Display Devices*. Wiley, Chicester, p. 307ff.

[3] Yang, D.K., and Wu, S.T. 2006. *Fundamentals of Liquid Crystal Display Devices*. Wiley, Chicester, p. 290ff.

[4] Armitage, D., Underwood, I., and Wu, S.T. 2006. *Introduction to Microdisplays*. Wiley, Chicester, pp. 12–64; Yang, D.K., and Wu, S.T. 2006. *Fundamentals of Liquid Crystal Display Devices*, Wiley, pp. 120ff. and 281ff.

[5] Drzaic, P. 2004. Application tutorial A-1, SID 04 application tutorial notes; den Boer, W. 2005. *Active Matrix Liquid Crystal Displays*. Newnes, Oxford, p. 211.
[6] Armitage, D., Underwood, I., and Wu, S.T. 2006. *Introduction to Microdisplays*. Wiley, Chicester, p. 263ff.
[7] Kikuchi, H. 2009. Potential and challenges of polymer-stabilized isotropic LCs for display applications 2. Seminar at Chimei Optoelectronics, Dec. 18, 2009.
[8] Lin, Y.H., et al. 2005. *J. Disp. Technol.*, **1**, 230.
[9] Maeda, T., et al. 1999. *J. SID.*, **7**, 9; Yang, D.K., and Wu, S.T. 2006. *Fundamentals of Liquid Crystal Display Devices*, Wiley, Chicester, p. 259ff.
[10] Wu, S.T., and Yang, D.K. 2001. *Reflective Liquid Crystal Displays*. Wiley, Chicester; Yang, D.K., and Wu, S.T. 2006. *Fundamentals of Liquid Crystal Display Devices*, Wiley, Chicester, p. 243ff.
[11] Chigrinov, V.G. 1999. *Liquid Crystal Devices: Physics and Applications*. Artech House, Boston, p. 264ff.
[12] Wu, S.T., Wu, C.S., and Kuo, C.L. 1996. *SPIE Dig.*, **2949**, 52ff.
[13] Lueder, E. 2001. *Liquid Crystal Displays: Addressing Schemes and Electro-Optical Effects*. Wiley, Chicester, p. 300.
[14] Armitage, D., Underwood, I., and Wu, S.T. 2006. *Introduction to Microdisplays*. Wiley, Chicester, p. 307ff.; Chigrinov, V.G. 1999. *Liquid Crystal Devices: Physics and Applications*, Artech House, Boston, p. 262ff.; Lueder, E. 2001. *Liquid Crystal Displays: Addressing Schemes and Electro-Optical Effects*, Wiley, Chicester, p. 297ff.
[15] Armitage, D., Underwood, I., and Wu, S.T. 2006. *Introduction to Microdisplays*. Wiley, Chicester, p. 40ff.; Lueder, E. 2001. *Liquid Crystal Displays: Addressing Schemes and Electro-Optical Effects*, Wiley, Chicester, p. 263ff.
[16] den Boer, W. 2005. *Active Matrix Liquid Crystal Displays*. Newnes, Oxford, p. 145ff.
[17] Blundell, B., and Schwarz, A. 2000. *Volumetric Three-Dimensional Display Systems*. Wiley, Chicester.

Index

Liquid Crystal Displays, First Edition. Robert H. Chen.